EXPLOITABLE MARINE ECOSYSTEMS: THEIR BEHAVIOUR AND MANAGEMENT

EXPLOITABLE MARINE ECOSYSTEMS: THEIR BEHAVIOUR AND MANAGEMENT

The nature and dynamics of marine ecosystems:
their productivity, bases for fisheries,
and management

TAIVO LAEVASTU
with contributions from
DAYTON L. ALVERSON and
RICHARD J. MARASCO

Fishing News Books

Copyright © 1996 by
Fishing News Books
A division of Blackwell Science Ltd
Editorial Offices:
Osney Mead, Oxford OX2 0EL
25 John Street, London WC1N 2BL
23 Ainslie Place, Edinburgh EH3 6AJ
238 Main Street, Cambridge,
  Massachusetts 02142, USA
54 University Street, Carlton,
  Victoria 3053, Australia

Other Editorial Offices:
Arnette Blackwell SA
  1, rue de Lille, 75007 Paris
  France

Blackwell Wissenschafts-Verlag GmbH
  Kurfürstendamm 57
  10707 Berlin, Germany

  Feldgasse 13, A-1238 Wien
  Austria

First published 1996

Set in 10pt Times Roman
by DP Photosetting, Aylesbury, Bucks
Printed and bound in Great Britain by
Hartnolls Ltd, Bodmin, Cornwall

DISTRIBUTORS

Marston Book Services Ltd
PO Box 87
Oxford OX2 0DT
(*Orders:* Tel: 01865 206206
      Fax: 01865 721205
      Telex: 83355 MEDBOK G)

USA
Blackwell Science, Inc
238 Main Street
Cambridge, MA 02142
(*Orders:* Tel: 800 215-1000
        617 876-7000)

Canada
Oxford University Press
70 Wynford Drive
Don Mills
Ontario M3C 1J9
(*Orders:* Tel: 416 441-2941)

Australia
Blackwell Science Pty Ltd
54 University Street
Carlton, Victoria 3053
(*Orders:* Tel: 03 347-0300
      Fax: 03 349-3016)

A catalogue record for this book is available from
the British Library

ISBN 0–85238–225–1

Contents

List of Contributors

Taivo Laevastu Senior Scientist, Alaska Fisheries Science Center, Seattle, Washington (retired). Dr Laevastu has published many books and numerous scientific and technical articles on fisheries, fisheries oceanography, marine meteorology and marine ecosystems.

Dayton L. Alverson President, Natural Resources Consultants, Seattle, Washington. Formerly Director, Northwest and Alaska Fisheries Center, Seattle, Washington. Dr Alverson is an advisor to both the US and Canadian Governments and to other national and international organizations on matters of fisheries research and management. He has published several books and numerous scientific articles on fisheries.

Richard J. Marasco Director, Resource Ecology and Fisheries Management Division, Alaska Fisheries Science Center, Seattle, Washington. Dr Marasco has been engaged in fisheries research and management in the North Pacific for the last 18 years. He has published numerous papers on fisheries resources, economics and management.

Preface

Any attempt to summarize the essentials of marine ecosystems is difficult, mainly because of the interwoven multiplicity of knowledge necessary for this task. Many basic and good descriptive works on marine ecosystems appeared in the 1930s, 1940s and 1950s. Since then a number of detailed and particular studies have been published, few of which have discussed their findings in the context of marine ecosystems as a whole. Recently several extensive summary works have been produced which emphasize specific aspects of the system (e.g. Longhurst 1981, Barnes & Mann 1991, Mann & Lazier 1991).

The main purpose of this book is to summarize the characteristics and dynamics of the numerous marine ecosystems that exist and to discuss man's effects on, and his attempts to manage, these systems. In particular, we aim to:

(1) Review the multitude of marine ecosystems that exist, with emphasis on their importance to man and the possible effects man can have on them.
(2) Review the great biodiversity of these systems, their natural variability and the environment-biota interactions that occur.
(3) Summarize the behaviour of fish ecosystems and their components, and review the ways in which fishing makes use of this behaviour.
(4) Summarize the effects of fishing on marine ecosystems, with particular emphasis on fish ecosystems.
(5) Review the past objectives, methods and effectiveness of marine ecosystem management.
(6) Outline the possibilities of comprehensive marine ecosystem management and its economic and bio-geo-political consequences and suggest procedures on which to base decisions for the exploitation and management of marine resources.

Gore (1992) recognized the need to educate the world's citizens about environmental issues. However, he underemphasized the effects of climatic changes and did not address the various local disturbances that often occur in marine ecosystems.

In Chapters 2 to 6 we give an abbreviated view of the numerous marine ecosystems and the processes that occur within them. This view is necessary to understand man's concern with these systems, their utilization and the harvesting of their resources. These descriptions provide a necessary background to our last two chapters on ecosystem management.

The Appendix of this book describes a basic ecosystem simulation programme with emphasis on fish ecosystems. Not all readers of this book will need it, but some will. It is included for these prospective users, because simulation models are usually documented in technical reports that are often not easily accessible.

The authors hope that this book will be a useful text for students of marine biology and ecology. Furthermore, we have attempted to make it understandable to conservationists concerned with the preservation and rational utilization of marine ecosystems.

Acknowledgements

The authors express their thanks to Mrs Marge Gregory and Mr Michael Hatch for typing the manuscript, to Mr Bernard Goiney for preparation of figures and to Mr Geary Ducker and Dr James Mason for editing the manuscript. Furthermore, the authors thank Dr Frank Fukuhara and other colleagues for valuable discussions and advice, and Mr Richard Miles of Fishing News Books for encouraging them to write this book.

Chapter 1
Marine Ecosystems and Man – An Introduction

1.1 Concern about the marine ecosystems

World-wide concern about the 'health' of the Earth's ecosystems has intensified in the last decade. This concern extends also to the marine ecosystems, only a small part of which is visible to man, and is mainly directed toward the problems of marine pollution, habitat degradation and too heavy exploitation of renewable living marine resources affecting their reproduction potential. This increased public concern about the environmental quality has resulted in a 'greening' of the value judgements made by the public and has affected natural resource exploitation policy.

Man's concern with the marine ecosystems and the 'state of the sea' focuses on four areas:

(1) The 'health' of marine food resources and their rational exploitation and management.
(2) The ocean as a depository of various wastes and the effects of pollution.
(3) The aesthetic value of clean sea-shores.
(4) The impact of marine shipping.

It is currently recognized that many traditional fishery resources might be fully exploited from the economic point of view (Alverson & Larkin 1992). There is also a growing mariculture along many coasts. Consequently, we need to ask how intensive fisheries have affected main ecosystems, what will be the effects of mariculture on the nearshore ecosystems and on the aesthetic value of our sea-shores. We must also know how marine waste disposal is affecting marine ecosystems and determine the safest way for disposing of waste in the ocean.

Terrestrial ecosystems are often easily visible and accessible to man. There is a natural tendency by some to consider marine ecosystems in terms similar to those used for terrestrial ecosystems. Such considerations can, however, lead to misunderstandings, as there are great and basic differences in nature and behaviour between terrestrial and marine ecosystems; some of these essential differences are shown in Table 1.1.

The slow progress in marine ecological studies and in understanding marine ecosystems in the past was partly caused by their complexities and the difficulties involved in conducting realistic controlled experiments. Furthermore, the scientific community has not been able to agree on a definition of marine ecosystems as such.

The current environmental activism and politics is apparently an outgrowth of the 1960s political and social activism and involves people with different ideologies and backgrounds. There are professional people with a biocentric world view who accept

Table 1.1   Generalized comparison of essential properties of terrestrial and marine ecosystems.

Terrestrial ecosystems	Marine ecosystems
(1) Easily visible and accessible	Invisible and difficult to access
(2) Basic organic production by sessile higher plants	Basic organic production mainly by microorganisms in water mass
(3) Production often limited by water	Production limited by some nutrient salts
(4) Grazing dominates over predation	Predation dominates over grazing
(5) Moderate complexity and biodiversity	Highly complex and biodiverse
(6) Advection rare and migration generally limited	Advection and migration dominant processes
(7) Habitat and ecosystem alteration by man easy	Habitat and ecosystem alteration by man difficult
(8) Cultivation dominates in exploitation	Wild stock catching and gathering dominates

the coexistence of man with other species, but advocate strong management of the natural resources by authorities. Many environmental activists are, however, amateur crusaders who sense that something is wrong, and tend to reject the management by governments. There are also people with radical ideologies who consider all human activity as destructive to the environment and advocate resistance at all costs.

Many of the diverse views about the need and manner to manage marine ecosystems are caused by differences in goals and by lack of knowledge of these ecosystems and processes in them, which this book attempts to alleviate.

1.2.   Complexities of marine ecosystems

The marine ecosystems and living resources in them are often invisible to the human eye in their natural environment. Most of the marine living resources utilized by man are also mobile and may be widely distributed over considerable area and depth. Most of the distributions are also very patchy. Therefore, obtaining information about these marine living resources is difficult and costly.

In describing biotic marine ecosystems and the processes within them, we consider the synecology of the assemblages or communities of species. Furthermore, we take into account biological and ecological characteristics of the species in relation to ecosystems, such as their 'ecological potentialities', distribution and abundance in the habitat, life span, reproductive habits, species as predator and as prey, fecundity and environmental tolerances.

Processes in marine ecosystems are in several ways more important than in terrestrial systems. On the other hand, perturbations, especially those caused by man, are much more profound in terrestrial than marine ecosystems. To understand the effects of man or marine ecosystems, it is also necessary to understand the natural variability of these systems; that is, variability caused by environmental changes as well as biological processes occurring within the ecosystems. Literature abounds in descriptions of variability, but is short in explaining synecology and the cause-and-effect relations in ecosystem processes and their quantitative evaluations. As pointed out by Ursin (1982), marine biologists are inclined to write a paper if they discover that something changes or oscillates. If everything is as usual there is no incentive to write a paper. Recently a dark picture describing the possible extinction of species is frequently found in the literature. Alarming statements cause the general public to doubt the ability of scientists to explain the true picture of the effect of man on the ecosystems, and require that governments 'reduce the scale of human activities' (*New York Times*, 13 May 1993).

Recently statements about the precipitate decline of fish stocks and the claims of overfishing have been reported in newspapers and in the trade press, but the evidence to support the claims is not always at hand. Firstly, the term 'overfishing' can have different definitions and criteria, and secondly, the fluctuations of fish stocks are affected by factors other than fishing. The authors of this book therefore review the effects of fishing on stocks in the framework of marine ecosystems and explore the possible impact of many on the biodiversity of these systems.

The marine ecosystem and the essential processes in it have often been depicted as a multi-linked food chain (Fig. 1.1). This food chain includes man as top predator on selected living marine resources. Although living resources are renewable, they are finite in size and have a limited renewal rate. Therefore, the exploitation of the common property resource must be fairly allocated and its exploitation managed.

Regional marine fish ecosystems function as minicosms. A myriad of interactions occur within the biota and between it and the environment. Marine mammals, man and fish are the main apex predators in marine ecosystems, which they control to a large extent by predation. The ecosystems cannot be fully understood by considering their constituents by species and/or species group only; we must concurrently consider quantitatively all the processes and interactions within the systems.

Rational management of marine ecosystems and living resources in them requires the consideration of the entire ecosystem and the processes operating in it which affect the dynamics of the species. A considerable amount of knowledge about marine fish ecosystems has been gained during the past nine decades. It is now possible to synthesize this information and to construct objective numerical dynamic ecosystem simulations, which we can use to assess the abundance and behaviour of the fishery resources and the factors affecting them. An example of the ecosystem simulation is given in the Appendix.

1.3   Utilization of marine ecosystems by man and his effects on them

The human population of the Earth is increasing in excess of 2% per annum. Consequently, food production must increase at a somewhat greater rate in order to keep in step with this population increase and to abolish the undernourishment that exists in some parts of the world. To achieve effective national and industrial planning, the extent of the available food resources must be estimated as well as their potential for expansion, exploitation and conservation. The food resources of the sea are dispersed

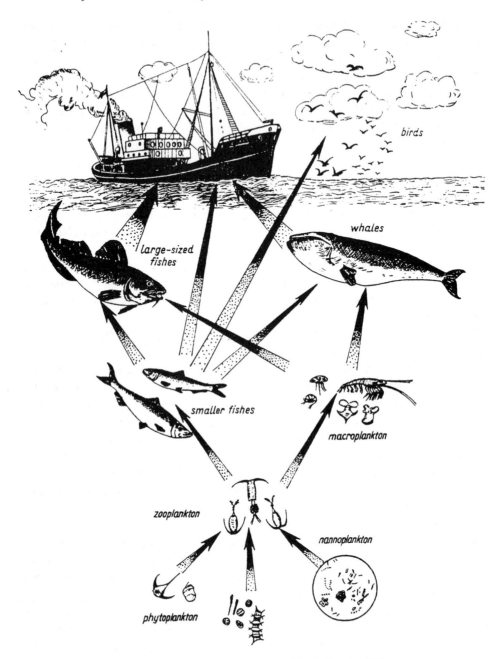

Fig. 1.1 Example of a five-link food chain in the sea (Demel & Kulikowsky 1965).

and their harvesting requires large amounts of energy, which is also in short supply and is expensive. The main problems in exploitation of the marine resources include locating of the resources, estimation of their abundance and response to exploitation, and consequently the designing of harvesting measures to ensure an optimum and sustainable yield.

The total catch for the world's fisheries has reached about 100 million t and seems to be stabilizing or even declining, and the world's fisheries at large are losing money. Rough calculations by FAO (1992) show that in 1991 the world's fishermen landed about $US70 billion worth of seafood with an operating cost of $US93 billion, not taking into consideration the capital costs of the world fleet, the replacement value of which is about $US300 billion. A billion here means 10^9. Thus, there is a serious over-capitalization of fisheries, which in return might affect the sustainable use of the resources.

Man is at present able to influence terrestrial production by cultivation, but his influence on the abundance of marine resources by cultivation is relatively small. Only a few species of marine organisms are actually cultivated to any extent, including oysters and shrimp, which can be cultivated in lagoons, ponds or estuaries, and salmon, which are mostly cultivated in pens. There also exists limited ocean ranching of salmon.

The aquaculture production in 1987 was about 10 million t (excluding algae), of which 6 million t was fresh-water production of fish and about 3 million t of molluscs. Marine fish production was about 0.5 million t and about an equal amount of crustaceans was produced. Most of the aquaculture production was from Asia (about 85%), with Europe being second (about 8.5%). By 1993 the aquaculture production has risen to 15 million t, including algae.

About one-half of the world's population lives within 100 km of the coast, the part of the sea most accessible to man. This nearshore ocean is also the most biologically productive part of the sea. The coastal areas of the sea are not pristine, having been affected by man for centuries, mainly in four respects, waste disposal, resulting in pollution and eutrophication, intensive harvesting of various resources of the sea, use of some coastal areas for mariculture, and using coast and nearshore waters for recreational purposes. Coastal marine ecosystems are also much more susceptible to catastrophic events than offshore ecosystems, both natural (e.g. storms) and man-made (e.g. pollution).

The utilization of marine resources by man is dependent not only on the total amounts of the resources present, but also on their concentration per unit of area. Fleming & Laevastu (1956) estimated that fisheries are profitable if the concentration of fish is limited in vertical extent and the concentration is at least 3–50 g/m^2, dependent on the value of the species.

The lack of adequate data to analyse and detect changes in marine ecosystems is a world-wide problem, and is often compounded by inaccessibility of data. Furthermore, the methods of data collection, as well as analysis, change too often, making the results difficult to compare. In the future more emphasis needs to be placed on the analysis of existing data from newly-raised points of view.

There is no doubt that conventional single-species assessment models, even if appropriate and applicable (which they are not), are inadequate for the assessment of the intensive fisheries of today. It does not require much perception to surmise that the removal of 1 million t of a given species is going to have substantial impact on its prey and predators and possibly on those stocks taken incidentally as by-catch.

Throwing into the mix management's need for forecasts to set catch quotas, the need for alternative assessment schemes is obvious. They are offered in this book.

1.4   Management of marine ecosystems

Management of marine ecosystems pertains to their rational use and utilization by man and to their 'fair and equitable' distribution (allocation). One of the objectives is to obtain optimum yields from the living marine resources (mainly fish) and sustain these yields over time. Connected to this objective is the maintenance of marine ecosystems in such a state that the biological production of the desired species is sustained. The second objective, pertaining largely to coastal areas, is to keep the marine environment clean and suitable for multiple uses by man. In some regions, an additional objective is to preserve some marine mammal populations. Thus the management objectives are essentially firstly economic, guided and restrained by biological–ecological conditions in the ecosystems, and secondly habitat–production orientated and aesthetic in respect to pollution, marine mammals and maintenance of biodiversity.

As soon as we start to harvest one species, abundance and age structure of the stock of this species changes. Changes are not limited to the target species, but affect other species also via predation and competition. That is, the whole ecosystem is usually affected by exploitation. In the past and at present, the state of the major fish stocks has been evaluated stock by stock. Alverson & Larkin (1992) considered 53% of major fish stocks of the world to be overexploited (mainly growth overfished), and thus mismanaged. They predicted that, in the twenty-first century, elements of the scientific community, environmental and conservation groups and politicians will unite in a call for more holistic ecosystem management of living aquatic resources.

Marine resources are usually considered common property resources and it is politically difficult for any government to deny access to any individual seeking to utilize the desirable fishery resources. This problem has been alleviated by the declaration of an exclusive economic zone (EEZ) that justifies management by governments on economic and social grounds.

Management includes fair and equitable allocation of resources to competing groups, which is a moral concept, while also considering economic efficiency and social impacts. It is desirable that fisheries management and its long-range planning establish certain future patterns for the wise use of the natural fisheries resources, on a regional basis considering the whole marine ecosystem and natural processes and the fluctuations in these. At present there is discrimination among fish species as food, especially in highly developed countries. Edible fish are often thrown back into the sea because there is no local market to them. This situation might change as many other, now easily marketable, resources become exploited to their fullest and edible, marketable fish products are made from fish which were earlier considered as 'scrap fish'.

The detection of human influence on marine ecosystems is difficult, mainly because of our inability to separate 'natural' changes from man-made changes. This separation requires long-term monitoring. However, many long-term monitoring programmes have been interrupted because of lack of funds and/or improperly set up initial monitoring programmes (Duarte *et al*. 1992).

One of the main obstacles to marine ecosystem management is the lack of adequate understanding of the current state and dynamics of the system we seek to manage, and the consequent oversimplification of the system and its natural behaviour as well as its response to actions by man. The main components of marine ecosystem management are:

(1) Describe the system, its integral parts and the interactions within it as well as with other ecosystems. The system must also be defined in space and time.

(2) Evaluate quantitatively the natural dynamics of the system and its essential components.

(3) Determine the goals of its utilization and the effects of the utilization under different scenarios and economic considerations.

(4) Determine the acceptable limits of change within the ecosystem under consideration (e.g. aesthetic, biological and economic limits), consider the instability of the system (especially in respect of recovery), and possible feedback systems affecting the instabilities (fluctuations) and their limits with optimum utilization and sustainability as main guiding principles.

(5) Determine the possible management and enforcement methods at hand and their effectiveness and legal status.

Chapter 2
Marine Ecosystems

2.1   General characteristics of marine ecosystems

There is no simple, generally accepted and understood definition of marine ecosystems. General statements and ecosystem definitions such as 'plants and animals living in a defined area and environment' or 'distinct and coherent ecological community of organisms and the physical environment with which they interact' as used for terrestrial ecosystems are not sufficient for describing the marine ecosystems. This is because of the pronounced temporal and spatial changes found in marine ecosystems, the variety of specific environmental conditions encountered in the sea, and the mostly unstable, advected and migrating biocoenoses and communities found there. Marine ecosystems have been defined on four grounds:

(1)   Mode of life and nature and functioning of dominant organisms (plankton, benthos and nekton ecosystems).
(2)   Specific environments (pelagic, demersal and brackish water ecosystems, etc.).
(3)   Geographic locations (North Sea, polar, etc.).
(4)   A combination of the three grounds listed above (e.g., Barents Sea benthic ecosystem).

There have been attempts to divide the ocean into 'large marine ecosystems' (e.g. Sherman *et al*. 1990), averaging the biological content and environmental conditions over large and diverse areas. Such large systems would mask fluctuations in the multitude of smaller, distinct systems, especially in the complex coastal marine systems. Furthermore, the descriptions of these large ecosystems tend to overlook the importance of predation by numerous possible predators on any single species.

Ecosystems contain plant and animal communities which are often specific to given ecosystems. These communities can be variable in space and time and difficult to define and delineate. The difficulties of defining marine communities have been reviewed by Nesis (1977) and Fager (1963). Owing to the great spatio-temporal variability in the sea we cannot clearly define marine communities as simple statistical units as is commonly done in terrestrial ecology. Marine communities are, however, often defined as ecotypes and by habitat (Table 2.1). Thus marine ecosystems and communities are not uniformly defined and one must rely on additional descriptive information for their characterization.

The long prevailing view in marine ecology, and especially in planktology, has been to characterize marine ecosystems from trophodynamic aspects. For example, Steele (1974) reviewed the marine ecosystems mainly from the point of view of matter and

Table 2.1 Criteria for characterization and classification of marine habitats, ecotypes and communities (some are also classified as ecosystems).

Element, habitat, criterion	Examples of ecotypes and communities
Depth	
in water column	Pelagic, demersal
to bottom	Shallow water, deep water
Distance from the coast	Neritic, coastal, oceanic
Type of bottom	Rocky, soft, coral reef
Mobility and way of life	Plankton, nekton, benthic
Temperature regime	Polar, temperate, tropical
Salinity regime	Oceanic, brackish water
Substrata use	Infauna, epifauna
Size of organisms	Nannoplankton, macroplankton
Dominance of species	Gadoid, clupeoid
Dynamics of environment	Intertidal, tidal, gyral
Life cycle characteristics	Migratory, sessile, holoplanktonic

energy transfer through the ecosystem at large – i.e. a 'food chain ecosystem'. He also attempted to construct a simple mathematical model of the transfer of matter in terms of carbon. However, lately this simple way of considering marine ecosystems has been challenged owing to the very complex and varied food and feeding relationships found in these ecosystems.

The 'eat and get eaten' condition is a considerably more important condition in marine than in terrestrial ecosystems. Terrestrial systems are in this respect much 'kinder' and based more on herbivory than marine systems. Most animals in a marine ecosystem die prematurely owing to predation and few are left to die from 'natural causes' such as old age or disease. Nearly everything in marine ecosystems is linked through predator–prey relations. Life-threatening competition causing starvation, environmental extremes, diseases and parasitism are but minor causes for death in marine ecosystems.

Fights are unusual (except in a few marine mammals). Thus the survival of the fittest also plays little part. However, size of the prey and the speed of escape are of great importance in the cruel marine ecosystem. Therefore marine animals produce many offspring, which are distributed widely with currents, many of them being eaten before they reach adulthood. Cannibalism is also quite common among fishes.

It is a rare ecosystem indeed that can be investigated without due consideration of transport and migration across the defined geographical boundaries. Inevitably the definition of the marine ecosystem depends upon the objectives of the researcher.

Great mobility, either by active migration or passive advection, is a special characteristic of marine ecosystems. No clear and regular food chains, such as herbivore to carnivore, exist in marine ecosystems, although this concept has been abundantly used in the past in describing quantitatively the energy transfer in aquatic systems. The composition of the food of most carnivores changes throughout their life cycle, and the size of the prey in relation to predator plays an important role.

The distribution of species in marine ecosystems depends primarily on the prevailing environmental conditions, especially the quasi-permanent current systems. Sharp gradients and boundaries of environmental conditions in the sea are rare, and ecosystems merge gradually into each other or change continuously, e.g. with depth.

Communities exist in the ecosystem in the sense that organized systems of organisms live together in defined environments. A fundamental definition of community is not possible in marine ecosystems. Living together depends on the mobility of organisms, their sphere of interaction with other organisms and a multitude of environmental factors. An assemblage of organisms can often be only a point in a continuum. Communities are therefore often defined on the basis of recurrent groups, taking into account the abundance of species in relation to other species in the community. This often leads also to the consideration of the dominance of the species and their position, especially in respect to trophic relations in the community.

For the more sessile organisms in the benthos the different communities might be characterized by the most numerous and typical species in them. Such characterization of level sea bottom communities by species composition and quantitative occurrence of them has been reviewed by Thorsen (1957).

The concept of 'key species' has also been applied in marine ecosystems. Key species is usually a larger predator whose influence and/or biomass dominates in a given ecosystem. For example, a starfish as a key species can be a predator on mussels in tidal zone and can reduce the amount and variety of barnacles and mussels in this zone.

The great differences between terrestrial and marine ecosystems, their behaviour and relationship to man, become apparent when comparing the conditions of life for plants on land with aquatic plants (Table 2.2).

The most common way to review the multiplicity in marine ecosystems is to list the classifications of the marine ecosystems by their mode of life:

(1) *Pelagic* ecosystems – ecosystems suspended in the water mass.
     Bioseston – living particulate matter in the water mass.
     (1.1) *Plankton* – small organisms suspended in the water mass; with very limited or no motility.
     (1.2) *Neuston* – organisms living in or on the surface film (of importance in limnology).
     (1.3) *Nekton* – free living, motile organisms in the water mass (e.g. fish ecosystems).
(2) *Benthic* or *demersal* communities – animals and plants living on the bottom or in close association with it.
     (2.1) *Infauna* – most of adult life spent buried in the sediment.
     (2.2) *Epifauna* – fauna on the surface of the sediment, mostly mobile.
     (2.3) *Sessile benthos* – fauna and flora, most of life spent fastened on substrate.

Another way to classify the marine ecosystems is by location and environment: e.g. coastal, neritic and oceanic; polar, boreal and tropical; coral reef, brackish waters.

Table 2.2 Comparisons between land and marine plants.

Subject or condition	Land plants	Marine plants
Taxonomic groups	The higher plants are the main producers of organic matter	The one-celled algae and multi-celled weeds in shallow water are the main producers of organic matter. Few species of submerged higher plants exist in the littoral zone of the sea
Size	Relatively large in size. Supporting structure and water transport limit the size	Small in size (except a few seaweeds). Relative surface area to volume limits the size (re uptake of nutrients).
Support	Must be able to resist force of gravity and provide protection against outer forces (wind, snow, etc.)	Unnecessary because of small size and relatively uniform and dense surrounding medium (water). No root system necessary
Space	Essentially two-dimensional. The surface of the soil is a limiting factor	Three-dimensional. The thickness of photosynthetic layer depends on the amount of suspended matter, affecting the penetration of light into the water
Availability of water	Conservation of water and special vascular system for water transport necessary	Freedom from desiccation
Availability of sunlight	Exposed to direct sunlight. Tolerate relatively high intensities	Absorption of infra-red and ultraviolet radiant energy by water is high. The extinction of light in the water limits the vertical extension of the photosynthetic zone
Availability of gases	From the air	Abundant supply of carbon available in form of CO_2 in water. Calcareous support structures possible. In some places and times 0_2 may be relatively low
Availability of nutrients and means of uptake	Available in soils. Must be transported in solution into different parts of the plant	Available in sea water in small quantities and may become limiting factor. Uptake through surface from the surrounding water
Seasonal variations in availability of nutrient salts	Little or no seasonal changes of nutrient availability. Seasonality of growth	Seasonal variation of some nutrient salts in photosynthetic zone. Seasonal succession of dominant species
Temperature changes and heat capacity	Low heat capacity of the air and relatively large sudden temperature changes	High heat capacity of the water prevents extreme and sudden changes of the temperature
Biological factors	Grazing by animals a minor factor; some local effects significant	Grazing by animals (zooplankton and fish) an important factor, influencing standing crop and production
Cultivation and harvesting	Many plants subject to mass cultivation; most plants can be harvested	No mass cultivation possible. Only few seaweeds are harvested to limited extent
Management	All land plants can be subjected to management	No active management of marine plants possible.

In the following sections we review the essentials of marine ecosystems from an ecological point of view and consider the characteristics with regard to their direct and/or indirect importance to man and the possible effect of man on them.

The book is descriptive by nature, avoiding ecological theory. The review of these ecosystems is done along two traditional lines; firstly in respect of motility, support and specific life form, i.e. plankton, benthos and nekton; and secondly as related to specific environmental features, such as coastal and offshore, demersal and pelagic ecosystems.

Owing to the great variability in species composition of the communities, only a few examples of the species composition and specific variability can be given here. The determinants of the biodiversity of different ecosystems are given in [2.4] whereas the environment and biota interactions are described in [2.5].

2.2   Major components of the marine ecosystems

The different major components of marine ecosystems, characterized by mode of life and nature of functioning, such as plankton, benthos and nekton, are considered as separate ecosystems, although they are related to each other mainly through food relationships. These ecosystems must also be specified by geographic locations and by environmental regimes.

2.2.1   *Plankton*

Plankton in the sea consists of microscopic organisms which are mostly invisible to the naked eye. These have very little or no direct use for man, but they are the basis of life in the sea through the production of organic matter and as a major source of food for other ecosystems. To understand marine ecosystems and their functioning, we review first the plankton ecosystems and processes therein under three headings, phytoplankton, zooplankton and bacteria. We describe their nature and give examples of their composition, variability and production in relation to other ecosystems.

Phytoplankton

The phytoplankters are the producers of organic matter in the sea. Examples are shown, greatly enlarged, in Fig. 2.1. The production of organic matter on the land and in the ocean is, on average, about equal (5 g dry weight/m^2/day). Its distribution on land is two-dimensional, but in the sea three-dimensional.

Average plankton (phyto- and zooplankton) contains about 85% water. Of the dry matter about 30% is ash and 50% carbon. These values vary considerably from one type of plankton to another.

The size of phytoplankton can be given as weight in mg of 1 million cells. Average diatoms can be 30 mg and larger (e.g. *Coscinodiscus* sp.) up to 150 mg. Average nude flagellates weigh only 0.02 mg.

The composition of phytoplankton varies in space and time. Normally diatoms

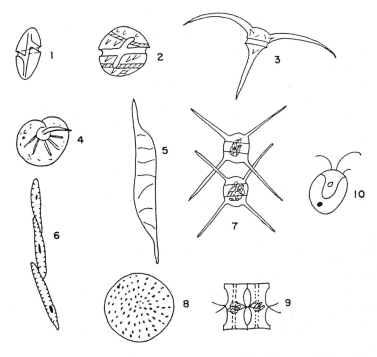

Fig. 2.1 Examples of some phytoplankton organisms (enlarged 100–400 times). (1) *Gyrodinium* sp.; (2) *Peridinium* sp.; (3) *Ceratium* sp.; (4) *Noctiluca* sp.; (5) *Rhizosolenia* sp.; (6) *Nitzschia* sp.; (7) *Chaetoceros* sp.; (8) *Coscinodiscus* sp.; (9) *Biddulphia* sp.; (10) *Chlamydomonas* sp.

dominate in high latitudes during the spring, whereas flagellates take over in late summer (Fig. 2.2). Microflagellates can also dominate phytoplankton in higher latitudes during the winter and early spring. Small organisms, such as flagellates, have relatively high metabolic rates. Therefore small populations (i.e. standing crop) can have high production.

In coastal areas and on the continental shelf neritic forms dominate (e.g. *Fragilaria* and *Hyalochaete*). The standing crop varies from 1000 to 84 300 cells per m^3 and is larger in neritic populations than in oceanic populations.

Some phytoplanktons are cosmopolitan (e.g. *Ceratium tripos* and *Chaetoceros atlanticus*), whereas some occur only in defined oceanographic conditions. Karohji (1972) found considerable differences in dominant phytoplankton species in different areas in the North Pacific. *Chaetoceros concavicornis* and *C. convolutus* were dominant in the western subarctic gyre and in the northern part of subarctic current, whereas south of the eastern Aleutian Islands *Denticula seminae* was dominant. In the Alaskan gyre *Nitzschia seriata* and *Phaeoceros* sp. were dominant. Karohji also found considerable differences in occurrence and distribution of species between two adjacent years. Furthermore, it was noted that the catch composition of plankton species varies with different methods of sampling, such as surface and vertical hauls with plankton nets of differing size.

A sub-surface chlorophyll maximum is a rather common phenomenon in the open

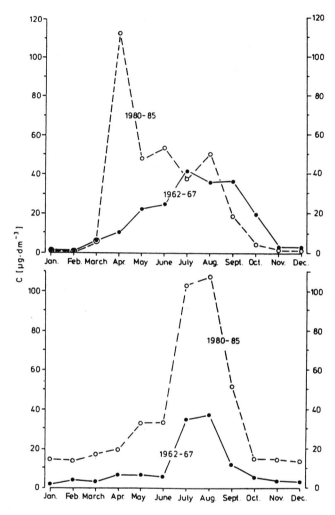

Fig. 2.2 Monthly means of (a) diatoms and (b) flagellates at Helgoland Roads (North Sea) in 1962–7 and 1980–5 (Gillbricht 1988).

ocean in low and middle latitudes. This is caused by photosynthetically active phytoplankton adapted to lower light intensities and living in depths where nutrients are not limiting the production (Anderson 1969) (Fig. 2.3). An example of the regional variation of phytoplankton standing crops in the surface layer in the North Pacific is given in Table 2.3.

Production of organic matter takes place during photosynthesis by phytoplankton, chemosynthesis being of less importance in the sea. Basic organic production or net primary production, i.e. photosynthesis minus respiration, is usually reported as organic production. The experimental measurement methods of basic organic production are not perfect, and large variations in space and time occur. The main methods of measurement are oxygen production and carbon-14 methods.

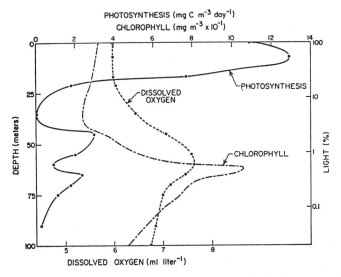

Fig. 2.3 Vertical distributions of chlorophyll a, photosynthesis, and dissolved oxygen at 45° 10′ N lat., 126° 56′ W long., 26 July 1968. The right ordinate represents depths to which specified amounts of subsurface light penetrate, expressed as percentage of light at the sea surface (Anderson 1969).

Table 2.3 Examples of the average standing crops of phytoplankton populations in North Pacific (Karohji 1972).

Area	Standing crops ($\times 10^3$ cell numbers/m^3)	Ratio to smallest observed crop
Western Aleutian waters	9800	200
Western Bering Sea	342	7
Eastern Bering Sea	48	1
Northern Bering Sea	8900	200
Alaskan coastal area	6800	140
Gulf of Alaska	200	4

The magnitude of the phytoplanktonic community and the rate of the photosynthetic process are determined by many factors. The limiting factors of the photosynthetic process can be classified as those factors which determine the available quantity of the materials and energy essential to the process, and controlling factors, those which determine the rate at which the process moves. The density of the plankton may increase to the point where it reduces light penetration, and the population itself then becomes a limiting factor.

The limiting factors of basic organic production in the sea are those that determine (1) the quantity and quality of transmitted light, (2) the quantity and kinds of nutrient salts and trace elements available to phytoplankton, (3) the quantity of CO_2, and (4) the magnitude of the phytoplankton populations and the level of their reproductive and metabolic processes. Thus the determinants for these limiting factors are:

(1) Determinants of light:
  (a) meteorological (e.g., intensity of cloudiness, wind, determining the sea-surface conditions and the penetration of light);
  (b) Hydrographical (turbidity): determining the fate of the light in the water.
(2) Determinants of nutrient salts and trace elements:
  (a) physical (currents and turbulent mixing transporting nutrients into euphotic layer);
  (b) chemical (solubility of a given element in sea water and the reactions with other elements affecting the removal);
  (c) biological (regeneration of organic matter releasing nutrients).
(3) Determinants of CO_2:
  (a) physical (mainly temperature – determining solubility of CO_2);
  (b) chemical (equilibrium in $CO_2 - H_2CO_3 - HCO_3 - CO_3$ system in sea water);
  (c) biological (consumption and production of carbon dioxide CO_2 and removal of carbonate ion CO_3^{--} by biological activity).
(4) Determinants of population size and physiology:
  (a) previous history of the population and its characteristics;
  (b) current biotic factors (grazing – determining survival, ectocrines affecting the physiological processes);
  (c) bacterial activity and diseases;
  (d) physical factors (currents, turbulence and mixing, transporting the population and dispersing it);
  (e) chemical factors (salinity, determining the osmotic pressure and density of the water – which affects buoyancy – and determining the species composition of population).

Available literature on phytoplankton contains numerous data on phytoplankton production measurements. These data are very variable in space and time. For example, McRoy *et al.* (1972) found integrated productivity of the southern Bering Sea to be 243 mgC/m²/day during the summer and 21 mgC/m²/day during the winter. The variation between the individual measurements was about 80 fold.

A summary of basic organic production in the world oceans is given in Fig. 2.4 (Hela & Laevastu 1962). Figure 2.5 shows the phytoplankton standing stock observed as colour of the sea by the National Aeronautics and Space Administration (NASA) for two different seasons (1978–81).

A special characteristic of phytoplankton production is the occurrence of 'blooms', i.e. intensive developments. The phytoplankton bloom, mainly of diatoms, in medium and higher latitudes usually starts in late March or early April in coastal areas and later moves progressively further offshore. The type of the phytoplankton also changes with time and distance from the coast, usually starting with small diatoms followed by medium size diatoms. In offshore waters flagellates and dinoflagellates usually dominate during the summer.

A usually harmless dinoflagellate, *Ceratium furca*, can also cause 'red tide' blooms. Walker & Pitcher (1991) described such a bloom, caused by accumulation of a slow-

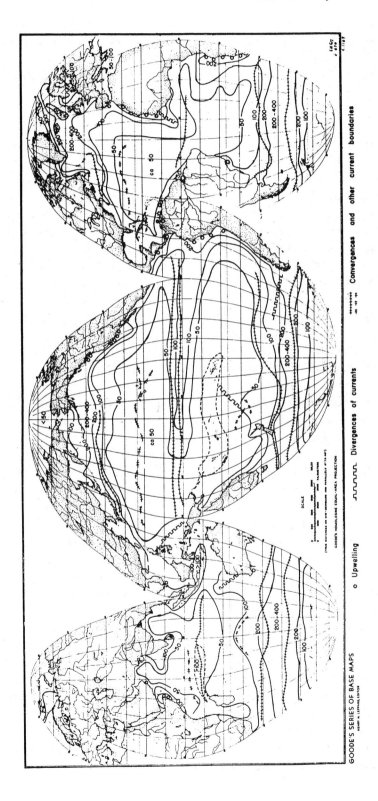

Fig. 2.4 Estimation of basic organic production in the oceans (gC/m²/year).

Fig. 2.5 Chlorophyll concentrations in surface layers for 1979 (January–March and July–September) (C.J. Tucker, NASA/GSFC).

growing ungrazed population subjected to concentration by physical entrainment and vertical migration.

Phytoplankton can produce certain antibiotics during intensive growth (Lucas 1947). These antibiotics prevent attack by bacteria and affect the grazing by zooplankton. It is known that zooplankton migrates into the patches of phytoplankton during the night to graze, and leaves the patches during daylight hours. This migration might be caused in part by the daytime production of antibiotics by phytoplankton. Furthermore, it is a kind of escape and production for the phytoplankton standing crop, because the crop is less grazed during the daytime assimilation period, when most biomass increase is being effected. Lucas (1947) has reviewed the evidence on the role of external metaboliates showing:

(1) The tendency for organisms to release metabolites into the external medium, including physiologically potent metabolites (ectocrines):

(2) The adaptation of 'successful' organisms to such metabolites in their environment, so that there is,

(3) The possibility that some elements of the community might appear to benefit from vitamins or be harmed by antibiotics.

Zooplankton

Zooplankton is of slightly more interest to man than phytoplankton as some of it (e.g. krill) is harvested directly, and it serves as direct food for fish, marine mammals and birds. Examples of typical zooplankton organisms are given in Fig. 2.6.

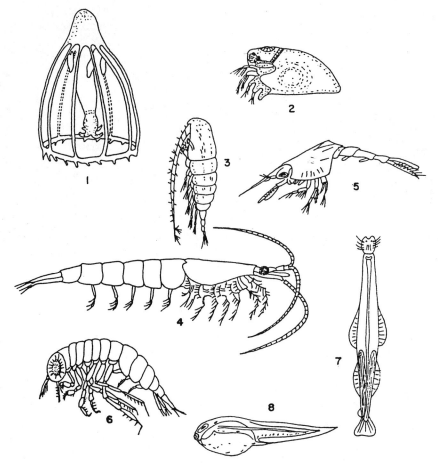

Fig. 2.6 Some of the most common zooplankton organisms (enlarged 10–100 times).
(1) *Aglantha digitalis* (O.F. Muller) (Hydrozoa); (2) *Evadne nordmanni* (Loven) (Cladocera); (3) *Calanus finmarchicus* (Gunner) (Copepoda); (4) *Meganyctiphanes norvegica* (M. Sars) (Euphausiacea); (5) *Eupagurus bernhardus* (L) (Decapoda) (larva); (6) *Thermisto abyssorcum* (Boeck) (Amphipoda); (7) *Sagitta elegans* (Verrill) (Chaetognatha); (8) *Pleuronectes flesus* (L) (Pisces) (larva)

Taxonomic identification of planktonic and benthoic organisms has been a difficult and time-consuming task in the past. It has, however, now been made easier with the use of computerized taxonomic keys (Estep 1989) (Fig. 2.7).

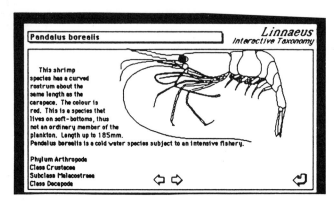

Fig. 2.7   The marine shrimp *Pandalus borealis* (Estep 1989).

The species composition of zooplankton communities is also variable in space and time. This is not surprising considering the great number of species; e.g. in the Bering Sea over 340 species, of which over 100 are copepods (Cooney 1981). Furthermore, the size of the zooplankton species and the mesh size of the nets for the collection of zooplankton varies, normally between 0.1 and 0.6 mm, which results in different observed pictures of the zooplankton community.

Large zooplankters can be about 20 mm long (e.g. *Euphausia pacifica*) and weigh about 40 mg. Average copepods are about 5 mm and weigh 5 mg. Euphausiids, which are important food items for other marine ecosystems, often dominate in deeper parts of the continental shelf and in offshore waters.

In medium and higher latitudes the average number of zooplankters under 1 m^2 of sea surface can range from 15 000 to 40 000 over the continental shelf and from 4000 to 10 000 offshore. An example of the standing stock of zooplankton at a Norwegian coastal location is shown in Fig. 2.8, whereas Fig. 2.9 illustrates the seasonal changes of species composition in an offshore location in Gulf of Alaska. There is usually a gradual change of zooplankton communities from one water mass to another, and so some zooplankton organisms have been used as indicators of water masses (Table 2.4).

Many zooplankton organisms make vertical migrations, usually towards the surface during the night. Morioka (1972) described pronounced diurnal vertical migrations of *Metridia lucens* between 400 m and surface. He furthermore divided the vertical distributions of zooplankton into three types, epipelagic, mesopelagic and bathypelagic. However, the division is not always consistent, as depth distribution and vertical migrations of zooplankton organisms vary considerably seasonally and spatially. Spawning in copepods usually starts earlier in lower than in higher latitudes and in neritic (coastal) than in oceanic areas.

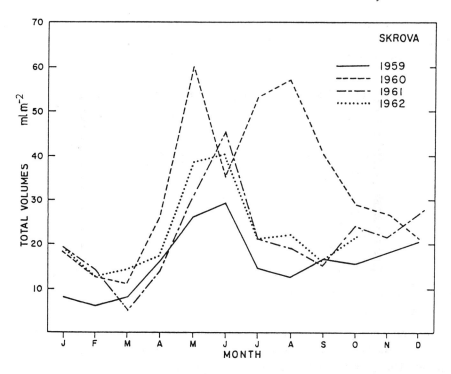

Fig. 2.8   Monthly mean total volumes in ml/m² of zooplankton at Skrova, Norway, 1959–62 (Lie 1965).

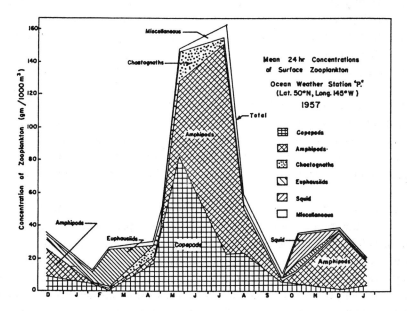

Fig. 2.9   Concentration and composition of surface zooplankton at station 'p' in the central Gulf of Alaska (McAllister 1961).

Table 2.4 The two groups of zooplankton species distinguished in the Canadian samples during NORWESTLANT surveys: (a) cold water forms; (b) warm water forms. (Bainbridge & Corlett 1968).

(a) Davis Strait and off SW Greenland	(b) North Atlantic
Copepoda	**Copepoda**
Microcalanus pygmaeus	*Scolecithricella ovata*
Euchaeta glacialis	*Euchaeta norvegica*
Acartia longiremis	*Heterorhabdus norvegicus*
	Eucalanus elongatus
Amphipoda	*Rhincalanus nastutus*
	Euchirella rostrata
Apherusa glacialis	*Pseudaetideus armatus*
Gammarus wilkitzki	*Metridia lucens*
Parathemisto abyssorum	
Parathemisto libellula	**Amphipoda**
	Lanceola clausi
Ostracoda	
Conchoecia borealis maxima	**Ostracoda**
	Conchoecia obtusata
Chaetognatha	*Conchoecia elegans*
Sagitta elegans	
	Euphausiacea
Pteropoda	*Thysanoessa longicaudata*
Spiratella helicina	
	Chaetognatha
	Sagitta maxima

The neuston layer (0–25 cm deep) can, during calm weather, contain more planktonic organisms than layers below (e.g. 100 000 specimens in $1m^3$ of water in the Bering Sea). Tintinnides, pteropods, eggs and nauplii of copepods, and some neustophile copepods such as *Oithona similis* usually dominate in the neuston. During calm weather some diatoms might also rise to the surface, covering it so that the sea surface looks oily.

The spatial distribution of zooplankton can vary considerably from year to year (Fig. 2.10) (Sherman & Jones 1980). The difference observed in the literature in the amount of zooplankton present in the sea is often caused by the differences in sampling methods used (e.g. mesh size of nets and depth of towing). Coarser zooplankton nets can miss the small metazoans, whose biomass can at times be more than ten times the biomass of copepods and other net plankton.

Examples of the relations between standing stock of phyto and zooplankton have been summarized in Table 2.5. Primary production measurements in the sea are made empirically, whereas all estimates of animal production are based on measurements of standing stocks combined with several theoretical assumptions, such as turnover rate, growth, consumption and mortality rates.

The production of zooplankton is not always proportional to the production of phytoplankton, for a number of reasons such as time and space shifts in maximum

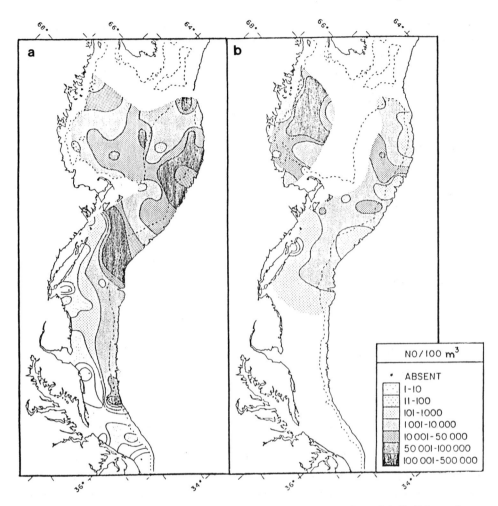

Fig. 2.10 Spatial trends in abundance of *Calanus finmarchicus* on the continental shelf off the north-east United States coast in April–May 1978 and March–April 1979 (Sherman & Jones 1980).

Table 2.5 Quantitative relation between phytoplankton and zooplankton (from various sources).

Locality	Time	Biomass in mg/m³		Phytoplankton
		Phytoplankton	Zooplankton	Zooplankton ratio
Arctic Seas	Biological winter	410	52	8
	Biological spring	2470	122	20
	Biological summer	560	230	2.5
Barents Sea	Summer	500–600	140	4
SW North Sea	February–March	27	241	0.1
Off Plymouth, UK	Yearly average	115–135	37–80	1.7–3.1
Coastal waters (gen.)	Yearly average	500–2000	100–800	2.5–5
Sargasso Sea	Yearly average	300	50	6

production of a low rate of grazing by zooplankton (Heinrich 1962a, Jones 1984). Only about 5–20% of phytoplankton production is utilized as food by zooplankton.

The zooplankton biomass in the southern Bering Sea is reported by various authors (e.g. Cooney (1981)) to be on average about 20–50 g wet weight per m^2, being lower over deep water than on the shelf and highest on the continental slope. The continental slope produces about 33 to 64 g C/m^2/year of zooplankton whereas the oceanic production is about 13 g C/m^2/year. The transfer rate from phyto- to zooplankton is on average about 15%, but variable in space, being lower in offshore and coastal areas than on the continental slope.

The production of copepods in the central and western Bering Sea is about 115 g wet weight/m^2, whereas in the northern Bering Sea it is about 10 g only. Productivity of net plankton and nannoplankton is about equal, but the nannoplankton can dominate in some areas and net plankton in another. The quantitative studies of nannoplankton are, however, limited.

Zooplankton does not only serve as food for fish and fish larvae (Fig. 2.11), but some zooplankters also prey on fish eggs and larvae. The probability of predation on fish eggs by carnivorous zooplankton is greater over deep water than over shallow water owing to the presence of a greater number of predators offshore (Bainbridge & Corlett 1968, Table 2.6).

The early larval stages of most fish feed on plankton and organic detritus derived from it, and thus the abundance of planktonic food is essential to the good condition of fish larvae. Some fish larvae, e.g. of plaice, start to feed on phytoplankton and then switch to zooplankton, especially on appendicularians. The positive correlation between the density of plankton and the density of the plankton-feeding fish population has been recorded in numerous places. For example, Einarson (1955) established a positive correlation between the amount of plankton present in the water and the catch of herring in the North Atlantic. This correlation does not, however, hold in many cases as both plankton and fish distributions are patchy and a negative correlation could also mean that fish might have grazed down the zooplankton patch.

Bacteria and heterotrophs

The basic biogeochemical function of the microflora in the sea is the oxidation of organic matter and creation of their own biomass. In this process oxygen is consumed and carbon dioxide liberated. Small flagellates consume bacteria and viruses, the latter being essentially organic particles in the colloidal size range (Gonzalez & Suttle 1993). It could be noted that 10–25% of the extracellular products in the form of soluble carbohydrates such as glycollic acid are lost from phytoplankton. These products can be utilized by bacteria.

Microflora, or bacterioplankton, consists of heterotrophic bacteria and fungi $< 10\,\mu m$ in size. In eutrophic regions its standing crop can be 0.2–2 g/m^3, in offshore waters in the temperate zone it is between 0.05 and 0.5 g/m^3 and in oligotrophic regions 5–50 mg/m^3. The production to biomass ratio varies from 0.3 to 1 (Sorokin

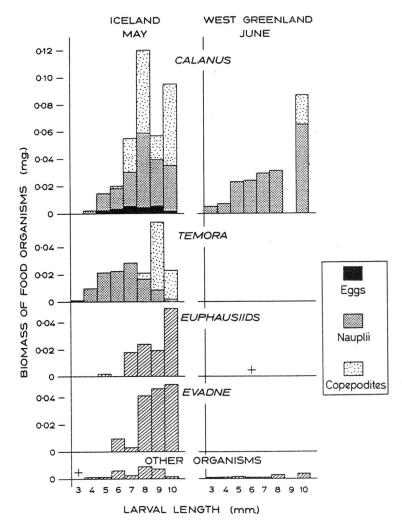

Fig. 2.11 The diet of cod larvae from the Faxa Bay area, Iceland, and in the Davis Strait off West Greenland during May and June (NORWESTLANT 2). The mean biomass of the various food organisms is shown for larvae of each length group from 3 to 10 mm. Food items have been separated into five categories with the histograms for two main copepod species subdivided into eggs, nauplii, and copepodites. The euphausiids present were nauplius and early calyptopis stages (Bainbridge & McKay 1968).

1977a). The bacterial biomass decreases rapidly in deep water below 200 m deep; there can, however, be a maximum below a permanent thermocline.

The total bacterial biomass in the sea can range from 1.2 to 100% of particulate organic carbon. Bacterial production can be of the same order of magnitude as phytoplankton production. Bacteria decompose organic matter, and without this decomposition and nutrient recycling there would be no plant or higher animal production.

Tsyban *et al.* (1992) found in the Bering Sea during the summer of 1988 671×10^3 bacteria cells per ml or 15.1 mg C/m^3. This was about twice as high a cell count as in

Table 2.6 Average number of predators per 30 min oblique haul (50–0m) with the 2 m stramin net off West Greenland during NORTHWESTLANT surveys. The averages refer to shallow (< 300 m) and deep (> 300 m) stations with cod larvae (Brainbridge & Corlett 1968).

Predators on cod larvae	Shallow	Deep
Aglantha	49.4	187.8
Other medusae	17.2	3.5
Siphonophora	2.3	5.6
Chaetognatha	10.0	1560.4
Tomopteris	+	51.8
Clione	2.7	3.0
Cephalopods	0.3	1.4
Euchaeta	0.9	248.6
Hyperiids	18.6	21.7

1981 and 1984. The Barents Sea bacteria population has been found to be between 10×10^3 and 500×10^3 cells per ml. The bacteria population decreases from surface to bottom.

Bacteria form clumps in the sea, which become available as auxiliary food source for microzooplankton (Seki 1970). Microflora is consumed by many filter feeding microzooplankters, especially by nauplii but also by bivalves and polychaetes on the sea bed. An important intermediate trophic link in the utilization of bacterioplankton are protozoa'ciliates and colourless flagellates. Protozoans consume bacteria and themselves serve as food for zooplankters. The biomass of protoza may reach 1 g/m^3 (Mamaeva 1992).

Most of the inorganic nutrients, such as phosphorus and nitrogen, that disappear from surface layers in springtime are not necessarily incorporated into organisms and/ or detritus, but are transformed into soluble organic substances (Gillbricht 1988).

Numerous investigators studying the culture of marine algae have demonstrated the influence of bacteria on planktonic algae. It has been shown that marine bacteria can settle on living phytoplankton organisms, especially at low light intensities when the phytoplankton are not actively photosynthesizing and are then not producing antibiotics; this direct settlement can kill phytoplankton. Furthermore, it is known that a large bacterial population can compete with phytoplankton for the uptake of phosphate.

2.2.2   *Benthos*

The benthic ecosystem consists of animals and plants that spend most of their life on or in the bottom substrata. This ecosystem is often divided into two subsystems:

- *Infauna* or endofauna – animals which are buried or buried in the bottom substrata during most of their adult life.
- *Epifauna* – animals when occur entirely above the bottom surface or have only their organ for anchoring sunk in it.

Two additional terms are often used to describe benthos:

- *Sessile benthos* – animals which are attached to the bottom and cannot move about.
- *Demersal* fish – fish which spend most of their post-larval life close to the bottom or on the bottom substrata and feed mainly on benthic animals.

Less than half of the world's commercial catch consists of benthic animals and demersal fish. Consequently, the magnitude and composition of the benthos is a factor affecting the standing crop and production of human marine food resources. Furthermore, most pelagic fish species are restricted to coastal waters during some periods of their life, especially during spawning, and the conditions of the bottom may have an important influence on this phase of the life of the pelagic species. Life in the neritic environment is influenced by benthos. The pelagic larvae of benthic organisms occur in the neritic environment in great quantities and serve as food for young demersal fish and small pelagic fish.

Earlier investigations of the benthos, especially those conducted in the first part of the twentieth century, were mainly concerned with the taxonomy and zoogeographic distribution of benthic communities. C.G.J. Petersen and co-workers started the first large-scale quantitative investigations on the marine benthos at the beginning of the twentieth century. They formulated quantitative methods for the study and classification of benthic communities and showed close connections between the characteristics and abundance of the benthos and of commercial fisheries. C.G.J. Petersen based his evaluation of the demersal fish resources of the sea (Petersen 1918) on estimates of benthos biomass. In this section we review the factors known to be important in influencing the abundance and composition of the benthos, emphasizing the quantitative approach.

The zone under the influence of the tides, is called the *intertidal zone*. It may be important for commercial shellfish production and it might also be locally important for subsistence fisheries (mainly sedentary fisheries). The *eulittoral zone* ranges from the low tide mark down to a depth of about 40–60 m (the average depth at which sessile seaweeds are found). This zone is considered important as a nursery ground for many young fish. The basic organic production on a suitable bottom in this zone, where attached vegetation exists, is generally high. The standing crop of *Laminaria*, and *Fucus* may reach $10 \, kg/m^2$ and more. Some of the organic matter produced in the eulittoral zone is carried by currents into deeper water, into the elittoral (sublittoral) where it is used as detrital food by benthic animals living here. This is an example of the continuous interactions that occur between different zones in the sea.

The elittoral (sublittoral) zone extends down to a depth of about 200 m, which is the average depth of the edge of the continental slope. The feeding grounds of most demersal fish and the spawning grounds of pelagic fish are within this zone and the main commercial offshore fisheries exist here. Below the elittoral, down to about 1000 m, is the archibenthic zone and the deepest zone below it is abyssal-benthic. Although the standing crop of the benthos in the upper part of the archibenthic zone on some continental slopes might have a sufficient biomass to serve as food for demersal fish, in general this deeper zone has little importance for commercial demersal fisheries.

Some of the general principles used for estimating the influence of environmental factors affecting the benthic biomass and its production are as follows:

- Temperature:
  (1) Biological turnover rate increases with increasing temperature. Therefore, with the same quantity of organic matter produced per unit time and unit area of bottom, the benthic standing crop is lower in high temperature areas than it is in low temperature areas, conditions being otherwise equal.
  (2) The standing crop is relatively greater when the temperature is low, especially below 0°C, because longer-lived individuals and populations lead to successive accumulation of generations and also because of decreased grazing at lower temperatures.
- Depth of water:
  (1) Standing crop decreases with increasing depth, especially below the euphotic zone.
  (2) Epifaunal communities are dominant in shallow water; infaunal communities are dominant in deeper water.
- Distance from the coast and configuration of the bottom:
  (1) The standing crop of benthos decreases with increasing distance from the coast.
  (2) Offshore ridges, banks and submarine slopes have higher standing crops than the surrounding deeper areas.
- Bottom substrata
  (1) The epifauna and flora dominate on hard bottoms.
  (2) The infauna increases with increasing content of organic matter in the sediments until it reaches an optimal value at a carbon content of about 3% of the sediment and then it decreases as the organic matter content increases further.

It is possible to group benthic fauna by sediment type and depth. It is, however, not always clear how the sediment type determines the faunal group. Furthermore, faunal discontinuities can occur over uniform sediments. Benthic biomass on a similar bottom type is correlated to food supply, mostly in the form of detritus. The available food supply varies with type of bottom, which is often a function of currents near the bottom and depth of the water. Intense movement of water close to the bottom increases the benthos, especially epifauna. Furthermore, the standing crop of benthos decreases with decreasing content of dissolved oxygen in the bottom water.

Benthos contains a great number of species, some of which are harvested by man (Fig. 2.12). Stocker (1981) found 472 species of benthos in the northern Bering Sea, encompassing 292 genera and 16 phyla. However, only 11 species accounted for 50% of both density and biomass and 37 species accounted for 75%. The species index of diversity varied from 0.09 to 1.47 (i.e. over one order of magnitude). He used cluster

Fig. 2.12 Examples of some benthic animals of economic importance (mussels, octopuses, clams, lobsters, shrimps and crabs, after Rass 1979).

analysis, dividing the communities into eight cluster groups, which would well correspond to community classification of Petersen.

The classification of benthos into communities and into infauna and epifauna used to be common practice in benthos research. This method originated with Petersen & Jensen (1911) and their co-workers and students (e.g. Spärck 1935). Examples of the benthos communities are shown in Fig. 2.13 and some examples of their quantitative and qualitative composition are given in Tables 2.7 and 2.8. The data in the tables come mainly from investigations in the North Atlantic and North Pacific oceans. Knowledge of the benthos in tropical waters is limited. Available data from the other oceans indicate that the conditions of the life and the corresponding communities are very similar in different areas of the world.

Thorsen (1955) writes: 'The same types of level bottom substrata at similar depths are quantitatively dominated by invertebrates belonging to the same genera but to different species.' Longhurst (1957) confirmed this theory in investigations of benthos in the tropics, showing the presence of communities which are the counterparts of those occurring at similar deposits in colder areas. The benthos of the intertidal zone and on tidal flats has special characteristics. It is often dominated by a few bivalves such as *Mya arenaria, Macoma balthica* and *Mytilus edulis* and the lugworm *Arenicola.*

The size of the standing crop of the benthos in a given location is not stable but varies considerably from season to season and from year to year. Changes in some environmental factors lead to changes in the standing crops in a rather short time;

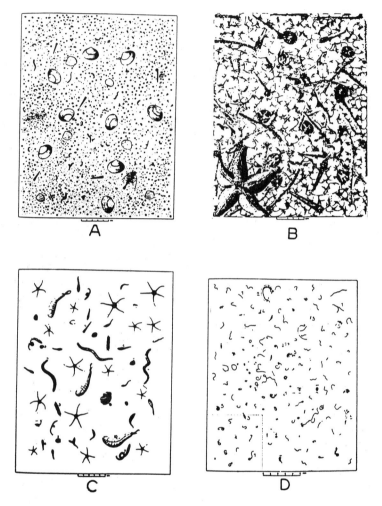

Fig. 2.13 Relative density and taxonomic composition of some benthos communities (0.1 m) (after Hagmeier 1951). (a) *Macoma baltica* community 270 g/m^2; (b) *Spisula subtruncata* community 5258 g/m^2; (c) *Venus gallina* community 191 g/m^2; and (d) Meiobenthos from *Venus gallina* community.

changes in others influence the standing crop of the next generation (e.g. severe winters, heavy storms that shift the bottom, prevailing oceanographical conditions during the larval stage and at settlement and suitable food in the plankton).

The number of different species groups (taxonomic groups) in benthos infauna can also vary considerably. Off the coast of Washington (North-east Pacific) Lie (1969) found by grouping a number of stations that the biomass comprised on average 20–64% polychaetes, 5–31% crustacea, 10–51% molluscs and 7–20% echinoderms. The average benthos biomass was about 22 g/m^2.

Stocker (1981) found about 300 g/m^2 of benthos in the northern Bering Sea, which is comparable in value to other seas in high latitudes. The benthos was dominated by benthos feeders. There is a considerable fishery for crabs (e.g. king crabs *Paralithodes*

Table 2.7   Examples of benthos communities.

Community (synonyms and variants)	Type of bottom	Depth of water m	Biomass g/m^2 and remarks	Subdominant species
Macoma baltica (*Tellina baltica*)	Sandy (muddy to hard)	1–100 (normally < 50)	100–300 (max 2 kg) Good 'fish food' community	Molluscs: Sandy bottom-*Cardium* sp; hard bottom-*Mya* sp;, *Tellina* sp.; muddy bottom-*Scobicularia* sp. Annelids: *Arenicola* sp.; *Ophelia* sp.; *Scoloplos* sp. *Nereis* sp.
Spisula elliptica (*Mactra elliptica*, S. *subtruncate* var.)	Sandy	Shallow water	10–350 (max 1.2 kg) Good 'fish food', variable	Molluscs: *Abra* sp. *Venus* sp., *Tellina* sp., *Cyprina* sp. Annelids: *Ophelia* sp., *Glycera* sp., Echinoderms: *Ophiura* sp., *Bisaster* sp.
Venus gallina (*Nucula* var., *Echinocardium* var.)	Soft bottom	Deeper water	Up to 500 Polychaetes and molluscs equal amounts	Molluscs: *Mactra* sp., *Spisula* sp. *Nucula* sp. Annelids: *Nephtys* sp., *Owenia* sp. Echinoderms and others: *Echinocardium* sp., *Bathyporeia* sp., *Ophiura* sp.

Table 2.8   Major benthic invertebrate groups reported from 1296 National Marine Fisheries Service exploratory fishing drags in the Western Gulf of Alasks (from National Marine Fisheries Service, undated).

Group	Occurrence in trawl drags	Maximum catch (kg/h)
Sea cucumbers	common	1360
Starfish	common	815
Sea urchins	common	755
Sea anemone	common	180
Basketstars	common	180
Sea pens	common	90
Mussels	occasional	725
Brittlestars	occasional	300
Crabs	occasional	25
Hermit crabs	occasional	30
Barnacles	occasional	35
Sponge	seldom	225
Whelks	seldom	15
Sand dollars	seldom	225
Clams	rare	90
Horse crab	rare	Trace
Sea fan	rare	Trace
Octopus	rare	Trace
Box crab	rare	Trace
Worms	rare	Trace

spp and snow crabs *Chionoecetes* spp) in the Bering Sea. The epifaunal biomass of the large organisms was about 3 g/m², of which *Asterias* spp and other sea stars made up about one-half in shallower water (< 40 m deep).

The amount of epibenthic invertebrates in the Gulf of Alaska is about 2.6 g/m², similar to the amount in the Bering Sea. Molluscs, crustaceans and echinoderms also dominate in the Gulf of Alaska (Jewett & Feder 1976), where snow crab (*Chionoecetes bairdi*) contributed 66% of the biomass (Table 2.9).

Table 2.9 Percentage composition by weight of leading invertebrate species collected during north-east Gulf of Alaska (NEGOA) trawling investigations, summer 1975 (Jewett & Feder 1976).

Phyla	Percentage of weight	Leading species	Average weight per individual (g)	Percentage weight within phylum	Percentage weight from all phyla
Arthropoda	71.4	*Chionoecetes bairdi*	454	92.6	66.2
		Pandalus borealis	8	4.0	2.9
		Lopholithodes foraminatus	420	0.6	0.4
				Total 97.2	69.5
Echinodermata	19.0	*Ophiura sarsi*	6	23.2	4.4
		Ctenodiscus crispatus	10	15.7	2.9
		Brisaster townsendi	10	11.2	2.1
		Pycnopodia helianthoides	482	10.3	2.0
				Total 60.4	11.4
Mollusca	4.6	*Pecten caurinus*	350	43.4	2.0
		Neptunea lyrata	180	12.5	0.6
		Fusitriton oregonensis	100	11.5	0.5
Total	95.0			Total 67.4	3.1

Some long-term changes in the benthos in the Arctic off Spitsbergen have been documented by Blacker (1965). He classified some benthic organisms as boreal and some as Arctic (the indicator species). Figure 2.14 shows the distribution of these indicator species in two time periods, 1878–1931 and 1949–59. Nothing can be said from his example about the speed of the change in this ecosystem. The quantitative estimates of benthic energy budgets are rough estimates. We do not know the turnover rate. The cross shelf exchange is also greatly variable spatially.

Many, if not most, of the larvae and juvenile stages of benthic organisms are pelagic. Most of the larvae pass through metamorphosis before settlement to the bottom. The recruitment of shellfish (clams) through settlement of larvae can vary greatly from year to year. Franz (1976) found that the surf clam stock off Long Island were age 9 years or older but mostly composed of age classes 1–3. The clam stocks off Rockway Point were younger. He concluded that the commercial fishery is dependent on massive settlements of larvae which occur irregularly and infrequently.

Most of the population in the benthos are detritus feeders, and are, therefore,

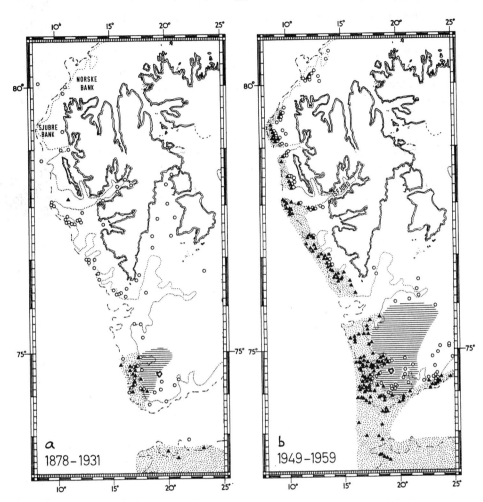

Fig. 2.14 Summary charts of the distribution of Atlantic benthic species (▲) and Arctic benthic species (○). (a) All past records, 1878–1931 and (b) present records, 1949–59. The stippled area indicates the bottom where Atlantic conditions predominate, as deduced from the occurrence of Atlantic species. Comparison of a and b shows how much the Atlantic influence has increased since 1931. The hatched area covers places where conditions may vary from Atlantic to extreme Arctic; similar areas may occur in shallow water off Spitsbergen (Blacker 1965).

largely dependent on the amount of organic matter present in the sediment. Idelson (1934) found a positive correlation in the decrease of bottom fauna with the decrease of organic matter in the sediments. Gorschkova (1938) also found a positive correlation between infauna and organic matter but a negative correlation between epifauna and organic matter in the sediments. In general, the population density is highest on a bottom which contains a medium amount of organic matter.

The amount of microflora (bacteria) in sediment varies with the amount of organic matter, which is usually also a function of depth.

According to Bader (1954) the lamellibranch population increases with an increase of organic matter in the sediment until bacterial decomposition of the organic material becomes the major limiting factor, when the population decreases. Bader also found the greatest lamellibranch densities on bottoms with particles of mean diameter 0.5 mm and an organic carbon content of 3%.

Because molluscs are the dominant group of benthic animals, the availability of calcium carbonate for their shells is a factor in the economy of these communities. The shells constitute on the average about 40% of the total benthos biomass in some areas.

Grazing by fish and other predators influences the standing crop of benthos to a large extent. Jensen (1919) estimated, for example, the standing crops of plaice food in the benthos in Thisted Breding to be $22.3–60.8 \, g/m^2$. The annual consumption of the same benthos was estimated at $31.8–84.3 \, g/m^2$. This means that the standing crop of benthos reproduced itself roughly once a year. Besides consumption by fish the benthos can be consumed by other predatory organisms contained in the benthos itself. Petersen & Jensen (1911) estimate that the whelks and starfish consume about twice as much benthos as do the demersal fish. On the other hand, Alton (1973) estimated that only about 17% of benthos in the Bering Sea is accessible as food for commercial demersal fish. Low temperatures in the northern part of this sea prevent commercial fish from feeding on benthos there.

Cramer & Daan (1986) compared the benthos consumption as obtained from cod and haddock stomach sampling in the North Sea with data on benthos production. They found that the consumption figures were too high in relation to benthos production. They concluded that the problem lies in incomplete quantitative sampling of benthos.

It is known that predators can be disastrous to the standing crops of some benthic species in shallow waters. For example, the grazing on *Mya arenaria* by *Simulus polyphemus* may cause sudden changes in the population of *Mya*. Little is known about the influence of predators on the standing crop of benthos in deeper water.

Parasites and diseases may in certain cases be as important as predators. The disastrous influence of the disease of eel grass in Danish waters is one of the extreme examples. The activity of boring snails (e.g. *Polynices duplicata*) on reducing the population of lamellibranchs in some areas is also well known.

Although the benthos is as important for fisheries as plankton are, relatively little attention has yet been given to it. Clearly the growth characteristics of a population of any particular area, in any particular period, is determined in large part by the availability of food. Quantitative study of benthic communities is therefore urgently required as part of the programme of fisheries biology to promote the evolution of carrying capacity of different areas in respect to demersal species.

2.2.3 *Nekton*

Nekton consists of organisms in the water mass which have the ability to swim and migrate. Nekton consists mainly of fishes and squids, although krill, a large zoo-plankter, is sometimes counted as a micronekton organism (organisms 1–20 cm long). In this book we consider nekton to be fish ecosystems.

Fish are divided by their way of life into pelagic and demersal. Pelagic fish (e.g. herring) spend most of their life in the water mass and use pelagic food; demersal fish (e.g. flatfishes) spend considerable time of their adult life near bottom and obtain much of their food from benthos. This division, though widely in use, is not strictly applicable because some species can be included into both categories. Furthermore, the mode of life of some species changes during their life cycle, being pelagic before maturation and changing thereafter to more demersal life (e.g. hake). Therefore, the terms semi-pelagic and semi-demersal are also used.

The nektonic (or fish) ecosystems are connected to other marine ecosystems through the food chain and also through their motility and migrations from one area and system to another. The fish ecosystems are the most important marine ecosystems to man and are intensively exploited. Intensive research on fish ecosystems is conducted by all countries bordering the sea and a voluminous literature exists in fisheries matters. In this book we summarize only a few aspects of fish ecosystems, mainly from the point of view of man's effects on this system, the natural behavioural effects and the nature and possible magnitudes of natural fluctuations in this system contrasted with the effects of man. The problems of biodiversity of these systems are dealt with in connection with by-catch problems.

Despite intensive and long-lasting research on fish stocks, our quantitative knowledge of fluctuations and their causes is incomplete. This is mainly caused by the difficulties of quantitative studies. Three chapters (4–6) in this book deal with the behaviour of fish and the utilization of this knowledge for exploitation and management. The temporal and especially spatial definition of these ecosystems is often uncertain because of the changes of behaviour of species during their life cycle and because of their great motility and resulting changes in communities.

Another, more general way of subdividing fish ecosystems is by ichthyofaunal regions, considering the species composition of ichthyofauna and the characteristics of dominant commercial pelagic and demersal species (Fig. 2.15). The boundaries of regions were selected by considering the distribution of the species, their relative importance in the commercial catch and some of the characteristics of the environment. It is, of course, obvious that no sharp boundaries exist in the sea and that the boundaries marked on the map can be considered as wide transition regions. In many cases, the boundaries of faunal regions coincide with the boundaries of current systems and of water masses and with geographical boundaries.

There are some ichthyofaunal regions in which the species composition of the catch is similar, though the regions are geographically far apart. These similarities of the species composition have been caused by similarity of environmental conditions. Among these similar regions the following four categories can be recognized:

(1) Boreal regions:
       Pacific Boreal, Atlantic Boreal.

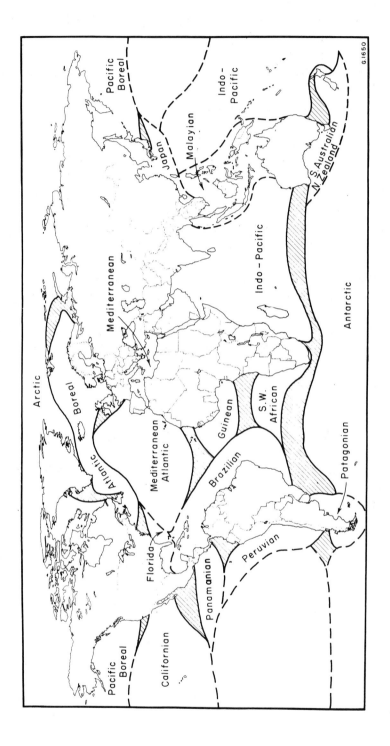

Fig. 2.15   Ichthyofaunal regions of the oceans.

(2) Upwelling regions:
     South-west African, Californian, Patagonian, Peruvian, South Australian, New Zealand.
(3) Subtropical transition regions:
     Florida, Mediterranean Atlantic, Mediterranean, Japan, Panamanian.
(4) Tropical regions:
     Brazilian, Guinean, Indo-Pacific.

This grouping of similar regions permits some conclusions to be drawn about the regions where little investigation has been made, by comparing them with another region in the same category which has been more investigated.

The ichthyofaunal regions can also be grouped according to the productivity of their communities per unit area of continental shelf. These differences in productivity of fish depend partly on the differences of basic organic or primary production, benthos, depth of the shelf, nature of the bottom and other factors:

(1) Highly productive for both pelagic and demersal species:
     Atlantic Boreal, Peruvian, South Australian, New Zealand, Patagonian, Japan:
(2) Highly productive for pelagic species, medium or low for demersal species:
     Pacific Boreal, Mediterranean Atlantic, South-west African, Californian, Florida, Panamanian, Antarctic.
(3) Low productivity:
     All tropical regions.

The polar regions and northern parts of boreal regions are dominated quantitatively by a few species in the nekton, whereas in the tropics there are many species but few dominant ones, except in upwelling regions where one or two pelagic species dominate. A summary of the numbers of commercially exploitable species in different latitudinal regions is given in Table 2.10.

Table 2.10 Number of commercial species in different latitudinal regions (from Rass 1979, modified).

Region	Number of commercial species			
	Total	Pelagic	Demersal	Diadromous (Migratory)
Arctic	20	6	6	8
Antarctic	15	5	10	—
Boreal				
Northern Hemisphere	50–75	12–15	35–50	5–40
Southern Hemisphere	40–50	14–18	23–30	—
Tropics	160–450	45–165	120–270	(5)

Quast & Hall (1972) list 432 fish species from Alaskan waters (Bering Sea and Gulf of Alaska) plus an additional 137 species from neighbouring waters which might occasionally occur off Alaska. Only 20 of these species might be considered commercial. In addition about 80–100 species are caught by survey gear and might occur occasionally as by-catch. Some species occur seldom as by-catch, not necessarily because they are rare but mainly because they either avoid the commercial gear or escape from it. Table 2.11 shows the rank order and abundance of species of a trawl survey as an example of the dominance of few species in demersal fish community. The abundance of demersal species is also stratified by depth strata (Table 2.12).

Pelagic fish and squids dominate quantitatively over demersal fish in the world catch of fish. Part of the reason is that they are lower in the food chain, utilizing zooplankton and some (e.g. anchoveta) even phytoplankton as food. The bulk of these small pelagic fish live in upwelling regions with high basic organic production. Furthermore, owing to their grater motility, some pelagic fish undertake extensive feeding migrations, thus grazing pelagic food over large areas.

Most fish species aggregate in shoals, thus making marine ecosystems patchy. Furthermore, if the individual shoals are considered on a smaller spatial scale, they represent single-species ecosystems. Most shoaling is for spawning, but many species

Table 2.11   Rank order of the 20 most abundant species encountered in the survey area during the 1980 cooperative US–Japan survey in the eastern Bering Sea and Gulf of Alaska (US National Marine Fisheries Service, unpublished data).

Species	CPUE kg/ha	Proportion of Total Fish or Invertebrates CPUE[a]
Walleye pollock	35.36	0.235
Giant grenadier	34.02	0.226
Pacific ocean perch	19.13	0.127
Pacific cod	16.53	0.115
Atka mackerel	12.88	0.085
Arrowtooth flounder[b]	5.37	0.036
Greenland turbot	4.12	0.027
Rock sole	3.96	0.026
Sablefish	2.85	0.019
Shortspine thornyhead	2.49	0.017
Squids	1.79	0.016
Rougheye rockfish	1.49	0.010
Shortraker rockfish	1.37	0.008
Pacific halibut	1.21	0.008
Skates	1.13	0.008
Great sculpin	1.18	0.008
Northern rockfish	1.05	0.007
Dover sole	0.95	0.006
Irish lords	0.95	0.006
Coryphaenoides sp.	0.76	0.005

[a] Total CPUE of all fish combined – 150.74 kg/ha.
  Total CPUE of squids, shrimps, and octopuses = 2.02 kg/ha.
[b] May include *Atherestes evermanni*.

Table 2.12 Average catch per area trawled (kg/ha) of major taxonomic groups encountered during the 1980 cooperative US – Japan survey in the eastern Bering Sea and Gulf of Alaska (US National Marine Fisheries Service, unpublished data).

Species groups	Depth (m)					
	1–100	101–200	201–300	301–500	501–900	1900
Gadidae (cods)	44.73	114.55	112.22	10.86	0.14	51.89
Macrouridae (grenadiers)	0	0	0	13.22	116.40	35.24
Scorpaenidae (rockfishes)	0.58	58.69	47.06	27.08	6.77	25.61
Pleuronectidae (flatfishes)	14.22	22.14	11.57	18.58	15.25	16.66
Hexagrammidae (greenlings)	5.70	49.62	0.71	0.04	0.00[a]	12.89
Anoplopomatidae (sablefish)	0.26	1.94	5.82	2.91	4.17	2.85
Cottidae (sculpins)	5.56	5.93	3.49	1.94	0.08	3.30
Rajidae (skates)	0.20	1.55	1.28	1.62	1.21	1.15
Zoarcidae (eelpouts)	0	0.01	0.01	0.16	1.50	0.46
Cyclopteridae (snailfishes)	0.00[a]	0.03	0.21	0.35	0.15	0.13
Agonidae (poachers)	0.02	0.08	0.05	0.02	0.04	0.04
Other fish	0.29	1.42	0.09	0.19	0.31	0.52
Total fish	71.56	255.96	182.51	76.97	146.02	150.74
Shrimps	0	0.02	0.06	0.02	0.02	0.02
Squids	0.03	0.44	8.09	3.24	0.67	1.79
Octopuses	0.43	0.26	0.28	0.15	0.08	0.23
Total invertebrates[b]	0.46	0.72	8.43	3.41	0.77	2.04
Total all species	72.02	256.68	190.94	80.38	146.79	152.78

[a] Less than 0.005 kg/ha.
[b] Includes only shrimps, squids and octopuses.

migrate and also feed in shoals. The size of shoals can vary from species to species, but also in space and time. A review of spatial distribution of pelagic fish and their shoaling behaviour is given by Harwood & Cushing (1972). For example, they note that the size of overwintering, non-feeding mackerel shoals observed during the day south-west of England measured 5 miles (8.0 km); by 1.5 miles (2.4 km) by 12 m thick, and contained 750 million fish. These shoals dispersed during the night. However, the size of shoals of fish can vary considerably; for example, the anchovy shoals off California normally have a diameter of only 10 to 30 m, rarely exceeding 60 m.

As mentioned earlier, any demersal fish can be demersal in the adult phase but pelagic as juveniles. An example of this is schematically shown in Fig. 2.16 for

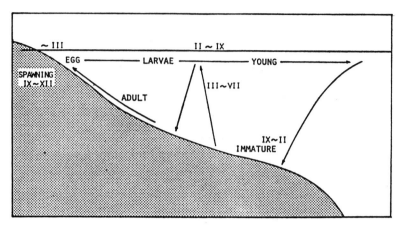

Fig. 2.16 Life cycle migrations of *Pleurogrammus azonus* (Yusa *et al.* 1977).

greenling. Several gadoid species, such as walleyes pollack and hake, are entirely pelagic during the 2–4 years of their juvenile life, but become semi-demersal later. Thus these species can belong to different fish ecosystems during their life cycle.

Squids form another abundant component of the nekton. A few species, such as *Illex* and *Loligo* can be found in continental shelf waters (Fig. 2.17). Most squid are, however, oceanic and predominantly tropical and subtropical. There can be aggregation of squids near the sub-arctic front which appear as thermal boundaries (Fig. 2.18).

Although we can deal with plankton and benthos as specific ecosystems, we cannot apply the same principles and consideration to nekton or fish ecosystems, which are the main subject of this book. Thus, we next consider environmental aspects and characterization of fish ecosystems.

2.3   Marine ecosystems as determined by the environment

Marine ecosystems can be classified by the characteristics of the environment. There are two major categories of these systems, the pelagic and/or offshore ecosystems and the neritic and coastal ecosystems where the demersal ecosystem predominates. There can also be a number of specific ecosystems (e.g. coral reef, brackish water, polar and tropical ecosystems).

Marine ecosystems are seldom well defined by their boundaries, as sharper boundaries are rare (e.g. coast, sharp change of bottom type) and relatively wide transition zones separate different, geographically-defined ecosystems.

Pelagic ecosystems have been described by different criteria, such as low latitude gyral regions (Blackburn 1981) and upwelling regions (Vinogradov 1981). Coastal and demersal ecosystems have been described as self- sea ecosystems (Walsh 1981), coastal upwelling ecosystems (Barber & Smith 1981) and deep-sea ecosystems (Rowe 1981).

In this section some of the essential properties of the environment-determined

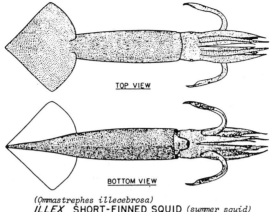

TOP VIEW

BOTTOM VIEW

(Ommastrephes illecebrosa)
ILLEX SHORT-FINNED SQUID *(summer squid)*

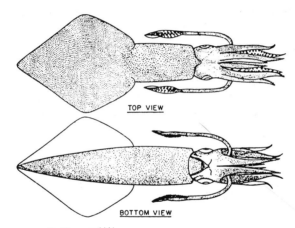

TOP VIEW

BOTTOM VIEW

(Loligo pealii)
LOLIGO LONG-FINNED SQUID *(Winter squid, common squid)*

Fig. 2.17 Some commercial squids from the north-west Atlantic (Lyles 1968); (a) short-finned squid *Ommastresphes illecebrosa* (summer squid); (b) long-finned squid *Loligo peallii* (winter squid, common squid).

ecosystems are described. The functioning of these ecosystems from a biological production point of view has been recently described by Barnes & Mann (1991) and Mann & Lazier (1991).

2.3.1   *Pelagic and offshore ecosystems*

Pelagic ecosystems contain mainly plankton and nekton (see [2.2.1] and [2.2.3]). There is also a specific ecosystem, neuston on the sea surface, which has been classified into several subsystems such as epineuston and hyponeuston (David 1967). This ecosystem consists of such organisms as *Velella*, *Physalia* and *Ianthia*, and even juvenile fish such as *Mupus maculatus* and *Mugil auratus* occur occasionally in neuston (Fig. 2.19).

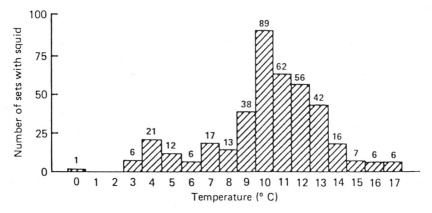

Fig. 2.18   Gillnet sets in the North Pacific in which squids were taken, related to surface water temperature (Fiscus & Mercer 1982).

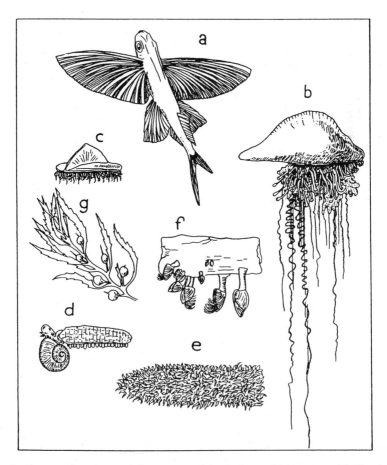

Fig. 2.19   Surface organisms (neuston) from warmer Atlantic Ocean; (a) *Exocoetus*; (b) *Physalia*; (c) *Velella*; (d) *Janthina*; (e) *Pyrosoma*; (f) *Lepas*; (g) *Sargassum* (Schott 1944).

Quantitatively the neuston ecosystem has a minor effect and minor importance in total marine ecosystems.

The basic organic production in the pelagic region occurs in the surface layers of the sea, usually above the seasonal thermocline. The annual production varies with latitudes and there is large seasonal variation in high latitudes. Many tropical and sub-tropical oceanic areas are characterized by low productivity and sub-surface chlorophyll (Fig. 2.3), indicating that basic organic production occurs at minimum light intensity where nutrients are available.

Most of the pelagic zooplankton and nekton communities are also limited to the upper layers of the ocean. The pelagic communities change with a seasonal rhythm. Episodic events, such as storms, can affect these ecosystems locally by advection and mixing. A specific seasonal succession in pelagic ecosystems occurs in higher latitudes. A somewhat similar succession occurs in regions of seasonal upwelling.

Pelagic systems are open systems, seldom reaching stability or maturity. There is a continuous transport and mixing of the communities and continuous migration of nekton, mainly in search of food.

A vertical (depth) structure of pelagic ecosystem is present in most oceanic areas. With increasing depth the biodiversity decreases. The deep-water pelagic ecosystem is biologically poor and species are specialized. Some species descend seasonally into the deeper layers (e.g. *Calanus hyperboreus* descends during the winter to depths greater than 500 m). Deep scattering layers occur usually at depths of 300–800 m. It can consist of myctophid fish, siphonophores, euphausiids and squids.

The pelagic zooplankton can be divided into communities which correspond to different surface water types as defined by oceanographic properties. Fraser (1961) recognized five types of plankton communities around the British Isles, mainly on the basis of water masses. He found that 'Each water mass, though labelled by a few indicator species that may be of greater or lesser importance in the ecosystem, has a characteristic fauna often typified by the abundance and condition of many species which are also found in other water masses but in different numbers or in different proportions.'

Fraser also described the various food interactions in pelagic communities. Pelagic fish are usually found during the feeding season in areas of plentiful food. However, salps, which are little used as food by fish, can develop rapidly in areas where phytoplankton is abundant and quickly graze it down, leaving little for other zooplankters which are the main food for pelagic fish. Ctenophores and chaetognaths can also denude great tracts of their zooplankton and young fish content. For example a mass development of the jelly comb *Mnemiopsis leidy* in the Black Sea in recent years has seriously affected the recruitment of sardines.

Some dinoflagellates (e.g., *Ichthyodinium chabelaudi*) can infect fish eggs and larvae and cause considerable mortality, thus affecting recruitment.

Bacteria are a functionally important component of pelagic marine ecosystem. Brown *et al.* (1991) estimated that bacteria in the water need to consumer 9–15% of the total primary production in the Benguela Current ecosystem to meet their requirements of carbon consumption. Bacteria biomass of 40 mg C/m^3 and mean growth rate of 50% of the biomass daily in the upper 30 m was 8–27% of the phytoplankton biomass and 26–44% of the phytoplankton production. Bacterial carbon consumption requirements were 86–147% of total primary production; i.e. most of the phytoplankton was decomposed by bacteria in the upper 30 m.

Similar results were observed by Nakajima & Nishizawa (1972) in the Bering Sea. They found that the amount of particulate organic carbon decreased nearly exponentially with depth in the upper 100 m. About one-third of the decrease was attributed to zooplankton grazing and the remainder to the possible utilization by ultraplankton and bacteria.

Pelagic ecosystems include pelagic fish, most of which spawn in coastal areas but undertake extensive seasonal and life-cycle migrations. Some semi-pelagic fish, such as hake, spend their juvenile years also as pelagic fish.

Two groups of large oceanic pelagic fish, salmon and tuna, are of great economic importance to man. In the North Pacific there are six species of *Oncorhynchus* and in the North Atlantic two species of *Salmo*. Salmon spawn in fresh water (i.e. in rivers, or a few species in lakes) and some species spend more than 1 year in fresh water before migrating to the ocean, where they feed for a few years and grow rapidly before returning to 'home' rivers for spawning. In the ocean salmon feed over large areas (see Fig. 2.20) and have been caught with various gear (gillnets, troll lines, longlines). Because these fish are of great local economic importance their stocks must now be supported by hatchery production of smolts, and their fishery is almost prohibited in offshore areas.

The food composition of the large pelagic fish, such as salmon, changes seasonally depending on the availability (and density) of food items (Fig. 2.21). LeBrasseur (1972) concluded that the pelagic organic production of the spring phytoplankton bloom is stored in herbivorous zooplankton such as euphausiids, and in smaller nekton such as squids and myctophids, which have a life span of a year or longer and thus permit the utilization of the spring production of herbivores by salmon and other larger pelagic species.

The other highly migratory, economically important species are the tunas and bonitos. Examples of the distribution of some tuna species are given in Figs. 2.22, 2.23 and 2.24. Several regional international bodies have been established for the management of tunas.

Squids are an important component of the offshore pelagic ecosystem, occurring mainly in offshore waters south of 40° N (Fiscus & Mercer (1982)). Some of them might occur during summer in the North Pacific north of 40° N in quantities sufficient for jigging (e.g. *Onychoteuthis borealijaponicus* and *Ommastrephes bartrami*).

The pelagic ecosystem is also home to mammals such as whales and dolphins. Pinnipeds, on the other hand, are partly terrestrial, because they mate in colonies and feed their young on land. Most marine mammal species undertake extensive seasonal migrations.

Pelagic ecosystems are often divided by current systems and/or temperature regimes. Temperature changes usually indicate changes in currents and upwelling as well as current boundaries. Such changes coincide with changes in the occurrence of

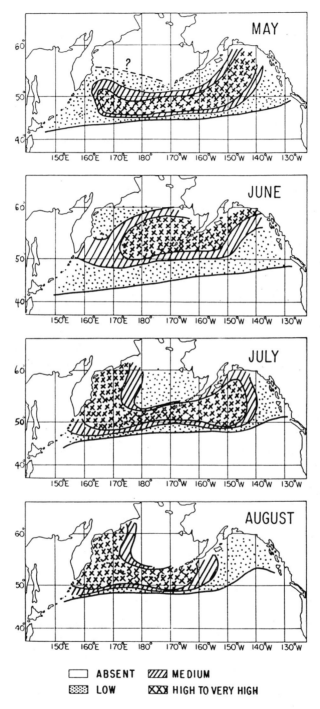

Fig. 2.20 Schematic diagram showing distribution and relative abundance of sockeye salmon in the North Pacific, May–August, maturing and immature fish combined (Manzer *et al.* 1965).

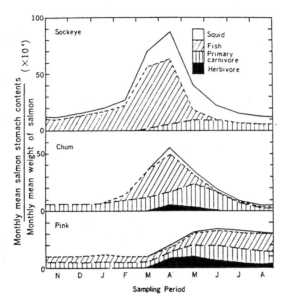

Fig. 2.21 Seasonal variations in the relative amounts of prey found in salmon stomachs in the North Pacific (LeBrasseur 1972).

shoals of pelagic species such as pilchard and anchovy, as demonstrated by Armstrong *et al.* (1991) off the east coast of South Africa.

2.3.2   *Neritic, coastal and demersal ecosystems*

Coastal ecosystems are the most important to man because they are visible and/or accessible to him. Coastal areas are densely settled by man and coastal waters are intensively used for recreation, local fisheries and mariculture, and some areas are also used for disposal of domestic and other waste. These coastal marine ecosystems are varied in space and largely dependent on the nature of the coast and sea bottom. Because of the great variability of these ecosystems, only some generally applicable ecological notes are given below.

The composition of littoral and shallow water communities depends greatly on the nature of the bottom sediment and on the presence of sessile plants (e.g. eelgrass *Zostera* and seaweeds). Macrophytes are usually well developed on rocky bottoms. *Laminaria* and *Fucus* are the most common and universal attached brown algae. Examples of common attached algae and seagrass are shown in Fig. 2.25. Zonation of different sessile algae occurs with depth, green algae being shallowest, followed by brown and red in deeper water. Most sessile algae occur in depths less than 10 m and very little in depths greater than 20 m. Daily production of *Laminaria* can be as high as $8 \, \text{g} \, C/m^2/day$ (or 65 g dry matter). Seaweeds are harvested in some areas for chemicals and in a few Asian countries for food supplement.

The soft bottom benthic ecosystems are reviewed in [2.2.2]. Some rocky shore communities are subject to intense shellfish exploitation, which might alter the eco-

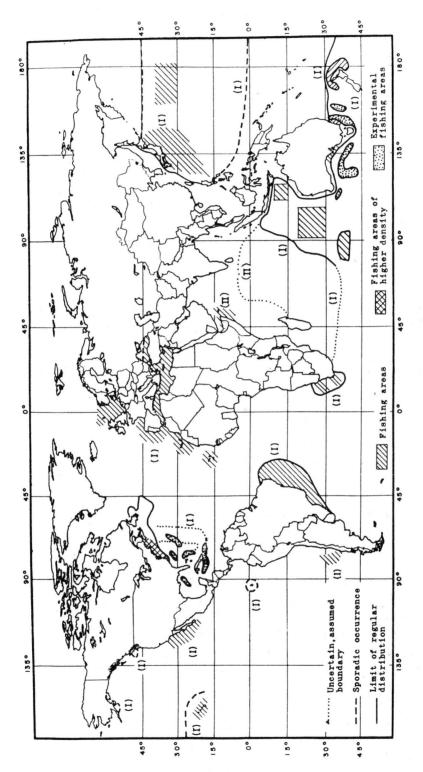

Fig. 2.22 Distributions and fishing areas of bluefin tunas (I) *Thunnus thynnus* and (II) *T. tonggol* (Laevastu & Rosa 1962).

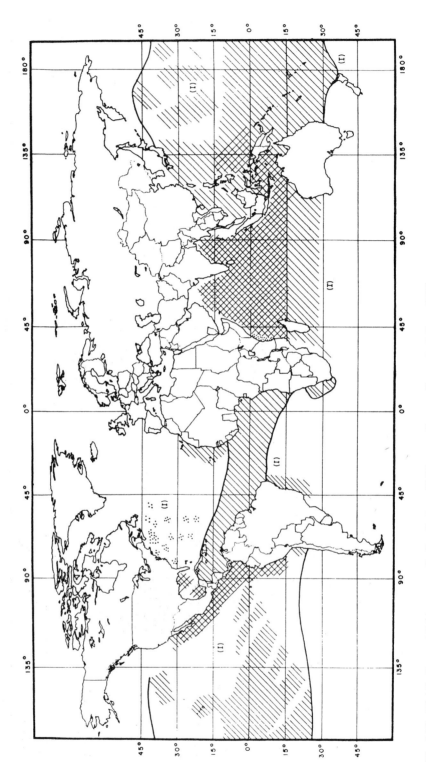

Fig. 2.23 Distribution and fishing areas of yellowfin tuna (I) *Thunnus albacares* (Laevastu & Rosa 1962).

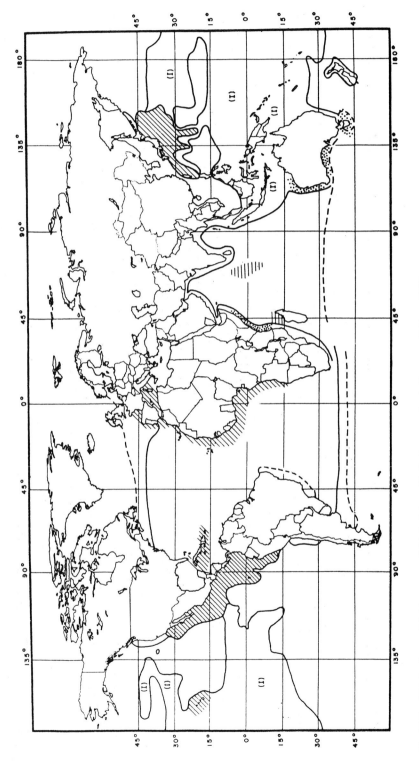

Fig. 2.24 Distribution and fishing areas of skipjack tuna (I) *Euthynnus pelamis* (Laevastu & Rosa 1962).

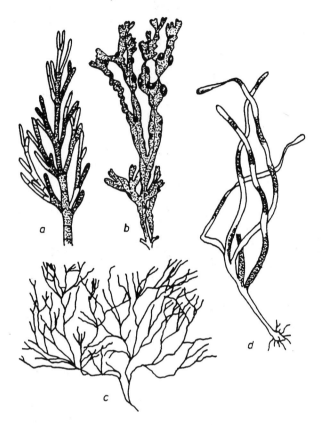

Fig. 2.25 Some common seaweeds and seagrass; (a) *Cladophora* (b) *Fucus vesiculosus*; (c) *Furcellaria lumbricalis*; and (d) *Zostera marina* (Pihu 1987).

system locally. Dye (1992) found that the restoration of such heavily harvested communities to a quasi-virgin state takes 8–9 years after cessation of harvesting.

Some of the epibenthic animals caught by man are protandric hermaphrodites. These include the deep-sea shrimp *Pandalus borealis* (Fig. 2.7) which occurs in the northern parts of the North Atlantic and North Pacific Oceans. This shrimp matures at about 1.5 years old at a length of 9 cm and is first male, and then after about 2.5 years or at a length of 12 cm it becomes female. The shrimp is usually caught between 75 and 350 m deep.

The abyssal benthic fauna live on a soft bottom. Its main components are glass sponges, some polychaetes, a few bivalves, sea urchins and sea stars (Porcellanasteridae). The total mean biomass is only 0.5–2 g/m^2. Most of the deep-water benthos are detritus feeders.

The variety of animals caught with commercial trawl gear from the sea bottom on the continental shelf is great. An example of such catches is given in Table 2.8. The quantitative composition of the demersal fish communities can vary from year to year. This variation is illustrated with the results of annual trawl surveys from the east coast of USA (Grosslein *et al.* 1980) (Fig. 2.26).

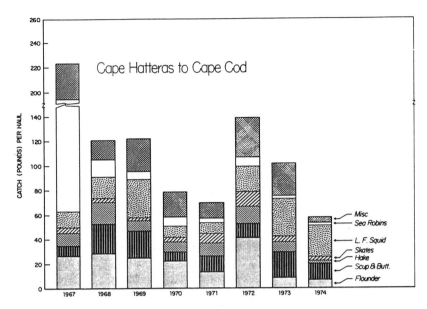

Fig. 2.26 Mean catch per standard 30-min haul of major species groups (spiny dogfish excluded) in autumn trawl surveys off the east coast of the USA. from Cape Hatteras to Cape Cod, 1967–74 (Grosslein *et al*. 1980).

The coastal areas are assumed to be nursery grounds for many fish. However, it is disputable whether they are the only nursery grounds for given species. It is much easier to catch young fish in shallower water, especially if they are near the bottom during daytime, than to catch them over deep water distributed over some depth range and where they can easily avoid capture by the towed gear.

Most commercial fish are found on the continental shelf at some stage of their life, especially during spawning. Many aspects of the fish ecosystems and their behaviour are dealt with in other chapters of this book.

Neritic zooplankton communities can be slightly different from offshore zooplankton communities in that neritic zooplankton contains larvae or benthic organisms. Furthermore, there are some zooplankton species whose overwintering eggs rest on the bottom and thus occur mainly over the continental shelf.

There has been a rapid expansion of shrimp culture in coastal lagoons, especially in Asia, where the 1989 harvest was 522 000 t (fresh and salt water culture). The commonly cultured species are *Penaeus* spp. Another intensive mariculture in coastal areas in higher latitudes is that of salmon. In 1991 the Norwegians farmed salmon harvest was 150 000 t. The mariculture activities have very limited and local effect on marine ecosystems at large. Attempts are being made in several countries to extend mariculture offshore using anchored pens.

In some subtropical areas artificial reefs have been created by dumping old cars and other solid waste off the coasts. These artificial reefs offer habitat to many species which are valued as sports fish. In the Gulf of Mexico good sports fishing occurs also around oil platforms, which act as huge vertical reef structures and attract numerous species of predator and prey. Anchored, floating fish attraction devices (FADs) are used in the tropics to attract and aggregate fish for capture.

2.3.3   *Specific marine ecosystems*

Coral reefs

Coral reef ecosystems are distinct coastal ecosystems in tropical seas. They have greater biological richness than the surrounding areas. An excellent review of reefs was given by Hughes (1991).

Coral reefs support biological communities rich in species, made up of sedentary colonial invertebrates, echinoderms, and a multitude of mostly small fish and other animals living in crevices and cavities. Coral reefs are also characterized by great stability, mainly due to climatic stability in the tropics. However, mass mortalities due to cold spells can occur in reefs near the southern boundary of the boreal region, (e.g. in Florida (Hughes 1991)). Hurricanes and other severe storms can also cause considerable damage to the coral reef and its communities.

Coral reefs have high basic organic production, mainly by attached algae. The animal life is not limited by this production and is controlled by predation which also contributes to high recycling of nutrients. Thus most of the carbon fixed by photosynthesis is respired by intermediate predators and little is left to support higher sustainable yield of large carnivores (i.e. to the desirable commercial yield).

A schematic section of a coral reef is given in Fig. 2.27. Reefs are built by calcareous

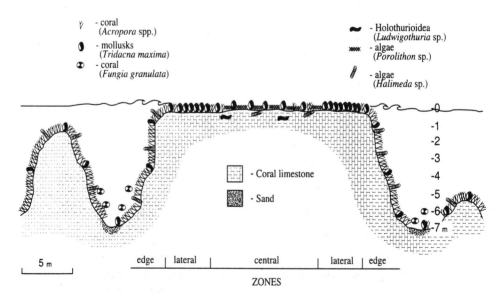

Fig. 2.27 Diagrammatic representation of a portion of the outstanding *Acropora-Tridacna* reefs connecting the islets Ana-Ana, Kimoa and Tridacna. Giant clams *Tridacna maxima*, thickly studding the reef, attain densities of 80 per m² (Kepler *et al.* 1992).

algae, calcareous sponges and corals of which there are 30–50 species. Living corals usually cover only a small portion of the reef.

Coral reef ecosystems have a high degree of maturity. The complex and intensively functioning coral reef ecosystem is very sensitive to anthropogenic effects, including eutrophication, pollution and intensive fishing.

The reef-building corals are also eaten by starfishes. Occasionally mass development of starfishes may destroy living corals, leaving the reef to be colonized thereafter by soft corals and algae.

Besides the local harvesting of fish from the coral reefs, the snail *Trochus* is harvested from the reef. Its meat is consumed by man and the colourful shells are used to make shiny buttons and jewellery all over the world. *Trochus* is also now transplanted and cultivated.

Brackish-water ecosystems

Brackish-water ecosystems are divided into two groups, the estuarine ecosystems and the brackish sea ecosystems (e.g. the Baltic Sea). Brackish-water ecosystems contain a mixture of fresh- and salt-water species. The number of species is usually low and dependent on salinity (Fig. 2.28). A rich scientific literature exists on the Baltic Sea

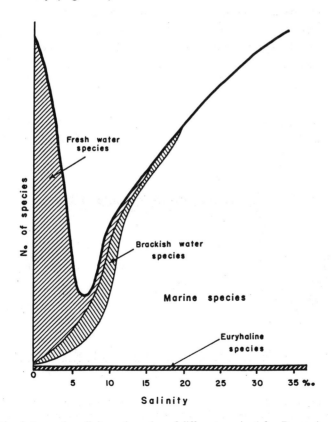

Fig. 2.28   Relation between the salinity and number of different species (after Remane).

and its ecosystems. This book does not deal with the specific brackish-water problems except for the eutrophication of the Baltic Sea, which is described in [7.2].

Polar ecosystems (Arctic and Antarctic)

Arctic and Antarctic marine ecosystems are characterized by low numbers of species, whereby few dominate, and there are high growth rates under favourable conditions and flexible reaction to unfavourable environmental changes.

Figure 2.29 shows the simplified ecosystem of the Antarctic and its essential food cycle. Owing to the remoteness of the Antarctic its exploitation by man is limited, except for whaling in earlier years.

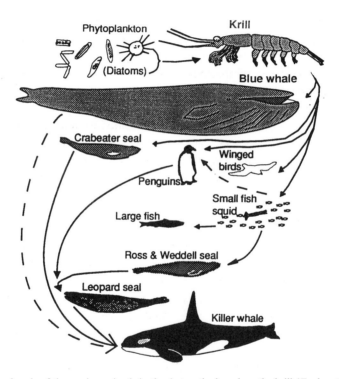

Fig. 2.29   Food cycle of the marine animals in the Antarctic, based on the krill (*Euphausia superba*).

Antarctic waters are frequented by five species of baleen whales (see Table 2.13). In addition, sperm whale and killer whale (toothed whales) are significant inhabitants and predators there. Commercial whaling has become an emotional issue world-wide and little if any whaling is expected in the Antarctic region in near future years.

There are six species of seals in the Antarctic. The Crabeater seal is the most numerous (about 30 million). The total abundance of the other species amounts to about 3 million. The main fish species in the Antarctic are Nototheniiformes, which are comprised of five families. They have been harvested in the past in subsidized fisheries. In addition there are southern blue whiting, Patagonian hake, rat-tails and skates.

Table 2.13 Baleen whales in the Antarctic (estimates by SCAR 1976).

Species	Number (thousands)	Mean weight (t)
Fin	84	48
Blue	10	83
Sei	40	17
Humpback	3	26
Minke	200	7

Great quantities of krill inhabit Antarctic waters, but its fishery has been possible only in subsidized fisheries. The total biomass of krill is estimated to exceed 1000 million t. The northern parts of Antarctic waters also contain considerable quantities of squids, which are harvested in Falkland and Argentinian waters.

The southern boundary of the Arctic is ill defined. There is a wide transition zone from northern boreal to Arctic fauna in the Barents and Bering Seas. Fewer mammals inhabit the Arctic ecosystem than the Antarctic, the main species being ice seals (i.e. ring and ribbon seals). The Arctic is also frequented by northern boreal mammals such as grey whale and grey seal. The Arctic fish fauna is also poor, the main species being the polar cod *Boreogadus saida*. Many northern boreal fish species (e.g. capelin, cod) use the southern Arctic as a feeding ground during summer.

2.4   Biodiversity in marine ecosystems and its determinants

Biological diversity has no precise definition. In terrestrial ecology it is associated with the extinction of species, which is a near impossibility in marine ecosystems (excluding some marine mammals and anadromous fishes). Species extinction is usually connected with environmental degradation. Degradations of the marine environment are temporary (including pollution, see [7.2]). Conservation of biodiversity was one of the focal points of the Earth Summit Conference held in Rio de Janeiro in 1992.

Biodiversity has been interpreted in the past as 'species richness'. Recently several new components have been added to this interpretation (Redford 1994), genetic diversity, species diversity, higher taxonomic diversity, community diversity and ecosystem diversity. The last two components encompass in essence the ecosystem. The main and noble aim of nature conservationists is to maintain biodiversity or species richness in all ecosystems on the earth, to guard against extinction of species and to warn when species are threatened or endangered.

We are unable to determine quantitatively the abundance of marine animals, including fish, because of their great motility and advection and the great temporal and spatial variability of marine animal communities. It is therefore not possible to determine with any degree of certainty whether any species of marine animal is threatened of endangered.

Biotic diversity in marine ecosystems is complex. It includes not only the community structure of species, but also processes that determine and modify it, e.g. turbulence and mixing, advection and migrations, and also the energy flow in terms of trophic relationships.

There are frequent reclassifications of marine species, and the sampling in marine ecosystem is quite haphazard. There is therefore no documented extinction of marine fish in historic times.

Marine ecosystems have no real equilibrium and no real stability, because their species composition fluctuates when disturbed by natural forces or by man, and normally does not return to an original state which in itself cannot be well defined. Therefore, we cannot define resilience in marine ecosystems except in very general terms. There are many other related terms in theoretical biology which have little application to marine ecosystems.

Many natural disturbances operate on marine ecosystems, the greatest and most universal of which is advection by currents. Man might also cause changes by fishing. The effects of man's action on marine ecosystems are the main subjects of this book and an earlier one by Laevastu & Favorite (1988).

The concept of biodiversity is related not only to the types of organisms in the community, but also to their abundance. It is difficult to find and define objectively any meaningful static index of biodiversity for marine ecosystems, especially because of the inadequate methods of sampling marine ecosystem. The world fish faunas (20 000–27 000 species) are not well known (Greenwood 1992).

Natural environmental disturbances are common in some marine ecosystems, whereas in others (e.g. the deep sea) great uniformity prevails. Local species diversity is often related to the role of key predators, their dominance over the rest of the community. There can also be cyclic changes in the communities, such as the grazing down of an algal community by sea urchins and a subsequent re-establishment of other algal communities.

Biodiversity in marine communities varies not only spatially and seasonally, but also with advection and with predation. Biodiversity of fish ecosystems has had large fluctuations for the last several hundred years, as abundance of fish scales in sediments indicate (Soutar & Isaacs 1974). Apparent changes in fish populations may occur, caused by temporal and spatial factors that are not taken into account in designing sampling programs. Pelagic fish especially show extreme variations of recruitment, before the start of extensive commercial fishing (Blaxter & Hunter 1982).

Great changes have occurred recently in fish ecosystems in high latitudes. An example is given in Fig. 2.30. Herring disappeared from the Barents Sea in 1970 and cod decreased from the mid-1970s to the end of the 1980s. Heavy fishing is assumed to have contributed to the decline in abundance of these species.

In contrast to species extinction and/or decrease in biodiversity is the discovery of new species in marine ecosystems, and the introduction and appearance of undesirable species from other areas. One example is the recent discovery (or new records) of 17 new benthos species off the Portuguese coast (Marques & Bellan-Santini 1993) (of the 113 species identified 17 (15%) were new records from the Portuguese coast).

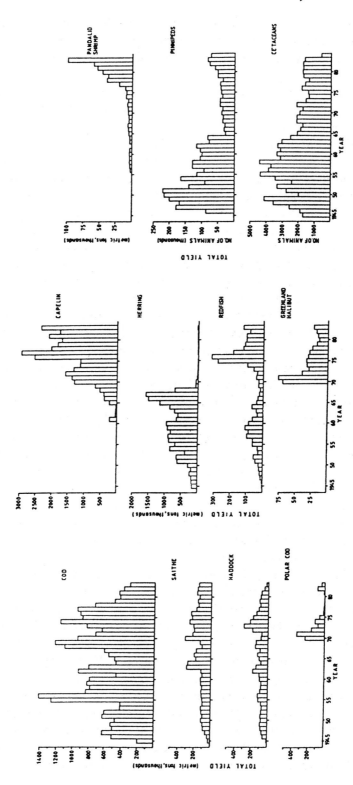

Fig. 2.30 Total yield from the main fish, shrimp and mammal resources from the Barents Sea and adjacent shelf areas. (ICES regions I, IIa and IIb) (Sources: Bull. Stat., ICES: fish, shrimp, mammals. Institute of Marine Research, Bergen).

Other examples are the accidental introduction of the Chinese 'wollhand' crab to European waters from ship's ballast water in the early twentieth century and the mass occurrence of comb jellies in the Black Sea in late 1980s. Transplantation of King crab and Pacific salmon to the Barents Sea are examples of intentional introduction of new species.

The biodiversity of tropical marine environments is great (Baker & Murphy 1991) and the number of species in tropical communities is high, but few if any species dominate, whereas in higher latitudes strong dominance by a few species is a rule. This dominance is also more pronounced in neritic than in oceanic communities. The biodiversity of benthic animals can be different only a short distance apart. The difference between the communities might be caused by the dominance of one or a few species, as shown by Branch *et al.* (1987) in communities off two small islands (Marcus and Malgas Islands) off the South African coast, about 4 km apart. The difference between the communities is demonstrated in Fig. 2.31 and 2.32. Off Malgas

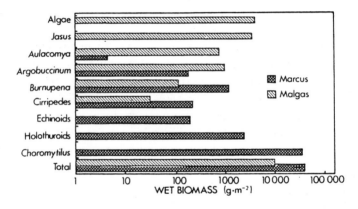

Fig. 2.31 Biomass of the dominant groups of organisms found in the subtidal zone at Marcus and Malgas Islands (Branch *et al.* 1987).

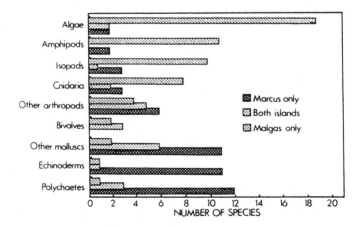

Fig. 2.32 Species richness of the major groups of organisms in the subtidal zone of Marcus and Malgas Islands (Branch *et al.* 1987).

Island, rock lobster and a variety of algae dominate. The average biomass of rock lobster is about $3800\,g/m^2$. Off Marcus Island few lobsters are present and the rich benthos fauna is dominated by black mussel ($35\,kg/m^2$). Associated with mussel beds are holothurians, ophiuroids, crinoids, polychaetes and whelks. Of the total number of species (156) only 34% are common to both islands.

Predation is a controlling factor not only in the fish ecosystem, but also in some subtidal benthic systems, as demonstrated experimentally by Branch *et al.* (1987). When rock lobsters were transplanted from Malgas Island to Marcus Island, the lobsters died owing to lack of proper food species. In contrast, predation by rock lobster at Malgas Island prevented the establishment of mussel beds there.

Biodiversity in marine communities varies with several environmental factors, including depth. However, the depth preferences of a given fish species may vary with the age of the fish and with the latitude, and many fish species are distributed through several geographically-defined ecosystems. An example of distribution by age is given in Figs 2.33 and 2.34. Salinity is another marine environmental factor which determines biodiversity, especially in brackish water. Figure 2.28 illustrates the relative abundance of species in relation to salinity.

For some time marine biologists have assumed that plant growth (phytoplankton production) and fish production might be quantitatively related. However, long and expensive studies have failed to demonstrate any clear direct relationship between either standing crop or production of phytoplankton and available standing stocks of harvestable fish, except possibly on a very large (regional) scale.

Examples of some of the factors which influence the diversity and stability of communities in two specific marine ecosystems are shown in Fig. 2.35 (Branch *et al.* 1987).

Lack of genetic diversity has been suggested as a possible problem in mariculture. However, no effects of changes of genetic diversity have been found which merit concern. Hybridization is a common phenomenon in marine ecosystem (e.g. between separate stocklets), and might even be a desirable phenomenon, ensuring survival of stocklets and survival of the genetically fittest. Genetic diversity has occurred naturally (e.g. in the form of different spring and autumn spawning stocks of Icelandic herring). Ojaveer (1981) found that the separation of spring and autumn spawning Baltic herring is effective; some hybridization might occur, but the percentage of hybrids in the stock is negligible.

Long-distance dispersal and immigration of emigration operate in marine ecosystems. Therefore genetic differentiation between stocks of the same species is rare. However, some genetic differentiation between, for example, Norwegian fjord herring stocks and offshore stocks has been found.

2.5   Environment and biota interactions

The marine environment, its physical and chemical properties and its dynamic processes vary over all space and time scales. These properties and processes affect the organisms in a multitude of ways. Furthermore, nearly all organisms interact with

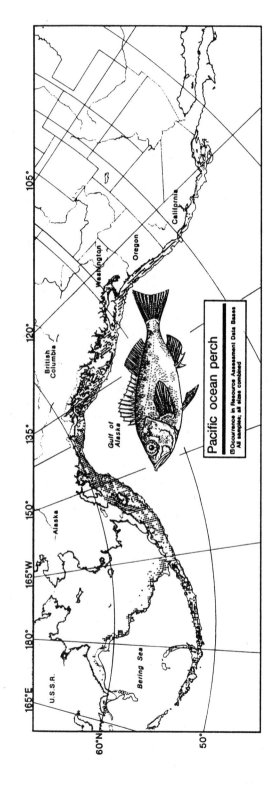

Fig. 2.33 The overall range of Pacific ocean perch off the west coast of North America based on an analysis of several resource assessment data bases for 1912–84 (Wolotira *et al.* 1993)

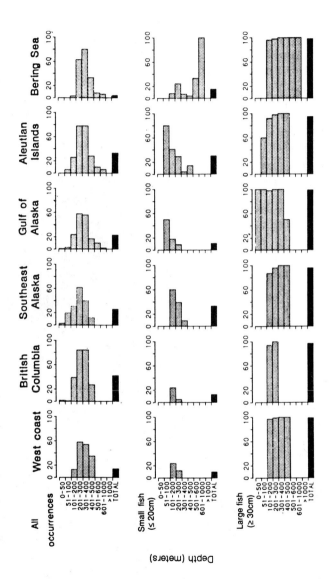

Fig. 2.34 Frequency of occurrence by depth interval by region for Pacific ocean perch off the west coast of North America based on presence in samples from resource assessment surveys during 1912–84 (Wolotira *et al.* 1993).

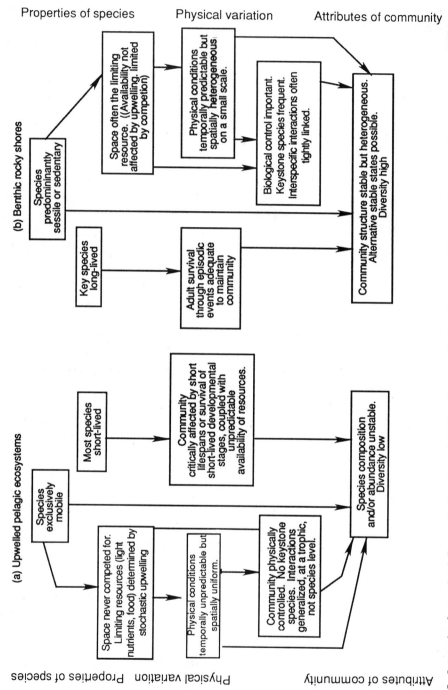

Fig. 2.35 Interacting factors that influence the diversity and stability of communities in (a) pelagic and (b) rocky-shore ecosystems (Branch *et al.* 1987).

each other within the ecosystems, the main interaction being to eat or to be eaten.

Every marine organism is to some extent adapted to its specific environment and to a specific ecosystem. This adaptation can have narrow or wide tolerances to various environmental and/or community factors. Usually one or a few of these factors can become limiting in respect either of living space or abundance in it. The life histories of species are also adapted to environment. For example, the spawning grounds of fish with pelagic eggs and larvae are usually located in upcurrent feeding areas and nursery grounds, and so the eggs and larvae are carried by currents to suitable feeding areas. Timing of spawning might also be adapted to the best availability of food for larvae. Fecundity might be adapted to counteract the possible mortality or loss of eggs and larvae. Community adaptation might be related to bottom substratum or to availability of other organisms as food.

The environmental factors which affect marine ecosystems can be divided into two groups; the constant and static factors (e.g. depth, type of bottom), and the variable and dynamic environmental factors (e.g. temperature and currents). The effects of various environmental factors on different marine ecosystems have been briefly described in [2.2] and [2.3]. The effects of the essential dynamic processes of fish ecosystems are summarized in Chapter 3.

The sea is a system with complicated and varied relationships and interactions between physical, dynamic and chemical components. These interactions may produce different effects on biota of the ecosystems. The discovery of the true biological dynamics and its regulation in the sea is hindered by the continuous and simultaneous changed in several dynamic processes in the sea that affect marine organisms. The intense fluctuations of an unknown number of factors affecting life in the sea mask the significance of single factors.

Countless studies have been published on the correlation between selected environmental properties and distribution and abundance of species in a static sense. Other studies have examined the potential correlation between changes in the environment and changes in ecosystems or in a single species in these systems. However, a good correlation may be purely formalistic and may have nothing to do with causality. This is especially so with a time series of data because many components may change quantitatively in one direction only, and be independent of each other over a long time span.

While plankton and most benthos are subjected to local environmental changes, and must adapt to such changes or suffer a multitude of consequences, nekton (mainly fish) and marine mammals can search for suitable environments through migration. The behaviour of fish in respect of environmental change is of considerable practical interest to man, especially with regard to their capture, and is reviewed in Chapter 4.

Sea water contains in solution nearly all the elements. Organisms take up these elements in various degrees and concentrate some of them in various organs, from a few tens to 100 000 times the concentration found in sea water (Krumholz & Goldberg 1957). Some of the metabolized organic compounds can be poisonous to other organisms. For example, the protozoan *Gymnodinium brevis* causes a red tide during occasional periods of mass development and has poisonous effects on fish and humans. Phytoplankton organisms excrete exocrenes which in dense patches appear to repel many fish.

Toxic algal blooms have been known for centuries. Recently more attention has been given to this phenomenon and to the accompanying paralytic shellfish poisoning, in an attempt, so far unsuccessful, to connect the higher frequency of red tide reports to increased pollution and even to global warming.

Phytoplankton production is largely controlled by light and its degree of penetration, the amount of nutrient salts present, and grazing. Views have been expressed in the press and by some government authorities about the effect of ultraviolet (UV) radiation on plankton production, especially in the Antarctic in connection with fluctuations of the hole in the ozone layer. However, UV radiation is absorbed in near-surface layers, and most phytoplankton shows photoadaptivity (Helbing *et al.* 1992). Other alarmist views on changes in the environment and marine fish ecosystems have been expressed in relation to the well-publicized but less well-founded global warming (Bigford 1991).

Regional sea temperature fluctuations, often brought about by advection caused by temporary shifts in wind regimes, might cause some changes in marine ecosystems. Based on a long-term series of investigations off Plymouth, Southward & Boalch (1986) suggested 'that the changes in temperature may affect the biological cycle in the western English Channel by altering species distribution over a wider area, and also more locally, by triggering switches of an unstable ecosystem. The resulting changes in species composition and in relative abundance of the different trophic levels appear to be independent of the basic productivity of the system'.

Skud (1982) convincingly explained how species dominance can be affected by environmental, especially temperature, change and how this change can affect the abundance but not necessarily density, and that the mechanism controlling abundance within the population can be intraspecific competition. Changes in dominance can also explain why a species in one area responds positively to, for example, temperature change in many years and then responds negatively, whereas in other geographic areas no response is observed.

The effects of the environment, climate and weather on fish have been summarized by Laevastu & Hayes (1981) and Laevastu (1993). Only a few examples of the effects of environment on the marine ecosystem are given here.

In the second half of the 1960s the east Greenland ice covered larger than normal areas north and north-east of Iceland. These cold conditions caused a decrease in phytoplankton production as well as amounts of zooplankton (mainly *Calanus finmarchicus*) present, and drastically changed the feeding migrations of the Norwegian herring (Jakobsson 1978).

Relatively profound changes in environmental conditions can also occur in the brackish Baltic Sea and provoke changes in the ecosystem and its components. For example, during a salinity increase, *Pseudocalanus* increases whereas *Limnocalanus* decreases. The *Temora* stock increases during periods of positive temperature anomalies (Kalejs & Ojaveer 1989). Besides variations in stock abundance of herring, periodic variations in the percentage of spring and autumn spawners also occur in the Baltic Sea.

The environment can determine population structure and many species-specific characteristics of ecosystems. Norwegian coastal cod and Arcto-Norwegian cod are discrete populations with different growth rates, ages at first spawning and migration patterns. Godø & Moksness (1985) demonstrated that these differences are related to the different environmental conditions of these stocks, but these species-specific characteristics can be changed under laboratory conditions.

Most of the Arcto-Norwegian cod stock spawn in the Lofoten area, from where the larvae are carried by currents into the Barents Sea (Fig. 2.36, Nakken & Raknes 1985) where they descend toward the bottom. Later the older fish move westward into the warmer Atlantic water (Figs 2.37, 2.38).

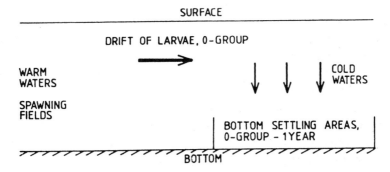

Fig. 2.36 Schematic representation of the transportation and distribution of Arctic cod (Nakken & Raknes 1985).

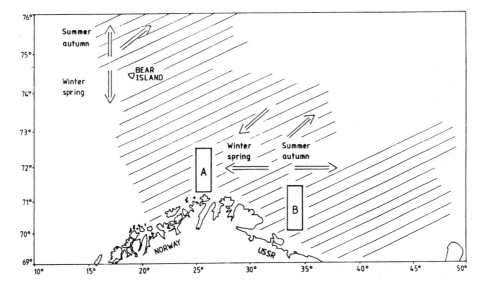

Fig. 2.37 Schematic representation of the distribution area of immature cod in the Barents Sea and the seasonal movements of the stock. A and B are used as reference localities for temperature (Nakken & Raknes 1985).

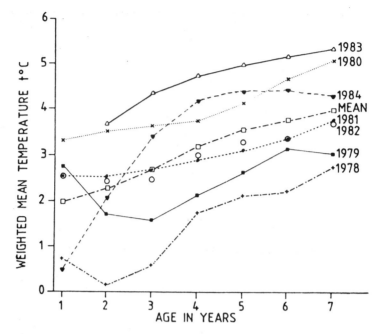

Fig. 2.38 Weighted mean values of bottom temperature for the various age groups of cod in February for the years 1978–84. The mean is the arithmetic mean for the whole period (Nakken & Raknes 1985).

Additional information on the effect of the environment on marine ecosystems can be found in Mann & Lazier (1991), who described in detail the essentials of biological–physical interactions in the oceans over different space and time scales. They deal mainly with plankton and sessile plant systems and with general production problems, but omit fisheries. Furthermore, they describe the processes that affect coastal ecosystems, including tides and estuaries.

Currents, especially wind-induced surface currents, carry marine populations and organisms from one area to another. Thus it can be concluded, and indeed observed, that closed or semi-closed current systems usually have the same pelagic ecosystems in their regions of distribution. The current systems also have their characteristic water masses and surface water types. Different water masses usually contain different zooplankton, some of which can be used as indicator species of these water masses. An example of the distribution of *Sagitta* species as an indicator of water masses and types is shown in Fig. 2.39 (Fraser 1952).

Corlett (1965) found some relationship between prevailing wind direction and wind-driven surface current in spring and summer zooplankton and recruitment of cod in the Barents Sea. This relation was further investigated by Saetersdal & Loeng (1983), who found that good recruitment of cod in the Barents Sea occurs when the temperature anomaly changes from negative to positive or is positive (Fig. 2.40). A positive temperature anomaly in the Barents Sea indicates increased transport of Atlantic water into the Barents Sea and consequently better transport of cod larvae from the Lofoten area.

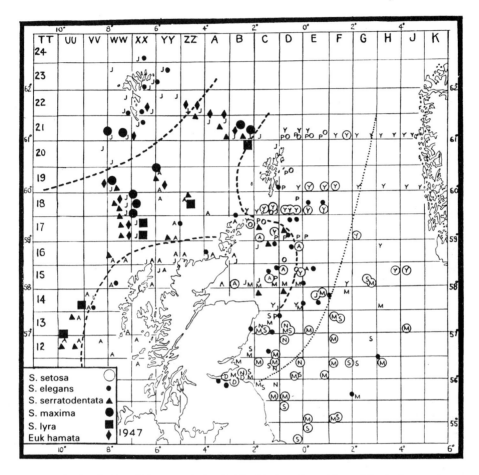

Fig. 2.39 The distribution of Chaetognatha in 1947. In this chart the positions of the plankton stations are shown by a letter indicating the month of the year: P = April, M = May, J = June, Y = July, A = August, S = September, O = October, N = November, D = December. The dotted line indicates the hypothetical northerly limit of distribution of *Sagitta setosa* in years of normal Atlantic inflow into the northern North Sea. The distribution of one species, *S. serratodentata*, in spring and summer is outlined by the broken line as an example of how the chart may be read (Fraser 1952).

The large-scale eddies of the currents might be places of formation of location specific marine ecosystems. The retention of fish larvae in eddies around banks has been pointed out by Fraser (1952).

The occurrence of low oxygen content in the ocean in the near bottom layers is rare and is usually caused by a 'catastrophic' event, a specific, lasting hydrographic or meteorological condition. Normally most of the oxygen uptake occurs in the water column rather than at the sediment surface (Thomas *et al.* 1976). Low oxygen content in the bottom waters can occur in some limited areas (e.g. in the German Bight, Dethlefsen & von Westernhagen 1983). Dead demersal fish and some benthic organisms are occasionally observed in these areas.

Sea ice causes some seasonal changes in marine ecosystems, although most high

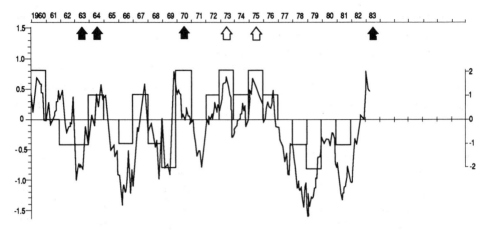

Fig. 2.40 Temperature anomalies in the Kola section in the Barents Sea in the period 1900–83 (continuous line). The histograms show the ice indices in the period 1900–83. Black arrows show year class with high abundance, and open arrows show year class with medium abundance of cod (Saetersdal & Loeng 1983).

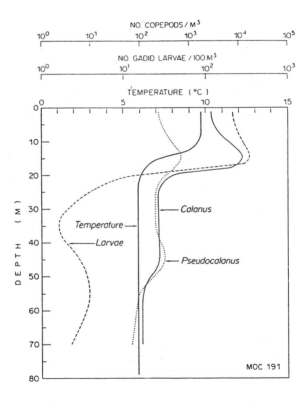

Fig. 2.41 Vertical distribution of gadoid (haddock and cod) larvae and dominant copepods *Calanus finmarchicus* and *Pseudocalanus sp.* in relation to the thermocline on the south-east part of Georges Bank before a storm (Lough & Laurence 1982). Note the different log scales used for copepods and gadoid larvae.

altitude marine ecosystems are adapted to seasonal occurrence of ice. Sea ice and ice margins can contain considerable phytoplankton blooms mainly diatoms (Alexander 1981), which can extent 50–100 km from the ice edge.

Many marine mammals are associated with sea ice in the Bering Sea, where about 1 million mammals use the ice for haul out and to bear their young. These mammals migrate to the Bering Sea with the ice in the autumn and leave when the ice retreats in the early summer.

One of the rare catastrophic events off the west coast of South America is El Niño. During an El Niño event (1982–3) one of the two surf clam species (*Mesodesma* sp.) disappeared from a sandy beach in Peru but had recolonized the beach after 3 years. However, another species (*Donax* sp.), which existed on the beach before El Niño, took over the space of the other, temporarily absent species. Arntz *et al.* (1987) found evidence (e.g. ancient mounds of *Mesodesma* shells) that such changes have occurred in the past and that recolonization might sometimes take decades and that the locations of recolonization might also change.

Storms affect the distribution of properties in the upper layer of the sea by turbulent mixing, which also alters the depth distribution of zooplankton and fish larvae (Figs 2.41, 2.42) (Lough & Laurence 1982). Pelagic larvae can be killed in the surface

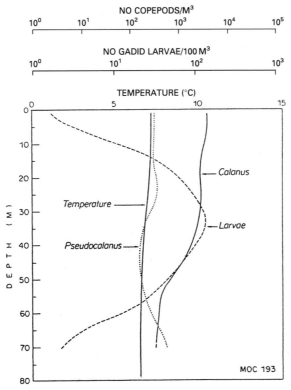

Fig. 2.42 Vertical distribution of gadoid (haddock and cod) larvae and dominant copepods *Calanus finmarchicus*, and *Pseudocalanus* sp. on the south-east part of Georges Bank after a storm (Lough & Laurence 1982). Note the different log scales used for copepods and gadoid larvae.

layers by the violent turbulence caused by storm-generated waves. Furthermore, strong wind-driven currents created by storms can cause anomalous advection of plankton and larvae and can cause fish to move from shallow to deeper water.

Recruitment failures are catastrophic events for fish populations. The causes of these failures usually remain unknown, although several hypotheses have been suggested (see summary in Laevastu & Favorite 1988). There is usually no relationship between stock and recruitment except at very low spawning biomass, and thus the effect of fishing rarely plays a decisive role in suppressing recruitment. For example, Winters *et al.* (1986) found that, for a herring stock off the east coast of Canada, egg production levels had no measurable impact on subsequent recruitment levels, and the relatively small stock had a remarkable degree of resilience. They concluded that the recruitment was controlled by density-dependent changes in larval abundance, which points to predation control. However, they also concluded that the environmental conditions during overwintering might affect spring spawning and might cause a change from spring to autumn spawning. However, no definite proof of this change was provided.

In summary, the environment is one of the determinants for an ecosystem. Consequently ecosystems are adapted to the environment (e.g. by latitudes and climatic zones, by distance from the coast and by depth of water). The environment can have anomalies and events (processes) which affect and influence the ecosystems and individual species in them. These influences are quite diverse and must be considered in relation to a specific species, ecosystem and event and questions asked pertaining to the changes in the ecosystem. Some essential processes affecting fish ecosystem are described in greater detail in Chapter 3.

Chapter 3
Processes in Marine Ecosystems and the Instability of These Systems

A number of physical, chemical and biological processes are simultaneously active in marine ecosystems. The intensities of these processes as well as their effects on the ecosystems vary in space and time. In this chapter we have attempted to summarize the essential processes that affect fish ecosystems and man's exploitation and management of them. The essential processes are predation, including the effects of top predators, migrations and other processes affecting distribution. The behaviour of fish, which is an integral part of the dynamic process in a fish ecosystem, is described in Chapter 4. All processes cause change in the systems; thus at the end of Chapter 3 we will be able to review the instability of the marine ecosystems.

3.1 Food interactions and predation in marine ecosystems

For over half a century the consideration of a simple food chain dominated the teaching of energy and matter transfer in the oceans. This simple food chain assumed that the primary production by phytoplankton is consumed by herbivorous zooplankton, which in turn is consumed by carnivorous fish of different stages, and use has been made of a simple transfer coefficient of 1 to 10 (i.e. 10% transfer) (Fig. 3.1). This concept also assumed that high phytoplankton production means a high yield of fish. Hundreds of papers and a few books have been published on this hypothesis and many simple mathematical models have been set up which compute the transfer of energy between the assumed, fixed trophic levels.

In the last few decades the complex and spatially and temporally variable food webs have been realized and assembled either in general terms for different sea areas (Fig. 3.2) or for different species (Figs 3.3, 3.4). Recently fallacies and shortcomings in the earlier simple food chain approach have been pointed out in some detail (e.g. Fransz & Gieskes 1984, Jones 1984). It has also been argued that there is an urgent need to understand the potential links between plankton production and fish recruitment and to evaluate whether environmental factors and fishing pressure limit the fish production (Brander & Dickson 1984).

One of the natural controlling principles in marine ecosystems is the concept of 'to eat or be eaten'. Predation in fish ecosystems starts on the first life stages of the system (i.e. eggs and larvae). Pelagic species, such as herring and sprat, can aggregate for feeding on spawning areas of demersal and semi-demersal species such as plaice and cod (Daan 1976, Daan *et al.* 1985). Herring eggs and larvae are also preyed upon by some carnivorous zooplankton organisms (Corten 1983) and the

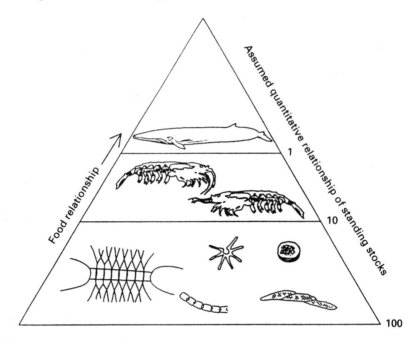

Fig. 3.1   Classic schematic presentation of pyramidal structure of life in the Antarctic seas.

sand eel (Christensen 1983). The copepod *Candacia armata* attacks herring larvae which are much bigger than the copepod itself. A gammarid amphipod *Calliopius laeviusculus* can consume about 16% of the total capelin egg deposition (DeBlois & Leggett 1993).

Fish larvae start to feed on small organisms, copepod eggs, copepodites, some phytoplankton and heterotrophs. With growth the main food becomes zooplankton, although a few small pelagic fish also feed partly on phytoplankton. Juveniles of demersal and semi-demersal species start to feed partly on benthos after settlement on the bottom, although planktonic and nektonic food is also taken, including larvae of their own and other species. Adult fish become more piscivorous and more cannibalistic with age and size.

Predation on benthos is dependent on prey size and availability (Arntz 1979) and may control the abundance of some benthic organisms. Predation intensity of benthos at a given location can vary from year to year with about one-half of the benthos production being consumed by commercial fish and the rest by non-commercial fish species and by predatory benthos.

The consumption of benthos varies considerably from species to species and in space and time (Fig. 3.5). Total consumption of benthos can be high in relation to estimated benthos production (e.g. 0.8 g C/m^2/year) in the southern North Sea by cod and haddock alone (Cramer & Daan 1986). The total production of benthos is usually not well known and underestimated.

The amount of fish in the diet of predatory fish increases with their age and size. The diet of fish also varies with season, being dependent on the availability of prey

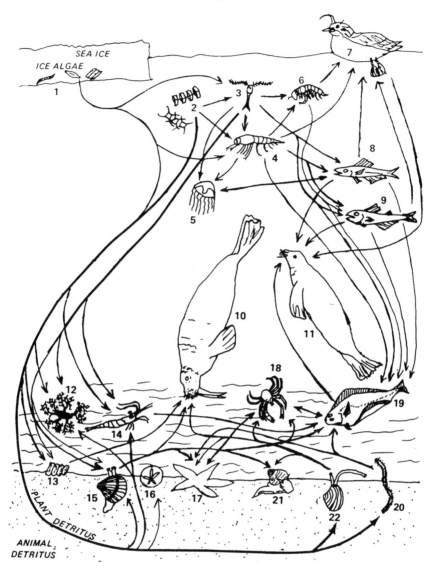

Fig. 3.2 The Bering Sea food web (a simplified version): (1) ice algae; (2) phytoplankton; (3) copepods; (4) mysids and euphausids; (5) medusae; (6) hyperid amphipods; (7) sea birds; (8, 9) pelagic fishes; (10) walrus; (11) seals; (12) basket stars; (13) ascideans; (14) shrimps; (15) filter-feeding bivalves; (16) sand dollars; (17) sea stars; (18) crabs; (19) bottom feeding fishes; (20) polychaetes; (21) predatory gastropods; (22) deposit-feeding bivalves (McConnaughey & McRoy 1976).

(Fig. 3.6). The food composition of ecologically similar species can be very different in different locations (Fig. 3.7). Fish predation on some fish species can be high, especially on smaller prey species. For example, in Icelandic waters the effect of predation by cod on capelin and fishing mortality of capelin are of a comparable order (Magnusson & Palsson 1989).

If a dominant pelagic species, such as the pelagic gadoid walleye pollack in the

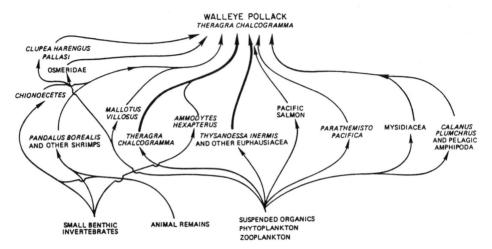

Fig. 3.3   A food web showing carbon flow to walleye pollack *Theragra chalcogramma* in the eastern Bering Sea. Bold lines indicate major food sources (Feder & Jewett 1981).

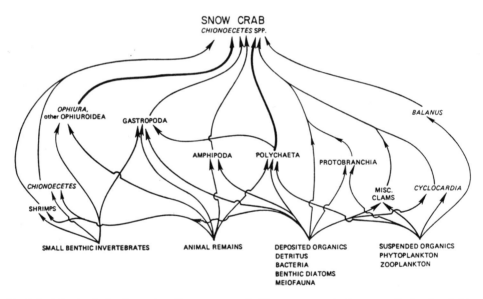

Fig. 3.4   A food web showing carbon flow to snow crab *Chionoecetes* spp in the eastern Bering Sea. Bold lines indicate major food sources (Feder & Jewett 1981).

North Pacific, is highly cannibalistic, cyclic changes of the stock size of this species can be caused by cannibalism: when the adult population is abundant, juveniles are under high cannibalistic predation. This results in reduced recruitment to the adult population in the near future and consequent decrease of adult population. In a few years the resulting smaller adult population exercises reduced predation pressure on juveniles, allowing higher recruitment and the future increase of adult biomass. Heavy fishing pressure on an adult population could also promote recruitment in

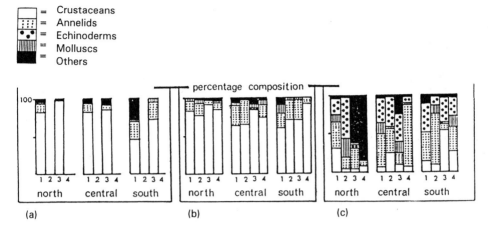

= Crustaceans
= Annelids
= Echinoderms
= Molluscs
= Others

Fig. 3.5 Feeding data for North Sea cod in (a) 1980 and (b) 1981 and haddock (c) in 1981 by area and quarter. (Cramer & Daan 1986). Percentage composition of benthic food by major taxa.

cannibalistic species, by removing older specimens, which are more cannibalistic and more piscivorous than younger specimens.

The food composition of highly migratory fish, such as tuna, can vary in space and time depending on the availability of food (Crawford *et al.* 1987). The food composition of pelagic Pacific salmon also changes with availability of food. This change is especially noticeable between offshore (pelagic) and coastal (neritic) ecosystems (Table 3.1, LeBrasseur 1972).

Only a small fraction of the available pelagic food in the open ocean is consumed by highly-migratory pelagic fish such as salmon; thus the oceanic food resource does not limit the carrying capacity of the pelagic fish ecosystem in the open ocean. Food density (availability in unit volume) might, however, be a limiting factor there for the growth rate (Favorite & Laevastu 1979).

Larvae and juveniles are mostly subjected to predation because of their small size and slow escape speed. Thus the survival of juveniles depends on the rapidity with which they (1) grow, i.e. pass through their most vulnerable prey size, (2) move into a more expansive environment with a reduced density of predators (e.g. to offshore waters), and (3) develop effective avoidance behaviour (i.e. greater size and swimming ability).

An 'escape size' (i.e. the size at which the mortality rate from predation has reached a recognizable 'levelling off') is graphically presented in Fig. 3.8, which also shows the relative change of natural mortality with age. The effect of the growth rate of juveniles on the time spent in the size range most vulnerable to high predation mortality is shown in Fig. 3.9. Finally, the relationship between the predation mortality coefficient and the time spent in the preferred prey size range is shown in Fig. 3.10.

Many, if not most fish, ecosystems in mid- and higher latitudes are predation-controlled, i.e. prey abundance and predator density-dependent predation on larvae and juveniles and the availability to the predator of alternative prey control

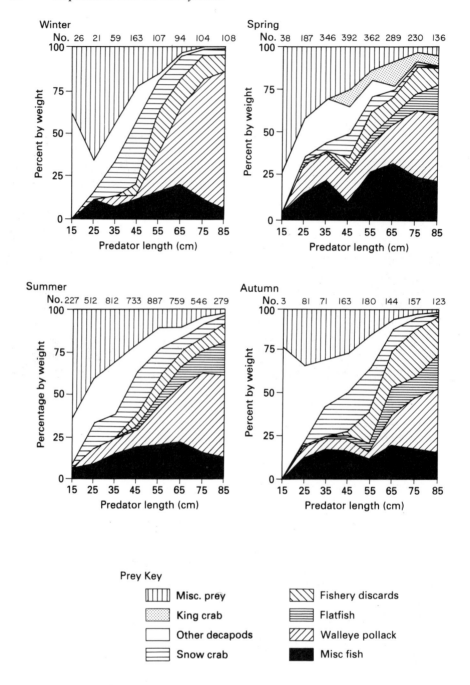

Fig. 3.6 Diet composition of Pacific cod, in terms of percentage by weight, season and predator size in the eastern Bering Sea (Livingston 1991). N = number of stomachs.

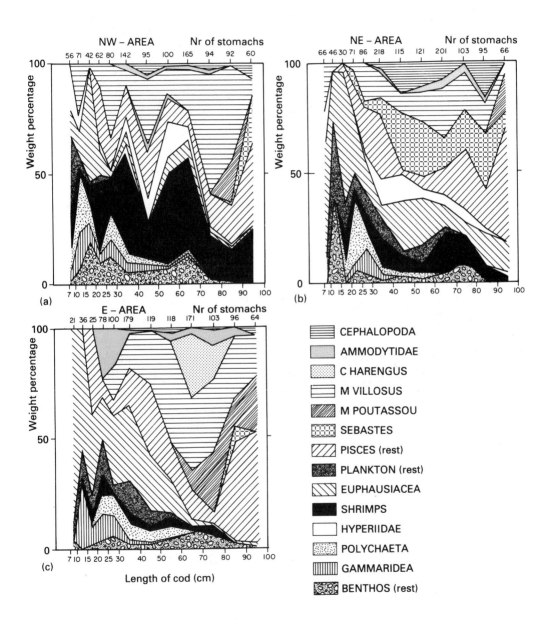

Fig. 3.7 The food composition of cod in Icelandic waters (weight percentages), in relation to predator length (cm). Based on material sampled in July, October–November 1980 and March 1981, which may be considered representative for the whole year (Palsson 1983). (a) The north-west, (b) the north-east and (c) the eastern area.

Table 3.1 Stomach contents of maturing salmon in the North Pacific by type of prey (% composition by weight) (LeBrasseur 1972).

	Herbivore	Primary carnivore	Secondary carnivore	Squid	Fish
Subarctic					
Sockeye	3	30	67	48	19
Chum	14	39	47	20	27
Pink	20	42	38	27	11
All species	12	37	51	32	19
Coastal					
Sockeye	6	82	12	3	9
Chum	4	60	36	1	35
Pink	12	33	55	2	53
All species	7	58	35	2	33

recruitment to exploitable stocks (Laevastu & Bax 1991, Sissenwine *et al.* 1982). Furthermore, predation influences most aquatic ecosystems. The predominance of predation over other causes of mortality is indicated by the observation that 74% of cod biomass production in the Bering Sea is removed by fish predation, the main predator being pollack, which is the most abundant species in the Bering Sea (Table 3.2). Mammals eat 9% of the annual production of cod. Cod is also one of the major piscivores in the eastern Bering Sea, with fish constituting 38% of its food.

Considering that predation is by far the greatest component of fish mortality and that it operates mainly on larvae and juveniles, it becomes obvious that predation will materially affect recruitment in most species. The predation rate depends on the relative sizes of predator the prey, their relative (size specific) abundance, and the availability of alternative prey. Thus predation is dependent on the relative density of prey items. There is ample empirical as well as experimental evidence for this (e.g. Gotceitas & Brown, 1993). Consequently, if recruitment were mainly affected by predation and predation is prey-density-dependent, recruitment would vary relatively little from year to year; long-term trends would be evident when the ecosystem composition changed.

Fish constitute about 22% of the food of fish in the eastern Bering Sea ecosystem (Fig. 3.11). Consequently the carrying capacity of a given ocean region with a high standing stock of fish depends not only on the production of basic food (zooplankton and benthos), but also on the availability and production of fish as food for other fish (i.e. the species composition and general trophic status of the fish ecosystem as well). If the level of a prey species such as capelin or herring in an ecosystem is low, the predator species might suffer partial starvation and the predation pressure on their larvae and juveniles would increase through cannibalism. Recruitment of a predominantly predator species (e.g. cod) would thus be reduced. This seems to have happened in the Barents Sea in the late 1980s (Hamre 1988, C.C.E. Hopkins, personal communication). The recovery of this piscivorous and other long-lived species in such

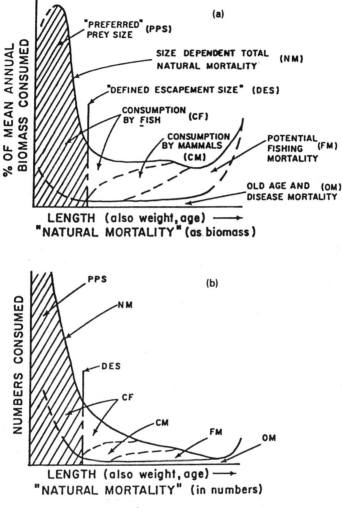

Fig. 3.8 Relative total annual mortality of fish. Preferred prey size and defined escape size are also indicated. (a) Biomass; (b) numbers.

a depressed ecosystem might be slow owing to the increased predation pressure on the larvae and juveniles of these fish, unless the recruitment of the main prey species (e.g. capelin) is favoured by some specific conditions. Growth rates of the piscivorous fish might also be affected.

Predation-controlled ecosystems such as the Bering and Barents Seas exist in other medium and high latitude regions. For example, Sherman *et al.* (1988) found that consumption of fish by all predators on Georges Bank is of the same order of magnitude as the total calculated production (Fig. 3.12). Usually predation of any commercial species in terms of biomass accounts for more than the harvest. In the Bering Sea the consumption by mammals and birds alone is about twice the total commercial harvest (see [3.4]). Similarly high fish consumption by the fish ecosystem

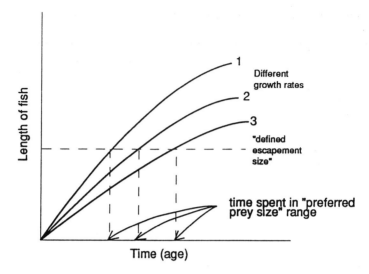

Fig. 3.9 Effect of growth rate of fish on time spent in 'preferred prey size' range.

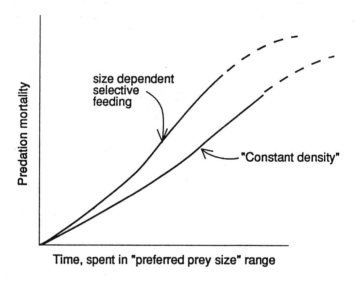

Fig. 3.10 Effects of time spent in 'preferred prey size' on predation (natural) mortality of fish.

has been computed for the North Sea, where saithe alone consume 434 000 t, of which 342 000 t is Norway pout and 60 000 t is haddock (Gislason 1983).

In most ecosystems the largest proportion of the fish biomass provides food for the fish themselves (Bax 1991) (Fig. 3.13). Intensive fishing, as in the North Sea, can change this balance somewhat. It should be noted that the composition of food can vary, not only by season but also annually (Fig. 3.14). In the example illustrated in Fig. 3.14 from the Barents Sea the reason for change of cod diet was the grazing down of the shrimp population and decrease of the capelin stock.

Table 3.2 Pacific cod biomass, its production and utilization in the eastern Bering Sea and Aleutian Region (area 900 000 km²).

	1000 t	%[a]
Total annual mean biomass (0.5 years to maximum age)	1470	
Exploitable biomass	956	65
Annual production	1352	92
Disease mortality	90	6.7
Catch (including discards)	150	11.1
Apex predation (mammals and birds)	115	8.5
Predation by fish	997	73.7
Predation on cod		
Pelagic fish	98	9.8
Semipelagic fish (pollock)	770	77.2
Demersal fish	130	13.0
Consumption by cod		
Zooplankton	1964	43.1
Fish	1577	34.6
Benthos	1016	22.3

[a] Percentage of exploitable biomass and annual production refers to annual mean biomass; percentages of mortality, predation, and consumption refer to the totals of these processes, respectively.

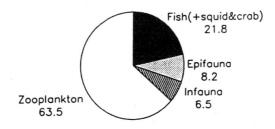

Fig. 3.11   Annual mean food composition of the fish component of the eastern Bering Sea ecosystem.

The knowledge of the food composition of species and the calculation of the consumption of various food items is necessary to make the biomass estimates for the different species within the ecosystem mutually compatible as indicated by their interactions through predation. For example, there must be sufficient pollack biomass in the eastern Bering Sea to sustain the estimated predation of 1.1 million t by marine mammals, predation by other fish (2.7 million), and cannibalism (up to 7.4 million t) in addition to an annual loss to the fishery of 1.1 million t.

3.2   Seasonal and life-cycle migrations

Plankton has no horizontal migrations but is advected by currents and has spatial and seasonal changes of species composition and abundance in any given location. The

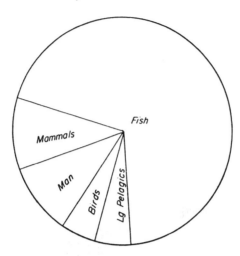

Fig. 3.12 Proportion of total Georges Bank fish production consumed by fish, marine mammals, man, birds and large pelagic (migratory) fish (Sissenwine 1990).

classical pictures of seasonal cycles of phyto- and zooplankton abundance (Fig. 3.15) hold only approximately, and great fluctuations in space and time can occur in these cycles (e.g. Adams 1987). Nellen & Schadt (1991) investigated the year-to-year variability of zooplankton in the north-west North Sea and found relatively great variations in zooplankton taxa as well as in its biomass from year to year, that appeared to be caused by advection. The only noticeable migration of zooplankton in some areas might be the seasonal vertical migration between deeper layers and the surface layers of the ocean.

Because of the great spatio-temporal variations and consequent patchiness it is difficult to detect year-to-year variations in the amount of zooplankton over larger areas. However, there are large annual differences in given fixed locations (Lie 1965) as well as differences between adjacent regions in the timing of peak abundance and spawning (egg production). Bainbridge & Corlett (1968) found that in 1963 the main spring spawning of *Calanus* had been completed by early June in the Davis Strait but was still proceeding in the adjacent Irminger Sea. Euphausiids showed the opposite trend.

Plankton development can show some long-term trends; e.g. phytoplankton increasing slightly in the 1970s and 1980s in Heliogoland waters (North Sea) (Fig. 3.16), whereas many zooplankton organisms in the North Sea showed a downward trend (Fig. 3.17). Colebrook (1985) found that the downward trend of the zoo-plankton occurred mainly in winter. Furthermore, he concluded that the zooplankton consists of an assemblage of species responding individually to environmental influences rather than reacting as an integrated plankton community. He also concluded that the link between primary and secondary production is weak.

Seasonal changes in most benthic communities are little known. There cannot be any seasonal migrations of the adults of sessile organisms and the possible seasonal migrations of epibenthic organisms is poorly known.

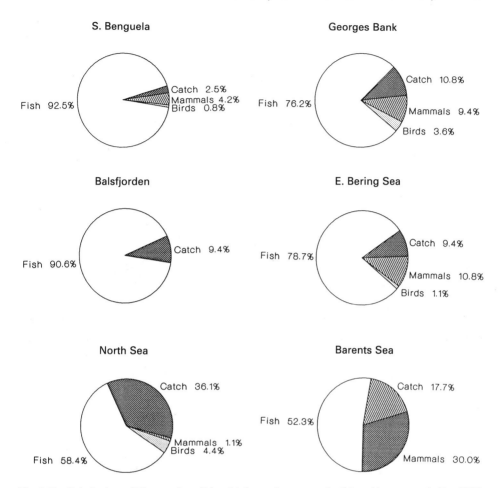

Fig. 3.13 Relative loss of biomass from fish to birds, marine mammals, fish and human catch (Bax 1991) (various data sources).

Fish larvae, which are part of plankton, are carried by surface currents, which are largely wind driven. There are year-to-year anomalies in surface winds that bring about differences in the dispersal of larvae between years. Saville (1965) concluded that there is a relationship between wind direction and strength and the distribution of larvae of Faroe haddock and Clyde herring. No such relationship was found in North Sea haddock.

Most fish migrate purposefully whether it be for feeding or spawning. Migrations can be seasonal or life-cycle migrations. These migrations result in changes in the species composition and relative abundance of any fish ecosystem in any given area, and make it difficult to define boundaries for marine fish ecosystem. Migrations make the identification of fish stocks difficult which in turn affects the management of fisheries on these stocks. The following example below shows some of the problems related to Norwegian coastal and Arcto-Norwegian cod stocks, as reported by Godø (1985):

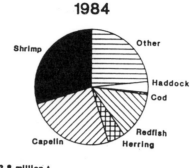

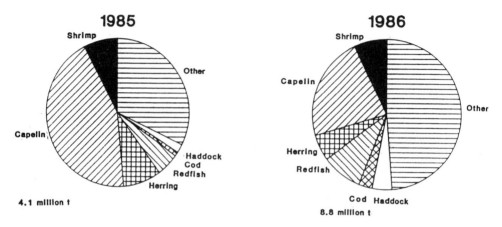

Fig. 3.14 Consumption by Atlantic cod in the Barents Sea from 1984 to 1986 (Bax 1991, reported by Torsvik 1987).

In early years the Norwegian coastal cod and Arcto-Norwegian cod stocks had been considered separate unit stocks on the bases of otolith patterns and analyses of haemoglobin polymorphism. These two stocks were also characterized by differences in migration. The coastal cod were less migratory in fjords and in coastal areas, whereas Arcto-Norwegian cod undertook the long migrations to the Barents Sea. Godø (1985) analysed the tagging experiments from the 1980s and found that there was a gradual change in the migration pattern of the non-migrating coastal cod in the fjords toward that of the migratory Arcto- Norwegian cod. No fjord cod was completely isolated from the cod stocks outside the fjords. Arcto-Norwegian cod also spawned on the spawning grounds of the local cod. The cod spawning off More and those spawning off Lofoten mingle in their feeding areas in the Barents Sea, but a great proportion of them return to their original spawning areas to spawn. Furthermore, cod spawning in the southern part of West Norway emigrated to the North Sea. Mingling of cod populations may also occur over very long distances (Harden Jones 1968).

There can be gradual life-cycle migrations of some species. Nedreaas (1987)

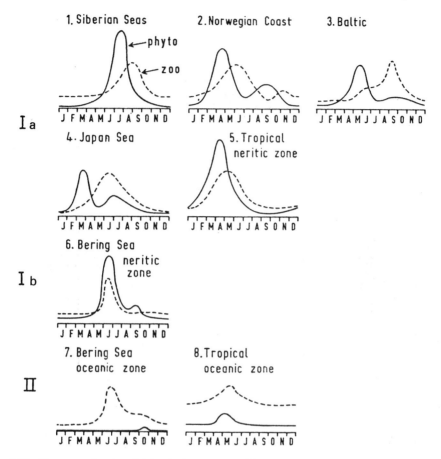

Fig. 3.15 The seasonal cycles of eight plankton communities. I. Cycles where the phytoplankton production and the animal grazing are unbalanced. In spring the phytoplankton biomass is greater than the zooplankton biomass. The seasonal fluctuations of the phytoplankton biomass are great.

(a) The spring maximum of the phytoplankton biomass precedes the maximum of the zooplankton biomass.

(b) The spring maximum of the phytoplankton biomass and the maximum of the zooplankton biomass occur simultaneously.

II. Cycles where the phytoplankton production and the animal grazing are balanced. The phytoplankton biomass is smaller than the zooplankton biomass. The seasonal fluctuations of the phytoplankton biomass are small.

The solid line is phytoplankton, the dotted line is zooplankton. The values of the zooplankton maxima are considered to be equal in all the graphs (Heinrich 1962a).

described a gradual migration of large two-year-old saithe away from the Norwegian coast, which did not, however, occur in all years and all locations. The main migration of three-year-old saithe away from the Norwegian coast occurred between February and June, when they feed mainly on krill and follow the krill offshore.

In many semi-demersal species there is a migration of older specimens into deeper water. For example, Sinclair (1992) showed that longline fishing effort was

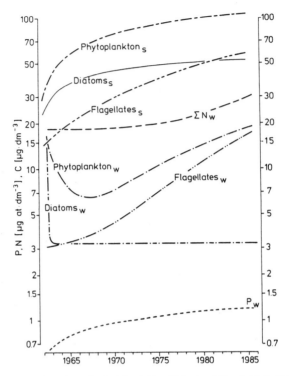

Fig. 3.16 Phytoplankton and nutrients in the North Sea, Helgoland Roads 1962–85: smoothed curves calculated from half-year means, s = summer (April–September), w = winter (October–March). Phosphate (P_w) and inorganic nitrogen ($N_w = NH_4 + NO_2 + NO_3$) near the surface are given in [µg at./dm^3], phytoplankton (C) as [µg/dm^3]. The curve for Diatoms$_s$ is not significant; the ordinate has a logarithmic scale (Gillbricht 1988).

concentrated in deeper areas where older fish were more abundant. An age segregation of cod by depth was observed on the Scotian Shelf, older fish moving into deeper water.

Seasonal migrations of fish in and out of marginal seas (e.g. the North Sea) and coastal areas is a common occurrence in medium and higher latitudes. Sparholt (1990) reported that between the first and third quarters of the year 0.5 million t of mackerel and 1.6 million t of horse mackerel move annually into the North Sea.

Other large-scale and year-to-year variable migrations of saithe, studied by tagging from 1954 to 1980, have been reported by Jakobsen & Olsen (1985). They observed mass migrations of saithe from northern Norway to Icelandic waters in some years (Fig. 3.18). From 1955 to 1959 the migration of immature saithe from about 63° N. was mainly northwards. After 1969 it was mainly southwards and recaptures were made in the North Sea. The unpredictable variations of the migrations between different defined saithe stocks have several consequences for fisheries management which have not been taken fully into account in the past, mainly owing to a lack of proper information.

Another example of a large-scale change in extensive migration routes has been

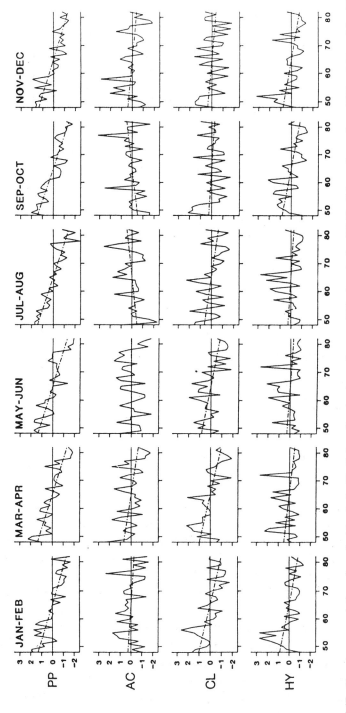

Fig. 3.17 Year-to-year variation of zooplankton in the North Sea. *Pseudocalanus elongatus* (PP), *Acartia clausi* (AC), *Calanus finmarchicus* (CL) and Hyperiidae (HY). Annual variations in abundance for each pair of months for the years 1948–82. The data in each plot (continuous lines) are reduced to zero mean-and-unit variance (abscissae) and the ordinates are in standard deviation units. Superimposed on each plot is a fitted straight line (dot-dashed line) (from Colebrook 1985).

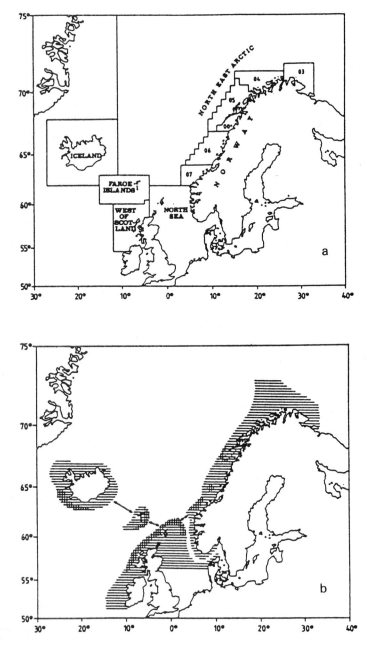

Fig. 3.18 Saithe in the north-east Atlantic. (a) Areas for the main stocks; (b) geographical distribution and main spawning areas. Arrows indicate probable migration routes (Jakobsen & Olsen 1985).

observed in the Atlanto-Scandian herring. Prior to 1969 it spawned along the Norwegian coast and migrated to the north and north-east of Iceland to feed and over-
winter. Its migration was greatly affected by hydrographical conditions and
availability of its main food, *Calanus* (Østvedt 1965). This migration pattern changed
drastically after 1969; herring did not then migrate to Iceland but remained in
Norwegian coastal waters, and some even moved to the southern Barents Sea.

While the migrations of most species in the North Atlantic are relatively well
known, the knowledge of migrations of North Pacific fish is meagre by comparison.
Best known are the migrations of the Pacific halibut (Skud 1977), based on a number
of tagging experiments. Considerable migration, both of juvenile and adult halibut,
occurred across the management areas (Fig. 3.19). Seasonal migrations of some
commercially-important species in the eastern Bering Sea were summarized by
Favorite & Laevastu (1981), based largely on commercial fishery data (Fig. 3.20).
Most of these migrations can be characterized as migrations into deeper water and to
continental slope in the autumn and back to shallower water in late spring.

The extensive feeding migrations of anadromous salmon have fascinated many
marine scientists in the past few decades. Examples of the ocean distribution of two
stocks of pink salmon with specific river origin are shown in Fig. 3.21. The navigation
of these species back to the river of their origin has given rise to many theories, none
of which can be fully tested. We have to accept the ability of the fish to navigate as an

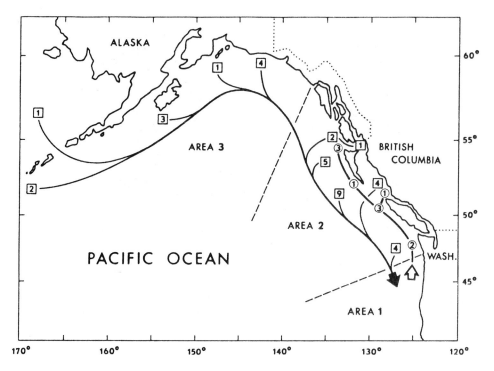

Fig. 3.19 Cross-boundary movements of Pacific halibut across the British Columbia–Washington border.
(Boxes show area of release for fish recovered south of Willapa Bay, Oregon; circles show recoveries of fish
that were released south of Willapa Bay) (Skud 1977).

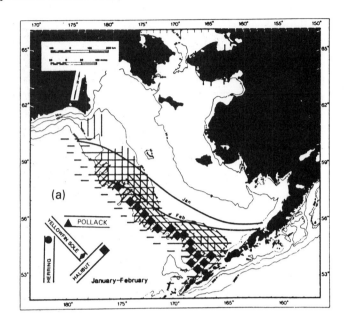

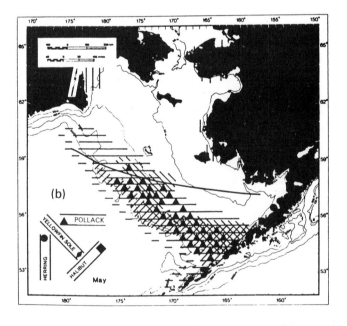

Fig. 3.20 Schematic diagrams of the distribution of halibut, herring, pollack and yellowfin sole in the eastern Bering Sea during various times of the year. (a) January–February (solid squares = spawning areas; long dashes = probable extension of shelf range and broad line = ice edge); (b) May (solid triangles = spawning areas; long dashes = probable extension of shelf range and broad line = ice edge). Some salmon present.

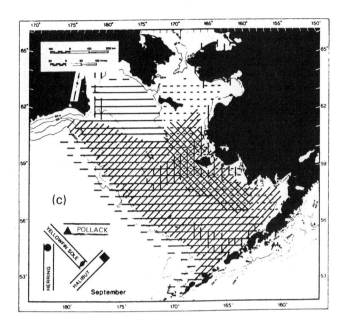

Fig. 3.20 *(Continued)* Schematic diagrams of the distribution of halibut, herring, pollack and yellowfin sole in the eastern Bering Sea during various times of the year. (c) September (long dashes = probable extension of shelf range; short dashes = juvenile distribution, where known to be different from adults). Salmon occur throughout the shelf area.

innate ability beyond the present comprehension of man. Some of these extensive migrations are also stock- specific within a given species (Kallio-Nyberg & Ikonen 1992).

Another example of transoceanic life-cycle migrations are the migrations of different species of tunas as known from tagging and commercial fisheries. Some of these migrations can also change with time, especially at their environmental boundaries. Examples of these changes are the occurrence of albacore tuna in some years off the Washington and British Columbia coasts and bluefin tuna in some years near Newfoundland. In the 1960s there was a bluefin fishery in the southern Norwegian Sea and the northern North Sea. Since that time bluefin tuna have not shown up in these areas.

Most of the fish migration must be considered as innate behaviour, developed over a considerable time span. Some environmental features, however, are known to affect the migration and aggregation of fish. Some of these features are the boundaries of oceanic currents, especially current convergences. More of these migration-influencing features are discussed in [4.3].

3.3   Effects of the environment on distribution

The effects of ocean environmental variables on the behaviour of fish have been summarized by Laevastu & Hayes (1981) and effects on distribution and abundance of fish caused by the weather and climate are described by Laevastu (1993). The

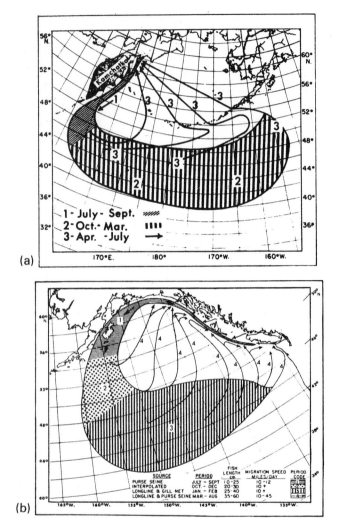

Fig. 3.21   (a) Ocean migrations of pink salmon stocks originating in south-eastern Alaska and British Columbia, and (b) probable migrations originating in East Kamchatka (Royce *et al.* 1968).

behaviour of fish as part of marine ecosystems, and also the aspects of fish behaviour pertinent to its capture, are described in Chapter 4. In this chapter we present only a few notes on the effect of environment on the marine ecosystems at large.

The distribution of benthos communities is largely determined by type of bottom and depth of water (see [2.2.2]). Large-scale climatological temperature changes might also determine the distribution of individual species in the benthos communities.

Plankton distribution is determined by the water type, its advection and in neritic plankters to some degree by distance from the coast. Many zooplankton organisms have diurnal and seasonal vertical migrations, some of which have considerable amplitude (e.g. 500 m) (Andersen *et al.* 1992). These migrations change the species composition and dominance in the plankton ecosystem at different depths.

Fish species have been classified according to their geographical distribution and temperature preference into arctic, boreal and tropical species. For example, polar cod and capelin in the Barents Sea are classed as arctic species, whereas cod and haddock are boreal species. However, Nilsson & Hopkins (1991) emphasize that this species–environment classification of communities cannot be applied in a strict sense, as discrete community boundaries are lacking and a given species which occurs in one predictable association is also likely to occur with another group of species under different conditions elsewhere.

The environmental requirement of any given species can vary throughout its life cycle and even seasonally. The environmental conditions also change seasonally and from year-to-year anomalies occur. The dynamic processes within the life cycle which are affected by the environment are schematically shown in Fig. 3.22. We can account for some seasonal changes in distribution and the corresponding relative regional abundance of some species, but year-to-year anomalies in distribution patterns are unpredictable. Sahrhage (1967) surveyed the distribution and density of the fish species in the North Sea during two summers (1959 and 1960) and two winters (1962 and 1963). He grouped the fish species in accordance with their distribution patterns. The following is extracted from Sahrhage's summary, illustrating the species-to-species differences in seasonal behaviour as well as response to environmental factors, and demonstrating the differences in seasonal distributions (Fig. 3.23), as well as changes in the fish ecosystem composition at any given location and time. Sahrhage classified the fish species in the North Sea according to their distribution into the following categories:

(1) Fish species with wide distribution, without strong seasonal variations: cod, haddock, whiting, long rough dab, dab, lemon sole, starry ray, bib and poor cod with rather wide distribution, and lesser silver smelt, Norway pout, hake, witch and megrim; the distribution areas of which extend into the North Sea from the north.

(2) Fish species with wide distribution and strong seasonal variations: grey gurnard, dragonet, plaice, sole, herring, sprat and spiny dogfish.

(3) Fish species with extreme seasonal variations in their distribution, summer immigrants: sardine, anchovy, red gurnard, horse mackerel and mackerel.

(4) Fish species with distribution clearly influenced by environmental factors:

   by water depth; lesser and greater silver smelt, blue whiting, coalfish, ling, blue ling, silvery pout (*Gadiculus thori*), redfish, black spiny dogfish, witch and halibut in deeper waters, and sprat, dragonet, plaice and sole in shallower waters:

   by bottom structure; lemon sole, witch and lesser weever:

   by water temperature; in particular sardine, anchovy, red gurnard and horse mackerel:

   the influence of salinity was not apparent.

(5) Fish species with long-term changes in distribution: haddock, witch, megrim and sole.

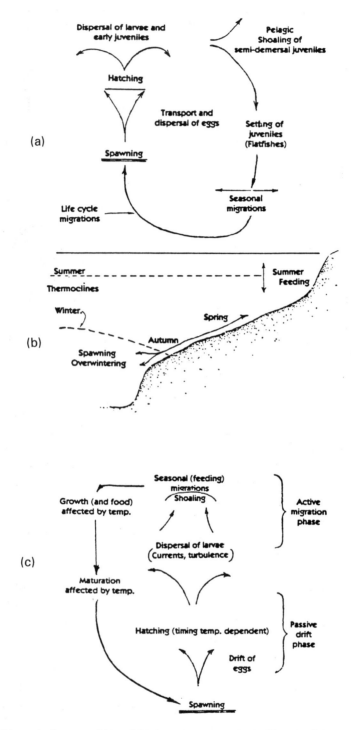

Fig. 3.22 Schematic diagrams of (a) and (b) dynamic processes in the life cycle of demersal fish which might be affected by the environment and (c) the effects of the environment (mainly current and temperature) on active and passive phases of the life history.

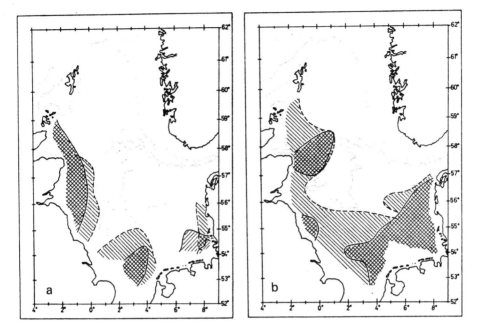

Fig. 3.23   Distribution of sprat (*Sprattus sprattus*) in the North Sea. (a) Summer 1959 and 1960, (b) winter 1962 and 1963 (Sahrhage 1967).

The effects of the environment on the abundance and behaviour of fish stocks has been summarized by Laevastu & Hayes (1981) and Laevastu & Favorite (1988). Numerous investigations have been conducted on the possible relationship between various environmental properties and the occurrence and distribution of fish. An example of the result of one such investigation is shown in Fig. 3.24, which depicts the relation between the occurrence of herring and sprat in the Baltic Sea in the autumn of 1968 to the thermocline and the more intensive mixing zone between intermediate and surface waters, where zooplankton abundance is higher than elsewhere. Such environment/fish distribution relations are species, location and season specific if they exist, and very few generalizations can be made in respect to ecosystems.

The influence of the environment on fish stocks can at times be gradual and difficult to ascertain, as was shown by Sharp & McLain (1993), who analysed the fluctuations of the Peruvian anchoveta stock and fishery and the environmental conditions before and during the collapse of this fishery during the 1982–83 El Niño. They found that the anchoveta collapse started as early as 1968 as part of a long-term ocean and atmosphere process, favouring the decrease of upwelling and coastal warming. Gradual changes occurred in the fish ecosystem from the dominance of coastal-upwelling species to more oceanic species. Sharp & McLain (1993) concluded that the ocean environment fluctuates relatively rapidly and can influence the populations to either increase or collapse. A fishery can influence these population fluctuations but cannot fully account for them.

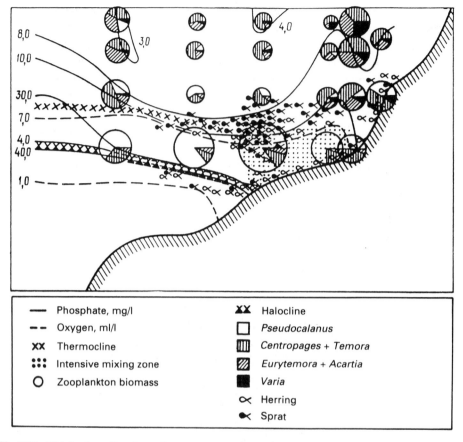

—	Phosphate, mg/l	XX	Halocline
--	Oxygen, ml/l	☐	*Pseudocalanus*
XX	Thermocline	▥	*Centropages + Temora*
⠿	Intensive mixing zone	▨	*Eurytemora + Acartia*
O	Zooplankton biomass	■	*Varia*
		∝	Herring
		✦	Sprat

Fig. 3.24 Distribution of herring and sprat concentrations in the central Baltic Sea in autumn 1968 as affected by biotic and abiotic environmental conditions (Ojaveer 1988).

3.4   Birds and mammals in marine ecosystems

Birds, marine mammals and humans are the top predators on the marine ecosystems. Table 3.3. summarizes the amount of fish biomass removed annually by these top predators from the North Sea and Bering Sea.

Table 3.3 Fish biomass and its utilization in the North Sea and the Bering Sea.

	Biomass (t/km^2)	
Category	North Sea	Bering Sea
Total finfish biomass	25.0	37.0
Catch (in 1980)	6.7	1.9
Consumption by mammals	0.1	3.1
Consumption by birds	0.3	1.1
Consumption by rest of the ecosystem (est)	(10.0)	(19.0)

All marine birds and many marine mammals (seals) breed on the coasts, whereas the breeding of whales and dolphins is independent of the coast. Marine birds are no longer utilized by man, but marine mammals have been harvested by man, possibly for thousands of years. Therefore it seems logical that these mammals should be considered and managed like any other living marine resource. However, some environmental groups have sentimental feelings for the protection of marine mammals, which are now unexploitable in many countries.

Marine birds, which are also top predators (Fig. 3.25), have so far escaped the attention of the protectors of nature as marine birds are not easily accessible to bird watchers. The following summary describes their effect on the marine ecosystem, mainly in respect of the quantities of fish consumed.

Fig. 3.25 Birds largely compete with man in exploiting the sea: cormorant swallowing a fish.

Wiens & Scott (1975) computed the predation of fish by sea birds off the Oregon coast to be 62 500 t annually. Forty-three per cent of the fish consumed was anchovy. A total of about 22% of the annual fish production in the neritic zone was consumed by birds.

The competition between fisheries and sea birds has been summarized in an excellent monograph by Furness (1982). Sanger (1972) estimated that in the sub-arctic Pacific region, sea birds consume up to 1.2 million t of food and void into this environment up to 240 000 t of faeces. Hunt *et al.* (1981) estimated that in the eastern Bering Sea alone birds consume 1.09 million t of food. The commercial catch of fish from this sea is about 2 million t. Thus there might be some competition with fisheries over the same resource. However, sea birds consume mainly larger zooplankton, fish larvae and juveniles and small pelagic fish.

Schaefer, quoted by Furness (1982), found that sea birds consumed 2.5 million t of Peruvian anchovy, or 17% of the stock annually, while the fishery took 7.5 million t.

In the North Sea region sea birds were harvested until about the end of the nine-teenth century. Many bird colonies have now shown an increase (e.g. the kittiwake populations between 1900 and 1963 increased by 3–4% per year and about 1% per year thereafter). Herring gulls have increased by 12–13% per year since 1930 (Furness 1982). Many sea bird colonies in Britain and Ireland are now showing signs of reaching their population ceiling.

Furness (1982) estimated that sea birds consume an equivalent of $4.0 \, kcal/m^2/year$ of pelagic fish, $0.1 \, kcal/m^2/year$ of offal and an equal amount of discards. He found that sea birds are not food-limited but mostly nest-site-limited, although food availability might determine breeding success in some colonies.

In South African waters, sea birds consume about 430 000 t of food per year (Crawford *et al.* 1991). There are about 2 million local breeding birds and 3 million migrants, with a total biomass of 5000 t. Of food consumed 31% is anchovy, 18% goby, 15% hake and 8% zooplankton. In the last 30 years there have been changes in the bird ecosystem, African penguins having decreased by about 400 000 birds and Cape cormorants having increased by some 500 000 birds.

Sea birds respond to changes in the marine ecosystem and their diet may be indi-cative of the collapse of some prey populations. However, data from sea birds cannot be used to indicate variation in the abundance of prey of particular fish populations on a regional scale (Adams *et al.* 1992).

Because man, marine mammals and birds prey on the same fish ecosystem, it is necessary to consider the nature of their respective predation on the same resource in greater detail in order to judge their effects on the fishery and *vice versa*.

A summary of a study on the trophic interactions of marine mammals, birds and fishery resources in the eastern Bering Sea by T. Laevastu & R. Marasco (MS) is given below to illustrate this interaction quantitatively.

The total annual fish consumption by marine mammals and birds in the Bering Sea is estimated to be 1.945 million t, which is close to the annual commercial catch in the eastern Bering Sea. The consumption of fish by birds is only about 9.7% of the total consumption by mammals and birds.

More than one-half of the fish consumed are semipelagic species, mainly pollack, the most abundant species in the Bering Sea. The highest consumption of pelagic and semi-pelagic fish occurs during summer, whereas the highest consumption of demersal fish occurs during winter.

The amount of fish consumed by mammals and birds is only about 4.4% of the mean standing stock of all fish in the eastern Bering Sea (about 44.7 million t), and consequently has no noticeable effect on fishery resources and fisheries. The main fish-consumers among mammals are the sperm and beluga whales, walrus, bearded and fur seals, and bowhead, grey and right whales (fish consumption of the last two species is disputable). The largest predators on returning salmon are the beluga whale, Steller sea lion, and older fur seals. This predation occurs during the summer spawning migration.

The nature of marine mammal predation on fish is variable from species to species and in space and time, and also depends on the availability of prey. Baleen whales in the Bering Sea feed largely on epibenthic amphipods and euphausiids, but also take fish, including herring, capelin, polar cod and pollack. Sperm whales, which migrate into the Bering Sea in summer, although feeding largely on cephalopods in the North Pacific, take a considerable amount of fish in the Bering Sea (about 20% of their total diet). Toothed whales feed largely on fish.

Some pinnipeds are exclusive fish and cephalopod feeders, whereas others (e.g. walrus) feed on larger benthic animals such as crabs. During the winter the benthos feeders also take demersal fish.

Many mammals feed on fish shoals, which they can locate by various means. Species selection is rare and feeding is subject to availability of prey and density-dependence. Thus both commercial and non-commercial species are eaten. Some mammals, e.g. sea lions and fur seals, have a tendency to follow fishing vessels, feeding on discards and attacking the catch in gillnets, on longlines and in trawls during hauling.

The bulk of the fish eaten by marine mammals are juveniles with an estimated median length of 15–20 cm. However, some mammals feed on larger, single fish (e.g. salmon, larger cod, pollack and halibut). Among the predators of large fish are the beluga whale, Stellar sea-lion and killer whale. Some marine mammals (e.g. the killer whale and rarely the walrus) even attack smaller mammals.

We could assume that of the 1.74 million t of fish eaten by mammals more than half (i.e. 0.87 million t) are commercial species. Most of the commercial species eaten by mammals, like those eaten by birds, are pre-fishery juveniles. For our relative magnitude estimates we can assume that 15%, i.e. 0.13 million t, of the commercial species eaten by mammals are of exploitable size. This amount is only 0.7–1.3% of the total commercial fish biomass in the Bering Sea (estimates vary between 10 million and 20 million t).

Some of the pre-fishery juveniles (0.74 million t), if not eaten by marine mammals, might grow and recruit to exploitable stocks. We assume that a relatively high proportion, 50%, of the biomass survives (i.e. 0.37 million t). Furthermore, the surviving fraction would grow before recruitment to the fishery; plausible assumption is a threefold increase in weight. This would add an additional 1.11 million t to the exploitable stocks, which would give a total of 1.24 million t of exploitable fish removed by marine mammals and birds or 6–12% of the existing exploitable stock. This prospective addition is well within the limits of accuracy of any resource assessment.

This 6–12% potential increase in the exploitable stock could be achieved only if all marine mammals and birds were exterminated and the carrying capacity of the Bering Sea would allow an increase of fish biomass. Even if this small increase of spawning stock occurred, it would not affect any future recruitment, which is largely controlled by fish predation. From a fisheries point of view, a 6–12% increase in exploitable stock would not translate into a noticeable increase of either catch per unit effort (CPUE) or total catch. The magnitude of fluctuation of stocks through natural causes (e.g. by various environmental anomalies in temperature) or man's fishing is expected to be larger than the total effect of predation by all mammals and birds.

A comparative view of the possible competition between mammal and bird ecosystems and the fish ecosystem can be obtained by comparing the annual consumption of zooplankton and benthos. The zooplankton consumption by mammals and birds is only 1.5% of the corresponding consumption by the fish ecosystem, and the consumption of benthos by mammals and birds is 5.5% of that by the fish ecosystem. Thus we can conclude that essentially there is no competition between mammals and fish for the basic food resource that exists in the Bering Sea, and that predation by mammals and birds would not affect the total carrying capacity of the Bering Sea.

Carrying capacity is also affected by the state of the fish ecosystem. If fish prey ('forage fish', larvae and juveniles) are low, the whole fish ecosystem would be lower and/or there would be a shift to higher zooplankton and benthos consumption. If drastic natural fluctuations of the whole fish ecosystem did occur (e.g. as in the Barents Sea in the late 1980s), some mammal species would seek food elsewhere.

The number of marine mammals fluctuates in time, and censuses of the Steller sea lions (Fig. 3.26) in the Gulf of Alaska and in the Bering Sea seem to suggest that populations have decreased from about 140 000 in 1960 to 65 000 in 1985 and to 22 000 in 1989. This decrease in seal lion numbers has caused some environmental groups to blame the commercial fisheries for the decline and they would like to shut down the whole fishery in this region (Loshbaugh 1992). Although changes have occurred in fish ecosystem in this region since 1950 (Alverson 1992), no single factor can be identified as the cause of the decline of the sea lion. Two factors cannot be ruled out, great uncertainties in the censuses themselves and eating of sea lion pups by the orca or killer whale. Fourteen sea lion pup tags were found in the stomach of a single orca by Lockyer (1993).

The increasing populations of marine mammals, especially seals, consume considerable quantities of marine fish in a number of regions. For example Canadian harp seal consume over 1 million t of fish each year (Harwood 1984), and especially valuable anadromous fish (salmon) during their homing near the estuaries (Fiscus

Fig. 3.26   The sea lion, the most terrestial of all pinnipeds.

1978). Because of the damage done to fisheries by marine mammals, especially seals, calls for culling of their populations fall foul of public opinion aroused by environmentalists, despite the fact potential benefits of seal culling have been well investigated and proved (Gulland 1987a, Butterworth *et al.* 1988).

Mainly because of the cessation of the international fur trade, the increase in seal abundance has been rapid in many parts of the world. Butterworth (1991) reported that numbers of the Cape fur seal doubled from about 1 million animals in 1970 to 2 million in 1990 and is expected to double by the end of this century. The 2 million animals consume about 2 million t of fish, which is the same quantity as taken by commercial fisheries in South African and Namibian waters.

In the north-east Atlantic Ocean there are 80 000 minke whales, 200 000 hooded seals and 750 000 harp seals. These species also consume 2 million t of fish, which is of the same order of magnitude as the total catch of the Norwegian fisheries.

The Antarctic ecosystem formerly contained the greatest number of marine mammals, especially whales (Fig. 3.27). In the early 1960s about 25 000 finbacks, 5000 sperm and sei whales, 1000 blue whales and 250 humpback whales were taken

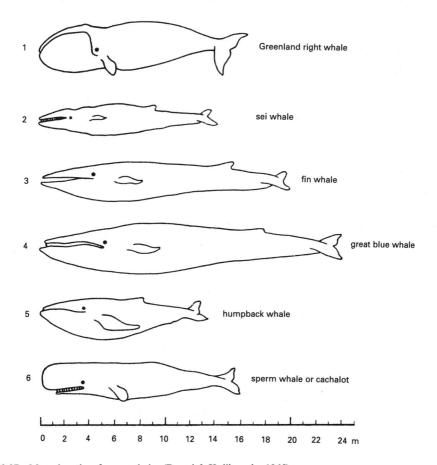

Fig. 3.27 Mean lengths of some whales (Demel & Kulikowsky 1965).

annually from Antarctic waters. Now that the whaling has stopped there are no reliable estimates of whale abundance in Antarctic waters.

The main food for the whales in the Antarctic has been krill. It is estimated that in the 1950s about 150 million t of krill were consumed by whales. With the decline of the whale population more krill has become available to birds and seals and their abundance has increased (Furness 1982). It seems likely that squids and fish have responded in a similar way. Present consumption by Antarctic sea birds, mainly penguins, is estimated at 15 to 20 million t of krill, 6 million to 8 million t of squid, and 6 million to 8 million t of fish per year (Everson, quoted by Furness 1982).

The consumption of zooplankton, fish and squids off the east coast of the USA has been estimated to be 0.17, 1.30 and 0.31, respectively (Scott *et al.* 1983). The gross annual cetacean biomass production on Georges Bank was estimated to represent about 14.5% of annual finfish production.

The consumption of fish and squids by sperm whale in the North Pacific is presented in Table 3.4, which suggests that consumption of fish by this cetacean alone exceeds the total catch by man for this region.

Table 3.4   Estimation of consumption of fish and squids by sperm whale in the North Pacific.

175 000[a] harvestable sperm whales in the North Pacific
30 t mean weight
= 5.25 million t biomass
5%[b] BWD (body weight daily), food requirement
= 18.25 times body weight annually
= 95.81 million t total food consumption

Food composition:
    85%[c] squids
    15% fish
Annual consumption by sperm whales in the North Pacific:
= 81.4 million t squids
   14.4 million t fish
Assuming $F_{max} \approx 20\%$[d], the minimum biomass of squids in the North Pacific $\approx$ 400 million t

[a] This is an absolute minimum estimate (Int. Whaling Comm. Spec. Issue 2, 1980). The total number of sperm whales in the North Pacific is estimated for 1977 as: females 411 000–525 000; males 376 000–474 000.
[b] The food consumption of whales is estimated in literature to 4–6% BWD. The minimum estimate is 2.5% BWD.
[c] Some estimates give up to 95% squids.
[d] This fishing coefficient of squids by sperm whale is probably too high; it corresponds roughly to F of pelagic fish.

3.5   Instability of marine ecosystems: its causes and consequences

The study of marine ecology, oceanography and other related marine sciences is in essence the study of changes of the marine environment and its living and non-living content, and their causes. Because marine ecosystems change in space and time, they are basically unstable. Thus two of the main points to be considered are the limits of these changes and whether there is a mean or average state of each of the systems around which they fluctuate. In this section we deal only with some specific aspects of the changes in marine ecosystems. The other chapters in the book also contain some information on changes and their causes.

The seasonal (Fig. 3.28) and regional variability of phytoplankton is a common phenomenon (see [2.2.1]). Large changes in phytoplankton and basic organic production can occur in upwelling regions when upwelling is brought to near-cessation (El Niño) during the shifts of large-scale wind systems (Blanchet *et al.* 1992).

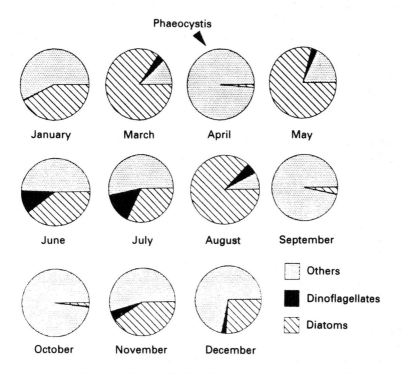

Fig. 3.28 Average monthly ratio of diatom, dinoflagellate and other species cell numbers at station Noordwijk 10 off the Dutch coast in 1990 (Colijn 1991).

Another phenomenon of phytoplankton, plankton blooms, has recently received attention because public awareness (especially of shellfish poisoning) has forced scientists and policy makers to investigate these blooms and give explanation for their occurrence. Intensive plankton blooms can occur owing to mass development of usually naked dinoflagellates such as *Gymnodinium flavum* ('yellow water'). *Gonyaulax (Alexandrium) polyhedra* ('red tide') and others.

Phytoplankton blooms and red tides were as impressive in the nineteenth century as they are today (Gillbricht 1988, Colijn 1991). Although eutrophication has been suggested as a major cause of the plankton blooms, there is no convincing evidence to that effect and their occurrence is not distinguishable from natural variations (Colijn 1991). Gillbricht (1988) wrote as follows:

'Bloom has nothing to do with an explosion of the population but is the consequence of an extended growth period (diatoms) and a relatively large initial stock (flagellates) with a slow-speed decline afterwards'.

Although the zooplankton component of plankton might have more moderate seasonal variations than phytoplankton (see [2.2.1]) and large-scale changes in general are moderate (Sherman *et al.* 1982a) some long-term and large-scale changes have been observed (Colebrook 1985), some of which have affected the feeding migrations of large stocks of pelagic fish (herring) (Jakobsson 1978).

The fluctuations of fish stocks and their possible causes have been summarized by Laevastu & Favorite (1988). There follow some notes pertaining mainly to ecosystem aspects of fish stock changes.

A characteristic of many pelagic, and a few semi-pelagic, fishes is that they have periodic outbreaks of highly successful recruitment upon which a fishery is often based for a number of years (Branch *et al.* 1987) (Fig. 3.29).

A common change in the fish ecosystem is the replacement of larger tertiary predators like gadoids or other abundant pelagic species, (e.g. herring and mackerel) by smaller, faster growing opportunistic plankton feeding fishes, (e.g. sand eel, Norway pout and sprat). Such changes in the fish ecosystem have been observed on both sides of the North Atlantic (Sherman *et al.* 1982b) (Fig. 3.30). The pelagic ecosystem is often largely controlled by the temporal variability of the physical environment. Species seldom compete for space or food in this system. Owing to the great mobility of the species, the localized depletion of food has little effect on the system.

While the pelagic ecosystem is relatively monotonous and physical variations can dominate the changes, rocky shores are regarded as areas of high variability, both spatially and temporally (Branch *et al.* 1987). In these areas it is often difficult to detect the cause of the change in community structure, whether it is caused by biological interactions or by changes and patchiness in the environment. Biological interactions in nearshore areas and an intertidal rocky coasts are considered to be the dominant cause of the change. The interacting factors controlling diversity in the pelagic upwelling system and in the benthic rocky shore ecosystem have been described by Branch *et al.* (1987) (see also Figs. 2.32 and 2.33).

In benthic ecosystems variability can be high, and episodic events such as storms affecting benthos in shallower water can occur. Much of the variability of the subtidal zone is controlled by biological interactions, especially predation. Furthermore, any particular species may act as 'keystone' species controlling the community. Natural fluctuations in biomass and recruitment in exposed benthic communities can be of considerable magnitude. Beukema (1982) studied annual variation in recruitment and biomass of five species in a tidal flat area of the Wadden Sea for 13 years. Some of his results are given in Table 3.5. Biomass tended to be more stable than numbers of recruits because of the buffering influence of several age classes present. The more age classes of adults there are in the total biomass, the more stable is the biomass. Successful and poor years for recruitment were roughly the same for the four bivalve species. Particularly heavy spatfall was found during the summer following the severe winter of 1978–9.

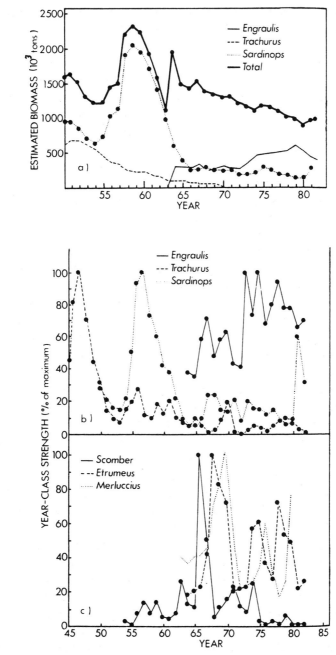

Fig. 3.29 (a) Estimated total biomass and (b) and (c) strength of recruitment of various species of neritic fish in the South African Western Cape sector of the Benguela (Branch *et al.* 1987).

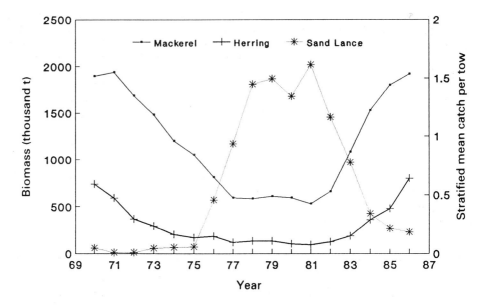

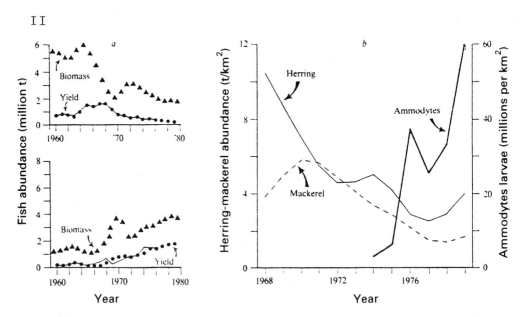

Fig. 3.30 (I) Trends in biomass of mackerel (age 1+) and herring (age 3+) in the north-west Atlantic derived from virtual population analysis, and trends in relative abundance (stratified mean catch in kg per tow) of sand lance (age 2+) based on research vessel surveys (Sherman *et al.* 1992).

(II) (a) Decline in herring and mackerel biomass (▲) and yield (●) (upper graphs), and subsequent rise in biomass (▲) and yield (●) (lower graphs) of small, fast-growing sprat, sand eels and Norway pout in the North Sea from 1960 to 1979. (b) Decline in herring (——) and mackerel (- - - -) biomass in t/km^2 and subsequent increase in larval *Ammodytes* (——) abundance in millions of larvae per km^2 for the north-west Atlantic shelf from Cape Hatteras to the Gulf of Maine (1974–9). Herring densities represent stocks ranging from southern New England to western Nova Scotia; mackerel densities include stocks ranging from Cape Hatteras to Labrador (1968–79), (Sherman *et al.* 1982).

Table 3.5 Annual mean values for biomass and recruitment, averaged from 15 sampling stations on Balgzand, Dutch Wadden Sea, for the five species contributing most to the macrobenthic biomass. Biomass values were obtained in late winter and numbers of recruits in late summer for the bivalves and in winter for *Arenicola* during the period 1969–81. Coefficients of variation (i.e. $100 \times$ standard deviation/mean) are all shown.

Species	Biomass (g/m^2)			Recruits (n/m^2)		
	Range	Mean	Coefficient of variation	Range	Mean	Coefficient of variation
Mya	2–11	5.4	53	2–933	114	222
Cerastoderma	1–16	4.4	102	16–507	138	111
Arenicola	3–5	4.1	18	1–8	4	60
Macoma	1–3	1.8	33	45–581	198	75
Mytilus	0–7	1.5	114	0–575	129	164
All other	2–7	3.6	41			
total	15–39	20.7	31			

Pronounced changes in the feeding migration of herring have been reported for Atlanto-Scandian herring by Jakobsson (1978) and for North Sea herring and mackerel by Corten & van de Kamp (1991). Whereas the change in the migration routes of Atlanto-Scandian herring was most probably caused by the changing feeding conditions (availability of *Calanus*), the changes in North Sea mackerel and herring are assumed to be caused by the changes in the flow of Atlantic water along the west coast of Scotland into the North Sea, but there is no physical oceanographic evidence that such changes indeed occurred.

Many changes difficult to explain have occurred in several fish ecosystems in different parts of the oceans, some of which have caused hardships to fisheries. A recent severe example is the drastic sudden decline of the adult northern cod stock off the Newfoundland and Labrador coasts (Young & Rose 1993). The disappearance of cod caused the announcement of a 2-year moratorium on fishing for northern cod. It is generally believed that the pronounced decline of northern cod was caused by heavy fishing and recruitment failures. However, there are six plausible causes, any one or more of which might have caused the disappearance of several hundred thousand tonnes of cod:

The first possibility is that the 'missing' cod never existed. The 1991 estimate of the stock was 1.1 million t, whereas in 1992 the size of the stock, consisting mainly of pre-fishery juveniles and pre-recruits, was estimated at only 600 000 t. However, an error of 50% in the stock assessment, especially over-assessment, is unlikely. The second possibility is overfishing. However, the reported Canadian catch was only 127 000 t and foreign unreported catch was estimated at 50 000 t. The third possibility is unusual natural mortality. However, no sign of mass mortalities, e.g. floating dead fish, has been observed and cod can tolerate sub-zero temperatures.

The fourth possibility is unusual emigration, but so far greater quantities of cod have not been found in Greenland, Grand Bank or in offshore waters. The fifth possibility, and a quite likely one, is heavy predation by seals. Harp seal populations have increased steadily and rapidly in recent years, and one of their main food items, capelin, has been decreasing. The sixth cause of decline might be some unobserved environmental change, either exceptional cooling or pronounced anomalies in the current system. Environmental anomalies are likely to affect recruitment rather than cause disappearance of adult fish. It seems that the causes of northern cod declines cannot be satisfactorily explained at present.

Sherman *et al.* (1988), discussing the changes in fish ecosystems which have occurred along the east coast of United States, pointed out that these are dynamic ecosystems and that wide fluctuations in abundance of the principal fish species are common, with significant changes in species composition and total biomass occurring from time to time.

The fish ecosystem is essentially a predator–prey system. Nearly all species can be at the same time a predator and a prey, and this relationship changes within the life cycle of the species, the juvenile stages being more subject to predation. If the biomass, either of the predator or of the prey, changes for any number of reasons including a different migration pattern for juveniles and adults, the total predation conditions of the ecosystem will change.

The various possible types of responses of predator and prey biomass are shown in Fig. 3.31, and although these responses can be classified into some defined types, there is considerable overlap in nature owing to various dynamic predator–prey processes (i.e. changes in space and time). Some of the types of response are as follows. When the predator biomass decreases (e.g. because of emigration), local prey abundance increases (type A); and when predator biomass increases, prey decreases (type B). When the secondary prey becomes predator to the primary prey (i.e. the secondary prey is a competitor to the primary predator), the predator biomass decreases (types C and F). If the prey is motile and its abundance decreases (e.g. by emigration), the predator biomass might decrease as well (e.g. forced emigration in search of food and/or because of starvation (type D) and if prey biomass increases then predator biomass might also increase (type E)).

Ursin (1982) reviewed the stability and variability of marine ecosystems, and found that, despite the variability of fish ecosystems, some resilience and some stability (or rather limits of fluctuations) can be observed in these systems. This stability is mostly controlled by one ill-defined process, density dependence. For example, a female cod spawns over 1 million eggs, which develop into larvae and juveniles and are reduced in about 5 years to two (say, a male and a female). The female again spawns 1 million eggs. If there were four cod left at the age of 5 years, the recruitment to the spawning stock is said to be very strong.

The process of reduction of the numbers of fish from eggs to adults, i.e. the recruitment process, varies from year to year in any given species. An example of recruitment variations is given in Table 3.6.

Ursin (1982) pointed out that if a species that is essentially a predator has one or more strong year classes, a nearly catastrophic change in the ecosystem would occur. However, the ecosystem is buffered in many ways against such catastrophe, e.g. by change in growth rate, changes in cannibalism, the appearance of otherwise sup-

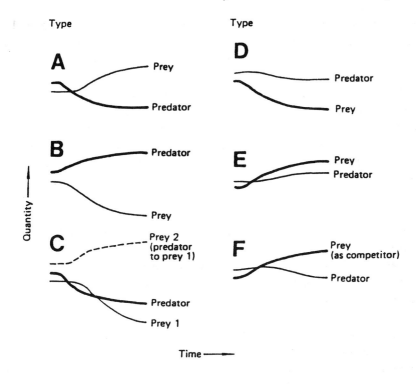

Fig. 3.31   Types of responses to biomass changes in predator–prey controlled ecosystem.

pressed predator in some stage of food chain and prey switching, discussed by Ursin (1982).

Some characteristics of fluctuations of marine biomass can be computed with holistic ecosystem simulation models. Some typical periods and magnitudes of fluctuations and annual rates of change are shown in Table 3.7. In addition, the following observations can be made:

(1)   The rates of change of biomass are dependent on the type of the species (e.g. flatfishes, semi-demersal fishes). Further, the magnitudes and periods of fluctuations are influenced by the life span and the rate of growth of the species.
(2)   The rates of decline tend to be more rapid than rates of increase.
(3)   The computed mean periods of fluctuations (4–7 years) correspond roughly to the periods observed in empirical data. The observed periods are irregular fluctuations due to irregularities in the occurrence of factors causing them, as discussed above. Increased growth rate, early maturation and prolonged period of spawning are population-stabilizing mechanisms.

Marine ecosystems as a whole fluctuate within some limits, which are difficult to ascertain as we cannot determine or define clearly an average condition of quanti-

Table 3.6 Recruitment variation in North Sea fishes 1963–75. Numbers adjusted to a value of 100 for the most outstanding year class of each species. 'Ratio' is the ratio of the numbers in the strongest and the weakest year classes (Ursin 1982).

| | | Gadoids | | | | Flatfishes | | Clupeids | | | Mackerel | |
| | | | | | Norway | | | | | | (two assessments) | |
Year class	Cod	Haddock	Whiting	Saithe	pout	Plaice	Sole	Herring	Sprat	Sand-eel		
1963	52	1	14	17	4	100	100	100			10	
1964	49	1	26	23	6	29	21	53			26	
1965	70	2	30	18	0.5	28	11	47			43	
1966	63	12	37	50	7	25	11	66	59		62	
1967	20	100	100	51		21	18	65	62		10	
1968	19	6	33	55	2	27	9	36	37		16	
1969	82	2	30	29		32	26	78	41		100	100
1970	100	14	33	29	45	25	6	62	20	100	9	12
1971	18	21	68	30	7	20	14	41	19	21		17
1972	35	4	90	40	16	62	19	19	46	47		4
1973	31	21	63	100	100	40	18	47	91	28		15
1974	51	40	92	27	38	25	7		100	86		11
1975	27	9	37	50	18	37	22		79	41		4
Ratio	6	100	7	6	200	5	17	5	5	5	11	25

Table 3.7 Some typical annual rates of change of biomass (with reference to mean equilibrium biomass B_e) and typical periods and magnitudes of fluctuations of biomass $(B_1-B_2)/B_e$). B_1 and B_2 are the biomass in January 1 year apart.

| | Annual change in percentage of B_e | | Fluctuations | |
Species/ecological group	Range	Mean	Mean period, years	Mean magnitude (in % of B_e)
Flatfishes	7.5–13	10.5	6–7	45
Sculpins and other non-commercial demersal	8–26	12	5	40
Cod	7–20	13	7	35
Pollock	11–34	19	5	75
Herring	20–50	38	5	80
Squids	35–70	50	4	100
Benthos	65–80	75	4	120

tative species composition. We can expect, based on past experiences, that relatively large changes in individual species biomass can occur, but we have not always been able to ascertain the cause of these changes. The apparent resilience of fish ecosystems varies from one system to another. Tropical demersal fish ecosystems seem to be either less resilient than pelagic systems or their fluctuation cycles are longer than those in higher latitudes.

Temporal and spatial changes of individual components occur in most ecosystems (see also [4.4]). There are few aquatic ecosystems where pronounced long-term changes of increased production have occurred. One of these is the Baltic ecosystem, which changed owing to eutrophication (Elmgren 1989) (see also [6.2]). Any possible effect of minor climatic change on marine ecosystems is highly speculative (Laevastu 1993).

Carrying capacity, though an imprecise term, usually refers to the maximum fish biomass that can be sustained by the basic aquatic fish resources, organic production, zooplankton and benthos, in a given region. However, the standing stock and production of fish biomass also depend on the species composition, ecological characteristics and especially on the principal food and feeding habits of the components of the fish ecosystem itself. Environmental physico-chemical properties and processes in the given region also affect the standing stocks of fish. The knowledge of these factors and processes is necessary before attempts are made to maximize the economic production from the sea. (See also Chapter 8.)

Overholtz & Tyler (1985) investigated the response of demersal fish assemblages of the Georges Bank during a period of heavy fishing pressure, 1965–74, and found that although the total biomass of targeted species declined while fishing effort increased several times, the assemblages maintained their temporal and spatial integrity and configuration.

Finally we must recognize one fundamental, though not necessarily overwhelming, cause of instability of fish ecosystems, namely the 'freedom of action' by fish, e.g. migrations, choice of food and other behavioural free choices. However, we cannot assume complete freedom of action in fish, as it is subjected to external environmental as well as internal innate influences. There seems to be no other secure and proven basis for the recognition of the degree of freedom in given species of fish, i.e. the degree of duration of its response to environmental variables from deterministic law, except adherence to principles of objectivity in interpretation of subjective impressions and equating them with objective facts. Thus, if we recognize that causes may be interlocked in a complicated way, forming a network of causes and objective evaluation in detail are impossible, we must partly enter the realm of metaphysics, where we still strive to retain objectivity by developing logical connections between sense impressions.

Chapter 4
Fish in Marine Ecosystems and the Natural Behaviour of Fish

4.1   The nature of pelagic and demersal fish ecosystems and their spawning migrations

Fish ecosystems, their species composition and the behaviour, interaction and dominance of species, change in space and time owing to the migrations of fish. Conveniently we tend to define fish ecosystems as either pelagic or demersal. However, there is no sharp boundary between these two kinds of systems. A given species can be pelagic for part of its life, especially in its juvenile phase, and be demersal for the rest. Commonly two criteria are used to designate a species as demersal or pelagic. The first criterion pertains to feeding habits (e.g. a fish is considered to be demersal if it feeds predominantly on sessile benthic animals). The second criterion is the method of capture (e.g. fish caught by bottom longlines, pots and demersal trawl are considered to be bottom fish or demersal). Furthermore, pelagic fish are migrating nearly all the time, whereas demersal fish have mainly seasonal and life-cycle migrations.

Fish migrations (e.g. spawning migration) can be seasonal and related to the life cycle, but most are feeding migrations, which cause interactions between marine ecosystems. Most pelagic fish migrate in shoals, searching for rich plankton (mainly zooplankton) patches or smaller forage fish upon which to feed, and when a richer patch is grazed down migration continues. Demersal fish graze on patches of suitable benthic organisms, mostly in loose formations. Many larger fish, e.g. salmon, tend to be more solitary in their feeding migrations, but migrating shoals of large fish, e.g. tuna, are not uncommon.

The distribution of organisms in marine ecosystems, and especially fish ecosystems, is patchy in both time and space. For example, there are changes in fish patchiness due to the dispersal of shoals at night or for feeding (e.g. cod). Patchiness is often linked to the structure of the environment, e.g. thermoclines and current boundary regions. Patches can be shoals with a relatively uniform pattern, but can also be without any definable structure. Patch structures are also dynamic. Patchy distribution and shoaling is more characteristic of pelagic and semi-pelagic species than of demersal fishes.

A given fishery resource can be large, but if it is dispersed it is not easily available to the fishery; i.e. its catch per unit effort is low and it is thus of little value to man. Some demersal fish belong to this dispersed category.

Fish cannot be characterized as pelagic or demersal by their spawning habits, because some pelagic species spawn on the bottom and some demersal species spawn in the water column. The Pacific herring, a pelagic species, fastens its eggs to sea weeds, whereas some Atlantic herring stocks are pelagic spawners. The Atlantic cod, a semi-demersal species is a pelagic spawner, whereas its counterpart, the Pacific cod, is a demersal spawner.

From a fisheries point of view spawning aggregations are one of the most important phenomena in fish behaviour. Spawning areas and periods when fish aggregate are important and are often the main fishing areas and periods for many species. For each species there is a centre of distribution where the main spawning grounds are located, and a number of fringe spawning areas which can disappear when the stock size decreases.

Each species and each stock has a defined spawning period and spawning area (ground) on which mature fish gather. To find the spawning ground is an innate ability of fish for which we do not have any good explanation. Fish homing to spawning grounds might be a function of learning and memory and subsequent utilization of various environmental, e.g. chemical, cues. Experimental proof of this hypothesis is difficult to obtain. Pitcher (1986) believed that shoaling might increase the accuracy of homing to spawning grounds.

Spawning grounds are usually located upcurrent from the feeding area. For example cod, haddock and saith feed in the Barents Sea, where they are also caught. However, their spawning areas are upcurrent along the Norwegian coast (Fig. 4.1), where spawning might last several months. The larvae then drift with the currents back to the Barents Sea, where their distribution can vary from year to year (Fig. 4.2), depending on wind-driven variation of currents. The above examples also demonstrate that many fish move from one defined large ecosystem to another, thus altering the nature of location-defined ecosystems.

Spawning grounds might appear to be downcurrent of the feeding grounds with reference to the surface current in some areas where sub-surface counter-currents exist which carry the larvae to feeding areas (e.g. sablefish along the west coast of North America). Furthermore, some fish might move in large current gyres in the oceans in which the spawning migrations are not against the current.

The timing of spawning can be influenced and spawning grounds can be displaced by temperature anomalies. It is known for some species that different age groups arrive at spawning grounds at different times. Little is known about the peculiarities of the behaviour of pre-spawning migrating shoals of most species, which might differ between stocks of the same species. Mohr (1971) stated that the faster-migrating Atlanto-Scandian herring shoals are more difficult to catch by seines and escape more easily, whereas North Sea and New England pre-spawning herring shoals are easier to catch.

Nearly all species leave the spawning areas after spawning and other species might move in to feed on the spawn and larvae. Spawning stress mortality is very large in some species (e.g. capelin) and in most species increases with age, averaging 10% of the spawning population per year.

Fish spawning can alter a given marine ecosystem in four different ways. Firstly there is the spawning migration, causing part of the species biomass to move out from

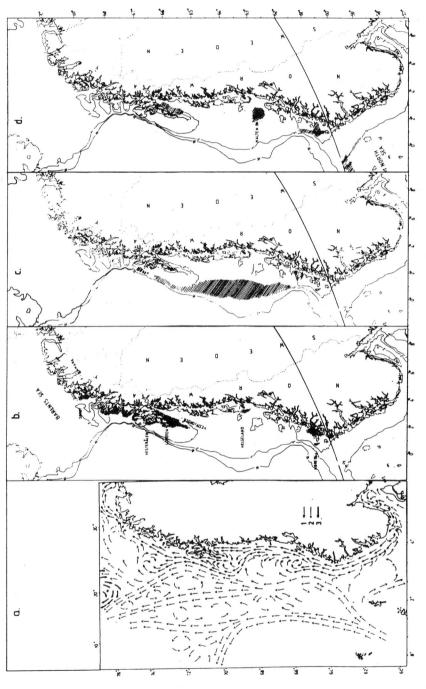

Fig. 4.1 (a) Surface currents in the eastern Norwegian shelf areas. Main spawning areas (hatched) of the boreal gadoids of the Barents Sea, (b) cod, (c) haddock and (d) saithe. (Bergstad *et al.* 1985).

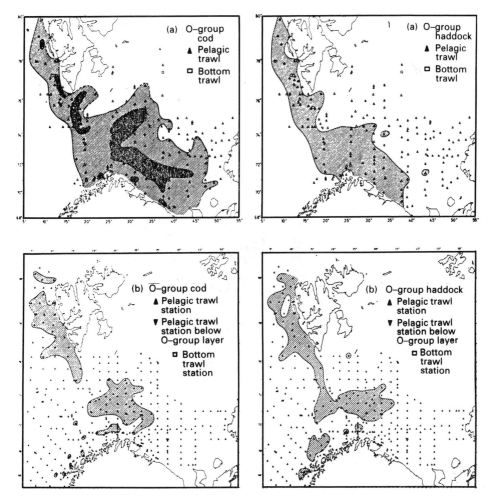

Fig. 4.2 Distribution of pelagic O-group cod (left panel) and haddock (right panel) based on the international O-group surveys in August–September 1970(a) and 1980(b) (Bergstad *et al.* 1985).

one local ecosystem to another. Secondly there is spawning aggregation in a specific ecosystem and usually a concentrated fishery on this aggregation. Thirdly 10–20% of the spawning biomass is released into the environment as milt and eggs; the eggs are subject to predation by other species that can move into the spawning area. Fourthly there is an increased mortality of spawners after the spawning process.

4.2   Fish as predator or prey: feeding behaviour and migrations

Fish ecosystems are the greatest predators on the other marine ecosystems and serve also as food for themselves (Fig. 3.11). The top predators of marine ecosystems are man through fishing, marine mammals and birds.

Most larvae of mid- and high-latitude fishes start to feed on nauplii of calanoid copepods, protozoa, flagellates and other small zooplankton. With an increase in the fish body size, the size of their prey also increases, as does the amount of fish larvae in their diet. The juveniles switch to larger zooplankton (e.g. krill) and demersal fish switch to benthic organisms. With increasing body size the adult diet consists of more fish (including cannibalism) (Fig. 4.3). Predation is one of the most important interactions in the marine ecosystems; it is spatiotemporaly variable and can cause changes in ecosystems.

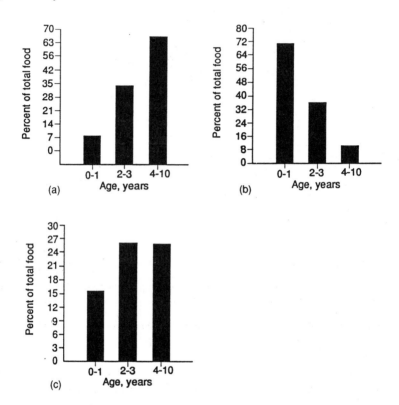

Fig. 4.3   Food composition of Pacific cod by age groups in eastern Bering Sea. (a) Percentage fish in food; (b) Percentage zooplankton in food; (c) Percentage benthos in food.

Figure 4.4 shows the disposition of the annual growth of Pacific cod, the predation on cod and mean food composition of cod in the Bering Sea. Predation mortality is the largest component of cod mortality (71%), whereas fishing mortality is only about 12% and consumption by marine mammals 9% of the annual biomass production. The diet of the cod is about equal parts zooplankton, benthos and fish. Fish need 3–6 times their body weight of food per year. Thus 1000 t of cod would require at least 3000 t of food per annum, of which 1000 t is fish.

Both light (vision) and smell (olfaction) are important in search for food and in the feeding process. Currents distribute the smell of prospective food items, such as bait on longlines, and fish approach the baits swimming against the current (Løkkeborg *et al.* 1989).

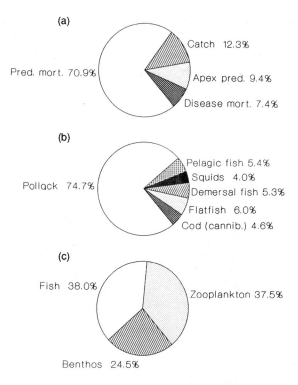

Fig. 4.4 Predation, biomass utilization and food sources of the Pacific cod as computed with holistic ecosystem simulation (inputs based on empirical data; outputs summarized over a period of 1 year). (a) disposition of annual increase of Pacific cod biomass (including ages from 0.5 year to maximum age) in the eastern Bering Sea (EBS) assuming equilibrium, i.e. removal = growth. (b) Predation on cod (larvae and juveniles) by other ecosystem components (apex predation by mammals is excluded). (c) General food composition of cod biomass. Cod biomass in the EBS = 1 130 000 t; annual increase = 1 220 000 t.

The complex predatory habits of fish have been revealed by stomach content analysis. Firstly, the food composition changes with the size/age of the species (Fig. 4.5). Secondly, the composition changes with time and depends on the availability of suitable prey items (Fig. 4.6). As the availability of suitable prey is different in different areas and at different depths, the food composition also changes spatially (Fig. 4.7). Some species also feed in different ecological regions and have different food preferences (Fig. 4.8). For example, the yellowfin sole is a selective feeder on benthic organisms, while the redfish is a pronounced pelagic feeder whose food composition also varies in space and time (Fig. 4.9).

Cannibalism on juveniles is a common phenomenon in the fish ecosystem, especially if one or few species are dominant in a given region (e.g. walleye pollack in the Bering Sea).

Fish feeding nearly always implies a search for food and a consequent migration. We can differentiate between seasonal and life-cycle feeding migration on the one hand and diurnal feeding behaviour and migrations on the other, with the latter being more important to commercial fisheries. Life-cycle feeding migrations include larval

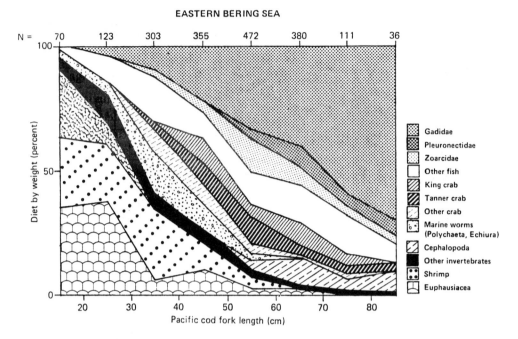

Fig. 4.5   Food composition of Pacific cod of different lengths in the eastern Bering Sea, 1984–6 (P.A. Livingston, personal communication).

drift from spawning grounds to feeding grounds and the active migration of spent fish to feeding areas. In many demersal species there are seasonal feeding migrations, to shallower water in summer and to deeper water in winter.

Vertical migrations are pronounced in many fishes in the spring and autumn, closer to the bottom during daytime and to mid-water during the night. Pelagic, near-surface fishes are mostly visual feeders and depend on light, whereas deep-water demersal fishes depend more on olfaction and mechanoreception for the location of food. Thus visual feeding depends on day length, water depth, turbidity (as affecting visibility), density of food, and volume searched. Although shoaling is considered to reduce predation mortality, shoals can also limit the access of individuals to food. Therefore hungry shoals are known to loosen up for feeding (Misund 1991).

When the stock abundance of a pelagic species decreases, shoal sizes usually decrease. Fish in smaller shoals have a slightly higher growth rate, possibly because of improved access to pelagic food. Higher temperatures and positive temperature anomalies prolong feeding periods and also result in faster growth (Dementeva & Mankovich 1966). Faster growth in turn leads to earlier maturation in many species, which results in higher age-specific spawning stress mortality and leads to rejuvenation of the population, i.e. a decrease of the mean age of the population.

Life-cycle and large-scale feeding migrations can change with time (e.g. Atlanto-Scandian herring), apparently triggered by the changing availability of pelagic food (Jakobsson 1978). Also, other behaviour patterns can change with time; e.g. Finn

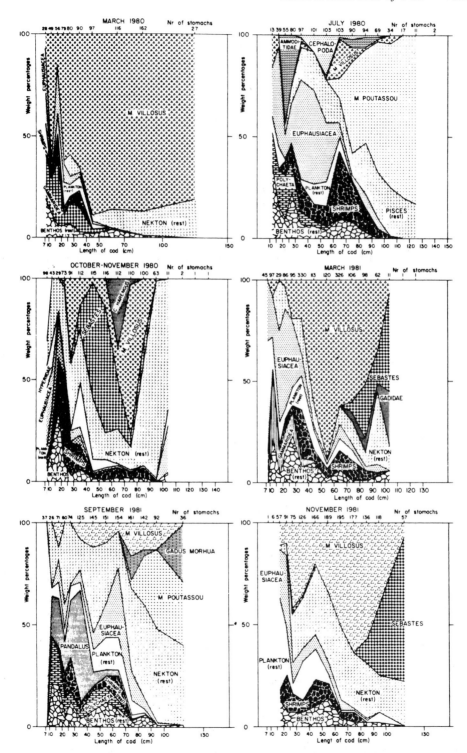

Fig. 4.6 Food composition of cod in Icelandic waters (weight percentages) in relation to predator length (cm) in March, July, October–November 1980 and March, September, November 1981 (Palsson 1983).

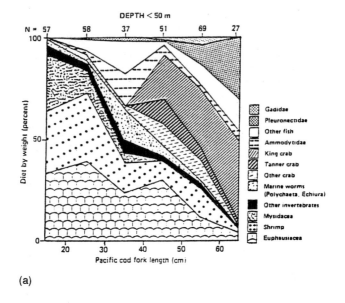

(a)

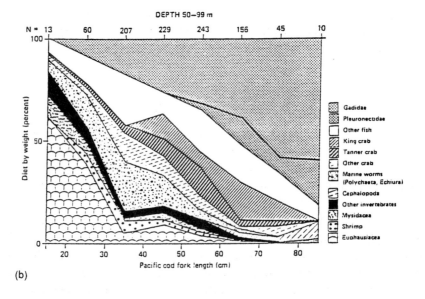

(b)

Fig. 4.7   Food composition of Pacific cod at different depths in the eastern Bering Sea, 1984–6 (P.A. Livingston, personal communication). (a) < 50 m; (b) 50–99 m.

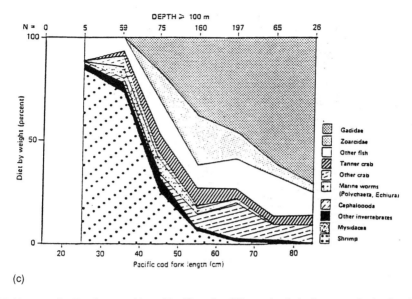

Fig. 4.7 *(Continued)*   Food composition of Pacific cod at different depths in the eastern Bering Sea, 1984–6 (P.A. Livingston, personal communication). (b) > 100 m.

Devold (cited by Mohr 1971) noted that the behaviour of Atlanto-Scandian herring had changed considerably since they were intensively fished.

Two generalities of feeding in the ecosystem, which can affect the productivity of that ecosystem, could be emphasized. Firstly the distribution of different sizes of organisms in a given ecosystem affects the trophic conditions, because the size-dependent feeding, i.e. size of the prey in relation to size of the predator, dominates in the marine ecosystems. Secondly the density of suitable prey items is an important determinant in the success of feeding. Both feeding-limiting conditions are alleviated by migrations in search for food and also by prey switching.

4.3   Natural behaviour and reactions of fish and stocks

Our knowledge of fish behaviour is now quite extensive, but its usefulness and importance in terms of contributing to social objectives such as fisheries management are in a period of transition. For thousands of years the behavioural aspects of fish have formed the basis of the technologies and strategies designed to harvest the living resources of the world's oceans and other aquatic environments.

Early studies emphasized fish behaviour as reactions to environmental stimuli, whereas much of the recent work tends to relate fish behaviour to gear performance. Many early studies also tended to generalize the observed behaviour, whereas recently it has been realized that each species and each size/age group may have some specific and distinct reactions. Even different stocks of the same species can react differently, especially in respect of seasonal response (e.g. migrations), and these responses can change with time (e.g. the general migration routes). Furthermore, it has now been

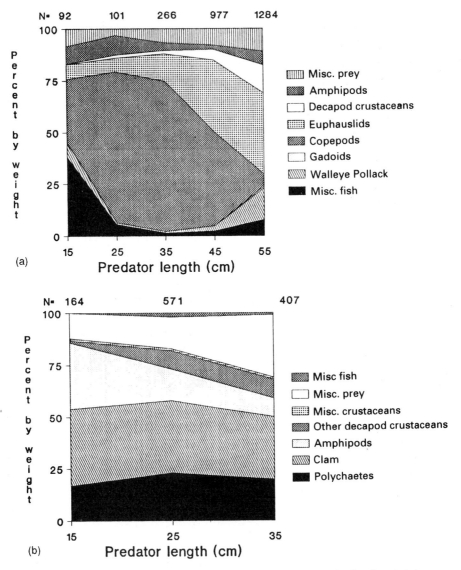

Fig. 4.8 Food composition of walleye pollock (*Theragra chalcogramma*) and yellowfin sole (*Pleuronectes asper*) in the eastern Bering Sea, 1984–1986 (Livingston 1991).

fully realized that individuals of a particular species react differently from shoals consisting of different life stages of the same species.

Much of the early work on fish behaviour in aquaria is difficult to relate to the behaviour of shoals and stocks in the open sea. Furthermore, the earlier studies, with few exceptions, have not made a significant impact on gear improvement and fisheries management (Harden Jones 1974).

Although we attempt to summarize fish behaviour from a cause and effect point of view, we must bear in mind that fish reactions can vary considerably and several

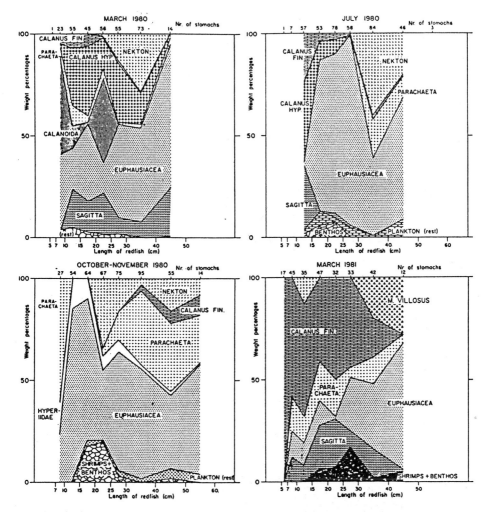

Fig. 4.9   The food composition of redfish in Icelandic waters (weight percentages) in relation to predator length (cm) in March, July, October–November 1980 and March 1981 (Palsson 1983).

stimuli might be present at any given time. Thus several cues might be involved simultaneously in fish behaviour. It seems that at times we must consider that a 'fish has its own mind', and many reactions cannot be explained from a human viewpoint.

4.3.1   *Natural (innate) behaviour*

Swimming speeds

Most natural (innate) behaviour of fish usually involves movement (i.e. swimming). It is therefore of interest to consider some aspects of swimming, especially speed. The review and summary of swimming speeds by Blaxter (1969) is still the best source of this information, though a recent summary of swimming speeds in relation to fishing gear by He (1993) is also valuable. Blaxter divided swimming speeds into two categories, (1) the burst speed, which is also referred to as escape speed and is of a short duration, measured in seconds, and (2) endurance speeds. Fish speed is usually given in terms of body lengths per second (BL/s). Burst speeds can be of the order of 10 BL/s, whereas endurance speeds for demersal species can reach about 4 BL/s, but are normally 2–3 BLs for longer distances, especially in shoals. Corresponding values for pelagic fish are 8 BL/s for shorter times and about 4 BLs over longer periods. It should be noted that swimming speeds are greatly affected by temperature (Inoue *et al.* 1993) being considerably slower in cold water. Endurance speeds for different species were given by Blaxter (1969), who also gave distances swum to exhaustion. Endurance decreases as swimming speed increases. Swimming speeds versus length are given in Fig. 4.10. Fish speeds are important in the capture process.

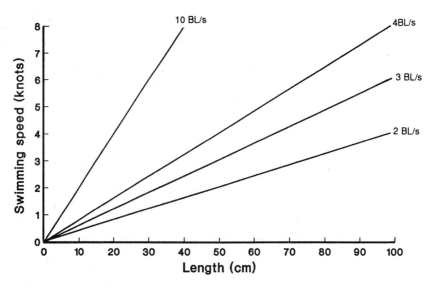

Fig. 4.10   Speeds as body lengths/second (BL/s) for fish of different length converted to knots for comparisons with towing speeds (Blaxter 1969).

The swimming speeds, as well as other fish behaviour, can be different for a single specimen from that of shoals, and therefore a brief review of shoaling behaviour is called for. Principal innate behaviours, their consequences and applications are summarized in Table 4.1.

Shoaling

Most commercial fishing occurs on fish aggregations, and the pelagic fishery is almost entirely based on catching shoals. Thus it is of value to review the nature of these aggregations, including their properties and behaviour. Shoaling behaviour and functions of shoals have recently been reviewed and summarized by Pitcher (1986).

Table 4.1 Innate behaviours of fish, their purposes, characteristics and peculiarities.

Behaviour aspects	Purposes	Characteristics	Peculiarities
Shoaling	Spawning Protection (Feeding) Homing	Polarized swimming; structure varies by species and size. Volume occupied is proportional to cube of fish length. More pronounced in pelagic fish.	25% of all fish species shoal. Shoals loosen and tighten diurnally. Hunger loosens, disturbance tightens shoals.
Aggregations	Spawning 'Resting' Feeding	Unorganized gathering on preferred ground and depths. 0.1–1 fish in m³. Move little.	
Spawning migrations	'Homing' Species aggregation	Each stock has own spawning grounds.	Main fishery on spawning grounds (most species).
Feeding migrations and behaviour	Feeding	Migration to shallow water in summer, deeper in winter. Migration speed decreases with abundant food. Smaller shoals – increased growth.	Diurnal vertical migration. Pelagic fish visual feeders, demersals use olfaction and mechanoreception.

Swimming speeds
    Burst (escape) speeds of short duration – 10 BL/s (Body/length/second)
    Sustained (endurance) speeds: pelagic species, normal BL/s, short duration 8 BL/s
                                     demersal spp, normal 2–3 BL/s, short duration 4 BL/s
    More tightly packed shoals move faster.

It is necessary to differentiate between shoaling and aggregation (Mohr 1971, Pitcher 1986). Shoaling is defined as synchronized (polarized) swimming behaviour at a relatively uniform speed. Aggregation is a disorganized gathering of random-oriented individual fish or small shoals, for example herring aggregations in the North Sea and off the Canadian west coast during their hibernation in deep water. The vertical and horizontal extent of aggregations and their density (e.g. 0.1–1 fish per m³) is more variable than in shoals.

There are several shoaling theories (Shaw 1962, Murphy 1980, Pitcher 1986), but we cannot fully verify any of them. Shoals are always in motion. It is therefore assumed that the near-field water movement, sensed by the lateral line where the rheotactic organ is located, is one of the most important mechanisms, together with optometric (visual) attraction and orientation, in shoaling and shoal behaviour.

Some of the peculiarities of shoaling behaviour that can be generalized and have been summarized by Misund (1991) are listed below. About 25% of adult fish species and juveniles of most species shoal. Shoaling might be considered as an innate behaviour with its main, but not sole, purpose to help decrease predation. Shoals seem to have no leaders. Shoal structure varies from species to species and by age/size of the fish. This might be largely influenced by the necessity of matching swimming speeds. Individual shoals tend to be spherical in mid-water and flattened near the surface and bottom.

The fish density in shoals can vary. Olsen (1981) found that a typical herring shoal contained 5–50 fish per m^3, which gave the highest value of about 10 kg per m^3.

Most shoal behaviour work has been done in tanks and it is difficult to extrapolate these results to the ocean. Thus a better knowledge of shoal behaviour is needed, which could be partly obtained either from fishermen or through well-planned field work (Misund 1991). Misund showed that acoustic estimates of shoals are uncertain, and that the volume occupied by a shoal is proportional to the number of individuals and inversely proportional to the cube of fish length. Furthermore, this relationship varies seasonally, and the shoals loosen and tighten diurnally. Hunger and feeding loosen the shoals whereas dangers tighten them. More tightly-packed shoals seem to swim faster. Shoals that detect predators compact, might split or do other reformation manoeuvres, which have been schematically summarized by Pitcher (1986).

Saetersdal (1969) pointed out that shoaling is less pronounced in gadoids than in truly pelagic fish (e.g. herring and mackerel). In some seasons, cod form irregular shoals in mid-water during day and single fish layers during darkness. Cod shoals near the bottom are usually tightly packed and move little. Konstantinov (1966) believed that contact between individual fish in these near-bottom shoals is lost, and that these shoals could be considered aggregations. However, tagging experiments on cod in the Barents Sea indicate that when cod of uniform size have joined into a given shoal, this association seems to last, sometimes for years. Mixed shoals of cod and haddock are also formed in which both species are of the same size and the distance between fish is about 0.5–1 m.

Long, narrow bands of shoals of Pacific hake, essentially a pelagic gadoid, have been described by Alverson (1967). These bands can be from a few hundred metres to 20 km long, 0.3–3 km side, and 6–20 m deep.

Aggregations of cod in the Barents Sea form at some preferred grounds (banks), where a trawl fishery is also concentrated (Konstantinov 1966). During spawning seasons aggregations of nearly all spawning fish seem to be sluggish and easy to catch. However, dispersed herring shoals of pre-spawners are agile, alert and difficult to catch (Mohr 1971).

Shoaling behaviour also changes seasonally. Williams & Pullen (1993) found that the shoal size of mackerel in the Tasmanian purse seine fishery increased significantly from spring to autumn. Surface shoals dominated during summer and sub-surface shoals in autumn. In summer shoals were composed almost exclusively of jack mackerel, but in other seasons mixed shoals of jack mackerel, redbait and blue mackerel were found.

Shoaling is important in fish capture, because most large-scale fisheries occur on fish shoals and other aggregations of fish.

Feeding behaviour and reactions

Fish are constantly searching for food. Most pelagic feeders feed visually, but in many demersal and semi-demersal species olfaction and chemical stimuli, as well as mechanoreception, are important for finding of food. Longline and baited trap fisheries depend entirely on luring the fish to bait and enticing the fish to swallow it. Therefore, a considerable amount of research has been done, especially in Bergen and in Aberdeen, on the suitability of different baits for different species.

The chemical sensing of baits and feeding behaviour of fishes has been summarized by Atema (1980). The smell from bait is transmitted to fish with soluble amino acid compounds which elicit neurophysiological and behavioural responses in fish. Fish chemoreception has been summarized by Hara (1984). Fish detect food by olfaction (smell) and gustation (taste). Although Hara assumed that the chemical stimulation by smell is non-directional, it has been observed that fish come to bait against the tidal current (He & Xu 1992). Pheromones, other exocrines and chemotaxis might play a large role in marine ecosystems, but research has been limited. Chemotactic communication is used for recognizing individuals of the same species and results in the formation of aggregations. Olfaction might also be used for homing to spawning grounds. Fish repellents (e.g. shark repellent) rely on olfaction.

The distribution of odours is affected by turbulence and currents, which are important in longline fishing strategy (Olsen & Laevastu 1983). Equally important is the leaching rate of baits, and bait distance in determining the 'smell field'. Bait must also be palatable to fish. Gustatory receptors in fish are found primarily in the mouth and lips, and barbels where present. Each predatory fish seems to recognize different mixtures of compounds during the olfactory luring and in gustatory uptake of bait. One of the aims in longline research is to find an artificial bait with long-lasting leaching and good palatability to different species.

Longline fishing is low-capital, relatively low-energy and species- and size-selective. Research on this fishing method conducted in Norway, Scotland and Japan in the last few decades has advanced our knowledge on attraction by smell, the effect, type, size and shape of bait and the fish's activity at the bait, which have all been found to be highly species-specific. One of the great problems in longlining is still feeding by bottom invertebrates on the baits.

Feeding reactions and the factors affecting them are species-specific and change with the age of the fish and with the season. Feeding frenzy is characteristic of some pelagic fish, such as tuna. In the fishery for tuna shoals, the feeding frenzy is stimulated by 'chumming' of bait before starting fishing with pole and line. Feeding reactions depend on the physiological state of the fish, (the 'hunger state'). The intensity of physiological processes are greatly affected by temperature, which also affects food requirements and growth. Furthermore, feeding reactions, the amount of food uptake, and growth depend on the availability of suitable food. For example, Khaldinova (1966) found that the haddock in the Barents Sea feeds more on capelin and krill when these two species are available, and grow faster than when it is forced by the absence of these species to feed on benthos.

Saetersdal (1969) pointed out that many species feed mostly at dawn and dusk and that feeding intensity varies with the physiological state, availability of food, and temperature. Zusser (1966) believed that light serves as the signal for feeding for most pelagic species and that the hungry fishes react more. Feeding reactions change in most species in the course of the year. This seasonal change in feeding affects dispersal of fish, shoal size and migrations, thus it affects the species composition and their relative abundance in the ecosystem.

Fright reactions

A fright reaction is defined as a quick reaction to suddenly-appearing or intermittent stimuli. Fright reactions usually subside after a short time, after which some timidity develops. The fright reaction of an individual fish is different from the fright reaction of a shoal; the latter seems to involve some kind of mass decision. There are two specific fright reactions in shoals, diving or rising of shoals, and splitting of shoals (e.g. as caused by ship noises).

   Induction of fright reaction is rarely used in fishing, except for driving fish into some stationary gear and preventing their escape from, say, beach and Danish purse-seines. Hashimoto & Maniwa (1971) described some experiments on the successful driving of jack mackerel and barracuda into stationary nets. In most purse-seine operations it is important not to frighten the shoals during setting of the seine and before pursing.

   Misund (1991) pointed out that vessel noises can induce a fright reaction in shoals, which results in increased swimming speed away from sound source, either sideways or in same direction as the moving sound source. Strange sounds appearing suddenly can cause fright reactions in shoals, but if the sound is continuous and does not move normal behaviour continues.

   Engås *et al.* (1991) described the reaction to vessel noises of a cod tagged with an ultrasonic tag:

   'The cod generally reacted to the vessel noise by an increase in swimming speed in front of the vessel, and, after a short distance (less than 100 m), by a sudden change in swimming direction out of the vessel path. On two occasions the cod were observed to perform a zig-zag swimming pattern in front of the vessel. The behavioural responses were always adequate to prevent the fish from being caught by trawl.'

   Parrish (1969) found that there was no panic response of fish in tanks to moving objects, and fish usually swam away about 3 m in front of the object. In darkness the effect and distance from the object decreased. Thus there can be some difference in fright reaction between day and night. Parrish also found that the behaviour of fish in the sea was essentially the same as in tanks.

   One of the commonest fright reactions is the predator avoidance reaction, which has been little studied in the natural environment. The extent of ship traffic noise as a cause of fright reaction in fish and possibly of fish migration has not yet been evaluated. We also do not know how fish learn to avoid ships and gear emitting frightening sounds.

4.3.2   *Reaction to stimuli*

The stimulus for fish to react and initiate a behaviour pattern can be either internal within the fish (innate) or external, perceived by the fish's various sensory organs, such as the eye and the lateral line. The main internal stimuli influencing behaviour are those concerned with preservation of life, e.g. hunger and reproduction. The combination of innate stimulil concerned with hunger and reproduction lead to a number of observed behavioural patterns in nature. The important external stimuli are light, sound, temperature and chemical constituents of the water mass, motion of the environment and the presence of dead and living objects in it, including prey and predators. An important external stimulus for fish is pressure, which we usually

Table 4.2   Reaction of fish to environmental stimuli and their application.

Environmental stimuli	Fish reactions	Remarks	Use, gear
Light	Feeding reaction, shoaling.	Threshold 0.1 lux; tuna feed in clear water.	In clear water aggregation for seining, jigging.
Sound	Mostly avoidance (ship and gear); breaking up shoals; swimming speed change.	Hearing frequency 50–500 Hz; threshold 20 dB. Rate of change important. Ship and trawl noise avoidance. Underestimate of resources.	Fish finding, resource assessment. Leading net noises used in set nets.
Currents	Rheotactic organ; orientation.	Fish use for migration and possibly for orientation.	Fishing tactics in set (and moving).
Turbulence	Avoidance, diving.	Affects depth of shoals.	Gear.
Temperature and thermal boundaries	Affects activity and food uptake.	'Mechanical' aggregation at temperature and current boundaries.	Finding of fish aggregations.
Food availability	Search for food Visual and olfactory.	Migration speed slowed in abundant food. Shoals loosen.	Fishing tactics and adaptation.
Response to bait	Olfactory stimuli for search.	Smell distribution affected by currents. Palatability of bait.	Luring to bait. Longline, baited traps.
Electric stimuli (electrotaxis)	Initial shock reaction. Movement to anode.	Use in brackish water and pulsating current in the sea.	Use forbidden in most countries.
Association to grounds	Spawning, feeding and 'resting' in eddies.	Association might be to different bottom, food availability, current conditions (e.g. eddies).	Major preferred fishing grounds.

interpret in terms of depth below the surface. Most of these external stimuli have species-specific reception thresholds and different responses. At any given time several stimuli can be present. Much of the fish's behaviour is difficult to interpret. Reactions of fish to stimuli and their applications are summarized in Table 4.2 and the relevance of observed responses to stimuli are provided in Table 4.3. The reaction of fish to food, bait, and food availability were summarized earlier in this chapter.

Table 4.3   Fish behaviour, stimuli, reactions and their applications.

Behaviour	Stimuli	Reactions (common)	Uses in fishery
Attraction	Light	Feeding	Aggregation for capture
	Sound	Fright	
	Chemical stimuli	Feeding	Luring to bait
	Food	(Feeding frenzy)	
	Presence of objects	Protection of predators	Aggregation (FADs)
Avoidance	Obstacles	Change direction, avoidance	Leading to traps
	Sound	Diving speed and direction	Noise reduction of ships and
	Turbulence	change	gear
	Turbidity	Diving	Depth of gear
	Predators	Moving out, deeper	Fishing strategy
		Fright reaction	
Fright	Sound	Avoidance, increased speed,	Driving into gear
	Moving obstacles	changing direction	Avoidance of escape
			Reduction of gear turbulence
			Visibility
	Predators	Fright	
Feeding	Food	Slowing speed,	Baited hooks and traps
	Hunger	feeding (including feeding	Chumming of bait
	(Migration)	frenzy)	

Reaction of fish to light

Light intensity affects the feeding of fish and the diurnal vertical migrations and orientation in shoals. Light is also a factor in a fish's reaction to gear. The minimum light intensity at which fish recognize their prey and obstacles is equivalent to about 0.1 lux (Blaxter 1980).

Light penetration in the sea is dependent on light intensity at the surface, which is largely determined by cloud cover and the sun's elevation, wave activity and the turbidity of the water. Different wave lengths of light have different extinction rates in water, so that the spectral composition of light changes with depth. It has been postulated that many visually-feeding fish, such as tuna, shun turbid waters. The maximum sighting distance of fish varies with the light condition in the water and with the turbidity. It also depends on the contrast of an object. The normal sighting distance is 5–10 m (Blaxter 1988).

Some species can be either positively or negatively phototactic. Advantage is taken of the positive phototaxis in some pelagic fishers to attract feeding aggregations of fish for purse-seining, dip netting or baited hook fishing. This attraction is economically effective only in relatively clear waters in the tropics and sub-tropics.

Attraction to light might occur for various reasons (e.g. phototaxis, feeding, curiosity, disorientation). The degree of positive phototaxis varies from species to species and also varies with season. The size of the area from which fish are collected by fishing light depends on the light intensity and especially on the turbidity of the water. Therefore a night-light fishery is effective mainly in clear oceanic waters.

Shoaling appears to be light-dependent. The main shoals seem to break up in the dark, although the fish may remain together as a group. If a shoal moves fast, some polarization in the dark may still exist, with the fish using other cues, such as tactile stimuli via the rheotactic organ. Table 4.4 (Blaxter 1965) gives the range of light intensities over which shoaling ceases.

Table 4.4   Range of light intensity over which schooling ceases (Blaxter 1965).

Species	Range of light intensity (lux)	Remarks
Clupea harengus	10^{-2}–10^3	$8\,m^3$ aquarium
Clupea harengus	10^0–10^{-1}	$225\,m^3$ aquarium
Sardinops caerulea	$< 10^{-1}$	aquarium
Alburnus alburnus *Atherina* 'Anchovy'	$< 10^{-1}$	
Engraulis encrasicholus *Atherina mochon pontica*	10^1–10^{-1}	
Hepsitia stipes *Bathystoma rimator*	5×10^{-1}	
Oncorhynchus spp	10^{-3}	
Phoxinus phoxinus	10^{-2}–10^{-3}	
Menidia	10^0–10^{-1}	

Vertical movements are part of a complex diurnal behaviour and might affect the rate of capture. Reasons for the species-specific vertical movements are related to food availability, either in the water column or on the bottom. There is also a 24-h endogenous cycle of activity in most fishes which is influenced by light. The eye of the fish is as sensitive as man's and might be more sensitive in some deep water fishes. Minimum vision threshold varies with the type of water, and in sunlight is 20–50 m but less in winter.

The day – night differences in trawl catches are complex, being related to endogenous fish behaviour, availability to the trawl, and trawl avoidance behaviour using vision, hearing and tactile sense. Woodhead (1965) summarized vertical distribution and diurnal behaviour of most important North Atlantic demersal species. Generalizations about these behaviour patterns are difficult, as they vary from species to species and within the same species. Vertical migrations and shoaling behaviour also vary with age, maturity state, season and geographic locality. The behaviour can also be affected by the environmental factors, including thermal structure and currents, and the availability of prey.

The avoidance of fishing gear and visual herding effects decrease at light intensities below 0.5–0.05 lux (Woodhead 1965). For moving gear, tactile and acoustic senses may be used by fish. Light and the vision of fish are two of the underlying mechanisms exploited by trap fisheries, especially in the design of the leading net. It is necessary to increase the visual contrast in the leading net with twine thickness as well as with colour. It has been found that yellow enhances the contrast (Nomura 1980). To decrease the visibility and contrasts of gill-nets and trawls, blue and green colours are often used.

The effect of visibility of a trawl on its effectiveness is variable. Parrish (1969) observed that during the night the orientation of fish to the trawl was less marked than during the day. In contrast, Engås & Ona (1990) found that trawling is equally efficient day and night and that the hearing of fish must play a predominant role in trawl avoidance. However, these authors also found slight differences between day and night in fish behaviour in front of the trawl. At night fish entered the trawl in the middle, close to bobbins. During the day, the fish entered more irregularly, using the whole opening of the trawl, but some haddock were lost over the headlines.

Table 4.5 shows the differences between day and night catches of haddock and the change of composition of their food between pelagic and benthic. Woodhead (1965) observed that the lower catches during the night were due to the greater number of smaller fish feeding exclusively on benthos (Table 4.5). This might also be too simplistic and general an explanation, as at the Faeroes the greatest catches of haddock are made at night. In most cases, however, the day and night differences in catchability of demersal species might depend on whether they leave the bottom during the night and are carried with currents.

Light affects feeding and the length of the daily feeding period in visually-feeding fish. In salmon culture in higher latitudes (e.g. Norway) artificial lighting is used over the pens during winter. This method increases the feeding period of salmon and increases the growth rate.

Table 4.5 Diurnal changes in the catch of haddock and feeding changes in relation to the size of fish (Woodhead 1965).

Length (cm)	$\dfrac{\text{Mean night catch}}{\text{Mean day catch}} \times 100$	Planktonic food %	Benthic food %
20–29	17	0	100
30–39	11	12	88
40–49	26	27	73
50–59	44	46	54
60–69	60	54	46
70+	75	65	35

Reaction of fish to sound

The sea is much more conductive of sound than of light. In the last few decades the study of fish reactions to sound, and the application of the results for improvement of gear, fishing tactics and resource assessment, have become among the most important factors in fisheries development.

The sea is a noisy environment and there are sounds from different sources present within the range of fish hearing. There are the natural noises, caused by breaking waves, which depend on the state of the sea, and there are also noises made by some fishes and mammals. Furthermore, man-made noises, the noises of ships' propellers and machinery, might dominate in many areas.

Sound absorption in the sea is dependent on the frequency of the sound. Low frequency sound, such as ships' sounds (10–600 Hz range) can be transmitted for long distances, whereas the higher frequency sounds, e.g. those from high-resolution echo sounders (100 kHz) are absorbed over short distances. Sound transmission in the sea is affected by changes in the speed of sound with depth, e.g. temperature profile and pressure effects, and by sound bouncing from the surface and bottom.

The hearing of fish is in the low frequency range (50 to 550 Hz). Engås & Ona (1990) suggest that fish might be sensitive also to infrasound (0.1–10 Hz). Enger *et al.* (1992) carried out experiments that showed that infrasound (10 Hz) produced spontaneous avoidance responses directed away from the sound source at an intensity of 35 dB above the fish's estimated hearing threshold. No such response could be seen at 150 Hz, even at 100 dB above the hearing threshold.

It is assumed that fish can detect sound using their ears, swim bladder and lateral line. The threshold for fish hearing lies in the range 10–25 dH, above the ambient noise level (between 50 and 1200 Hz frequency) (Olsen 1969, 1971), so it is possible that fish can perceive a noisy ship at a distance of 70–80 m. This distance might vary; it might be greater if the thermocline is deep and with small gradient and the depth is shallow or moderate, but can be closer with shallow and sharp thermoclines and in deeper water (for sound propagation see e.g. Laevastu 1966). The noise detection distance is also dependent on the background noise level.

In general, reaction to sound stimulus can be expected when the stimulus is at least 5–15% above the background level (Blaxter 1988). Sound can attract and repel, and these reactions depend on a number of properties of sound as well as the conditioning of fish to these properties.

As the noises emitted by vessels' propellers and machinery in the frequency range of 10–600 Hz are audible to fish, the fish display avoidance reaction, which varies from species to species (Misund 1991). Ona & Godø (1990) believed that propeller cavitation is the main source of noise affecting fish. Furthermore, these researchers suggested that the rate of change of sound pressure is more important for fish reaction than sound level.

Olsen (1971) pointed out that theoretically a vessel can be detected by fish in calm weather several nautical miles away. However, in practice the detection distance is much shorter and is limited by the level of ambient noise. The fish's reaction distance is also considerably shorter than its detection distance. Fish reaction varies also with the reason and age of fish. Pulsed signals seem to disturb fish more than continuous sound.

Different species react differently to ships' noises (Fig. 4.11). Haddock shoals in mid-water tend to swim sideways away from ships, whereas cod near the bottom tend to dive and pack closer to the bottom. Thus noisy ships are less effective in catching haddock with mid-water trawls, but more effective in catching cod with otter trawls. Olsen *et al.* (1982) observed that when a herring shoal was 75–100 m from the vessel, disturbance behaviour of the shoal occurred and fish started to descend and move away. The density of herring in the shoal was reduced by about 50% after the passage of the ship.

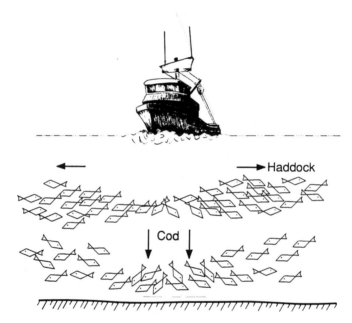

Fig. 4.11 Schematic presentation of the reaction of haddock and cod shoals to ships' noises.

Vessels' noises and fishes' reaction to them have been assumed to be responsible for breaking up fish shoals. This hypothesis has been carried even further by assuming that the increased growth rates of some fishes in smaller shoals might be attributed to the effect of ship traffic having caused the existence of smaller shoals, but there is no proof for these hypotheses. A.R. Margetts (cited by Mohr 1971) had noticed that fish avoided slow-moving vessels more than fast-moving vessels.

Moving trawls and ropes and nets in currents also emit low-frequency sound. Leader net vibration is of the order of 100 Hz (Nomura 1980), which provides audible guidance to traps. Vessel avoidance caused by vessel noises reduces the availability of fish to trawls. The vibration of trawls, trawl warps and netting will cause the splitting of the shoals in front of the trawls and the herding effects of wires, and influence the catching efficiency and selection of the trawls. Recently these aspects of fish behaviour have been described by Engås & Ona (1990), Ona & Godø (1990), and especially by Misund (1991). These authors also pointed out the present shortcomings in acoustic estimation of shoals and shoaling fish in general. The sources of error in hydroacoustic surveys (e.g. the downward and sidewise migrations, shadow effects in shoals, tilt angle change and bottom effect) have also been discussed and demonstrated by Olsen (1981).

Standard echo sounders, operating at 38 kHz, have also low-frequency components (500 Hz and less) associated with the carrier frequency pulse. These low- frequency pulses are audible by fish, which can react from a great distance before being detected by the echo sounders. These low-frequency noises radiated simultaneously to the carrier frequency can induce escape reactions in fish shoals and effect catches of tuna seiners, longliners and troll boats (Bercy & Bordeau 1985).

Reaction to electric stimulus

Fish react violently to electric stimuli, which can be used in fish capture mainly in fresh and brackish water and to a limited extent in sea water. However, most countries forbid electrical fishing.

Initial excitement with an electric current is followed by swimming towards the anode (electrotaxis), followed by loss of motility and equilibrium (electronarcosis). In a pulsating current the initial excitement is more vigorous than in steady current. This technique was used in an experimental Dutch beam trawl, where it served to stimulate the flatfish, mainly sole, to move out from the sandy sea bed.

Electrotaxis is increased by an increase in the pulse rate or by an increase of the pulse width. Electrotactic response varies with the size of the fish and from species to species in fish of the same size. Electrotaxis predominantly occurs in fishes which use their body-tail muscles to swim. The best results for electrotaxis are achieved with the use of a weaker continuous field with a superimposed pulsating field. Electrofishing was used in brackish waters in Russia (Daniulyte & Malukina 1969).

Reactions of fish to environmental properties and stimuli

The study of a fish's response to environmental stimuli, including the reaction to gear, has provided ideas for gear improvement, including increasing their selectivity. It is realized that generalization is impossible and that the response varies not only by species and by age group, but also with season and with prevailing environmental conditions. Response of fish to environmental stimuli and their application are summarized in Table 4.3.

Of the various environmental stimuli affecting fish, water temperature has received the most attention. Its effects on fish have been summarized by Laevastu & Hayes (1981). Of the 'active' temperature impacts on fish, the most important are its effect on activity (e.g. swimming speed) and food requirements, and possible effects of thermal gradients (thermocline and frontal zones) on the aggregation of fish. Temperature gradients in the sea are functionally related, vertically to turbulence (through the thermocline) and horizontally to currents (through fronts).

Violent turbulence from wave action causes fish to move into deeper water. It can be assumed that fish also want to avoid local turbulence caused by moving fishing gear. Although they are a mode of transportation, the effects of currents on fish behaviour are not fully understood. It is assumed that in the absence of visual cues for orientation fish can perceive current with their lateral line (rheotactic) organs. The threshold of perception is, however, unknown. It is assumed that fish orientate to currents by heading into them. Experiments by Greer Walker *et al.* (1978) clearly demonstrated that some flatfishes do purposefully use tidal currents for migration.

It is known that many fish species aggregate at thermal and current boundaries. It is unlikely that this aggregation is purposeful, with the possible exception of feeding on accumulations of pelagic food. The concentration of fish at fronts can, however, be explained by interaction of migration and the effects of currents and temperatures on migration speed, as demonstrated by Seckel (1972) and Laevastu (1992).

Many fish species aggregate at preferred locations, which become fishing grounds. These locations can change seasonally (e.g. during their spawning periods) and the species or groups present on the grounds can change as well. Often the fish aggregations are associated with bottom features (depth, slope, type of bottom). Harden Jones (1974) emphasized the need for knowledge of bottom features associated with fish concentrations, but few research data on this subject are available. Nomura (1980) emphasized the importance of bottom configuration for set-nets. He also pointed out that many fish tend to aggregate near shelters and that fish tend to swim along depth contours (slopes).

Concentration of fish at the preferred grounds are most often aggregations rather than shoals. However, aggregation and shoal structure and shape, which are often species characteristics, can vary with season and with environmental conditions (Shaw 1962).

The aggregation of fish in the centres of eddies is sometimes suggested. The reason for the aggregation in eddies and lees on the grounds might be the need for rest (i.e. not being advected by currents). The aggregation at slopes with stronger currents, which brings pelagic food past the quasi-stationary aggregations, might also be considered.

Attraction

Fish attraction techniques are used to create fishable aggregations or to attract fish to gear. Light, sound, smell and protection reaction (e.g. FADs) have been used as attractant stimuli. Attraction to light is used in night-light fishing for pelagic species in clear waters (e.g. sardines and mackerel). The attraction to light is mostly a feeding reaction and is dependent on the hunger state of the fish and varies seasonally (Zusser 1966).

There have been many attempts to attract fish with sound. Some angling through holes in the ice using suspended objects to produce sound has been limited in its use.

Attempts have been made to aggregate some species, e.g. cod, during spawning by recording and playing back some characteristic sounds made by spawning fish. These attempts have not led to any practical application. Hashimoto & Maniwa (1971) have described successful experiments on luring yellowtail and mackerel with underwater sound.

Attraction of fish to chemical stimuli (smell) is the main basis for longline and baited trap fishing. A voluminous literature exists, mainly from Norway, Scotland and Japan, on the suitability of different baits, including artificial baits as attractants, and the importance of the leaching rate of the baits and the distribution of the smell from baits by currents (Olsen & Laevastu 1983).

A suitable method for fish attraction and aggregation in the tropics is the anchoring of floating objects of various shapes, around and below which a number of different species aggregate (fish attraction devices, FADs) (Fig. 4.12). The FADs are usually anchored on known migratory paths of fishes (Dickson 1993). Artificial reefs also fall into this category. The reasons for fish aggregation at FADs and in artificial reefs are still uncertain, although the prevailing explanation is that fish search for protection under and around these devices.

There are other aspects of natural fish attraction, though less important from the applied point of view, which we do not fully understand. These include attraction

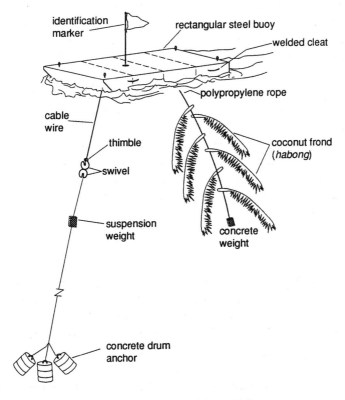

Fig. 4.12   Steel fish attraction device (FAD) or payaw (Dickson 1993).

between different sexes, spawning aggregations, and attraction to different grounds and specific types of bottoms.

An interesting approach to studying fish attraction is the conditioning of fish to light or sound signals during feeding. This practice is used in some freshwater fish cultures, and has been tested in the marine environment in arms of fjord where some fish, e.g. cod are held in large, natural enclosures (N. Balchen, personal communication).

Avoidance

Applied interest in fish avoidance reactions is directed towards efforts to minimize avoidance of moving gear, to direct fish to fixed gear (trap nets), or to make gear selective using the different avoidance reactions of different species and age groups. Avoidance can be classified by different criteria, such as avoidance of obstacles, avoidance of some stimuli (e.g. sound, currents, turbulence or moving objects), avoidance of discomfort caused by the environment (e.g. high turbidity), or natural and learned avoidance of capture by predators, including man.

Most gears (e.g. gill-nets) operate on a minimum avoidance reaction, which is achieved by lowering the visibility of the net by decreasing its contrast with the surrounding water. This is achieved by changing the thickness and colour of the material of the net. According to Parrish (1969) net avoidance by individuals of a species can be more pronounced than that by shoals. The shoal size, in terms of 'shoal pressure', is considered to be important.

Considerable attention has been paid to avoidance reactions triggered by low-frequency sounds, 20–600 Hz, which is the frequency range of fish hearing and also the frequency of ship noises and noises of the towed gear. Studying the reactions of sonically tagged cod, Engås *et al.* (1991) found that the fish reacted to vessel noise by an increase in swimming speed and by a sudden change of swimming direction out of the path of the vessel, returning and moving diagonally in the opposite direction. They assume that the fish might have been trapped between two lobes of the noise field of the ship.

Wardle (1993) concluded that, in trawl avoidance, the fish is affected first by low frequency sound, followed by visual stimuli. The manner of avoidance of a moving noise source, ship and/or gear, can be different in different species and age groups. Wardle (1993), for example, pointed out that haddock have a tendency to rise above the headline of the approaching trawl, whereas cod tend to dive or remain close to the bottom.

Among the environmentally-induced avoidance reactions are diving of fish into deeper layers during storms, apparently to avoid the high turbulence caused by wave motion near the surface, and the movement of some coastal, shallower-water fish into deeper water during storms, which might result in avoiding the turbidity near the bottom caused by a whirling up of sediment. This migration might be triggered by the sound of waves breaking on the coast.

4.4   Changes in marine ecosystems – the behaviour of ecosystems as a whole

Marine ecosystems are not stable, but fluctuate in space and time (see [3.5]). The studies of oceanography, marine fishers and marine ecology are essentially studies of changes in the marine ecosystems and of their causes and effects. The fluctuations of fish stocks and their plausible causes have been recently summarized by Laevastu & Favorite (1988). This section [4.4] contains notes and examples of the inevitable changes of marine ecosystems as the basic behavioural characteristic of these systems. In [7.4] we shall compare the natural changes in these systems with possible changes produced by man.

Changes at the ecosystem level include: (1) changes of abundance of difference species and changes of dominance in species, and (2) changes of overlap (coexistence) of species by migration and colonization. The changes can be regular, periodic (e.g. seasonal) or non-periodic (irregular) and long-term.

Considerable changes in species abundance have occurred in many places, where some species biomasses decrease while others increase (Fig. 4.13). These changes also cause changes in trophic flow (Fig. 4.14).

In an extensive monitoring of the German Bight from 1954 to 1981, Tiews (1983) found that some species increased in abundance, some decreased and most fluctuated, showing no definite trend. He suggested that the species which increased fitted into the ecological niche of species which declined. The important local fishery for shrimp

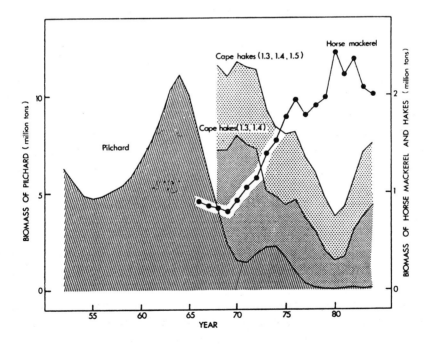

Fig. 4.13   Virtual population analysis (VPA) estimates of the biomass of pilchard *Sardinops ocellatus*, Cape horse mackerel *Trachurus capensis* and Cape hakes *Merluccius capensis* and *M. paradoxus* off Namibia, showing expansion of the horse mackerel resource following collapse of the pilchard resource and the inverse trends in biomasses of horse mackerel and Cape hakes (from Crawford *et al.* 1987).

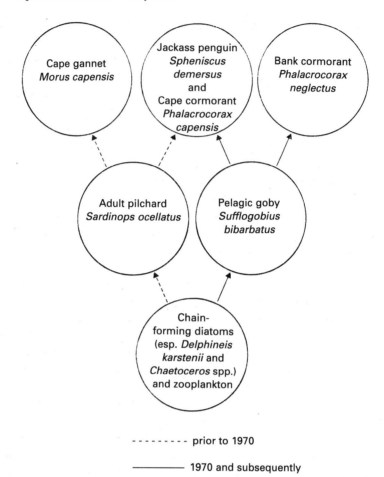

Fig. 4.14 Probable alteration in trophic flow through to have occurred in the intense perennial upwelling region north of Luderitz, South-west Africa (from Crawford *et al.* 1987).

could not have been responsible for these changes, as its effort was fairly constant and minor. Climatic cause was also an unlikely explanation for the observed changes.

The largest changes in marine ecosystems have occurred in the upwelling regions, off Peru and South-west Africa, and in the high latitudes (Greenland and Barents Sea), where the species composition of the ecosystem is relatively simple and many species live close to their environmental tolerance boundaries. The greatest fluctuations in fish ecosystem abundance have occurred in pelagic fish. Blaxter & Hunter (1982) described the fluctuations of clupeoid species. The profound changes which occurred in an upwelling region during El Niño have been well documented in a book by Arntz & Fahrbach (1991). Many regional fish stock fluctuations caused by fishing as well as by regional climatic fluctuations were summarized by Laevastu (1993).

When stock abundance changes, the general biology and ecology (e.g. distribution, spawning area and growth rate) of the stock may also change. Such changes in the

Atlanto-Scandian and Icelandic herring stocks have been summarized by Bakken (1983).

The causes for changes and fluctuations in marine ecosystems can seldom be determined with certainty. Usually several processes and factors affect the ecosystem simultaneously and the contribution of each cannot be ascertained. Some of the causes which have received considerable attention, include:

(1)   variation of recruitment caused by a multitude of factors;
(2)   variation in predation (which also affects recruitment);
(3)   intensity of exploitation of the targeted species, which also affects the abundance of non-target species, mainly via changes of predation conditions;
(4)   anomalies in several environmental factors and processes, and
(5)   change in migration patterns and other behavioural processes (e.g. changes in major spawning and feeding areas).

Catastrophic events can also impact the ecosystems (e.g. cessation of upwelling and severe winters with extensive ice cover in high latitudes).

We have learned to consider fluctuations and changes in marine ecosystems as unavoidable and normal events. However, the question arises whether there is a normal or mean state of any ecosystem to which a distributed system might return in time, and around which it fluctuates. This question pertains especially to attempts at recovery or rebuilding of commercial species biomasses, which are often attempted within fisheries management. However, unfortunately we cannot define the normal state of any marine ecosystem. Although recoveries of some stocks have been observed over varying time spans, other stocks have not fully recovered to their former states. In some cases the recovered stocks have also changed behaviour, e.g. migration routes, feeding areas and spawning ground. Thus, stock recovery is often wishful thinking.

Some aspects of the behaviour of species as well as ecosystems have received little attention. Firstly, there has been little quantification of fish behaviour, considering the marked difference in behaviour between individuals of a single species. Secondly, the ability of marine fish to learn and its consequences have rarely been studied (Fernø 1993, Soria *et al.* 1991). A few studies and the abundant anecdotal information indicate that a fish's ability to learn might play a major role in avoidance of capture and in reaction to environmental stimuli.

Man is interested in predicting changes in marine ecosystems by attempting to find causes for changes. Causality requires, however, real verifiable dependence of an occurrence of one event on another through repetitive observations or experiments. Even if we could eliminate all other simultaneously-acting factors but one, and could make many repetitive observations, we will find variability due to experimental errors in our results; i.e. chance enters into scientific activity. This issue is more complex when many factors affect a phenomenon and we can neither eliminate these factors nor know the quantitative relationship between them and the phenomenon under observation, especially when dealing with living things with their own freedom of action. Thus it is imperative that we be aware that the results of predictions in fisheries ecology are open to modification in the light of new experiences.

4.5   Fishing gear and strategies in relation to fish behaviour

A large variety of fishing gear is in use, especially in artisinal fisheries. Some basic properties of the primary types and their fishing methods are summarized in Table 4.6. For classification of fishing gear see von Brandt 1959, 1984. Here we consider only the basic problems of large, commercial gear, which yield the bulk of world catch, with some behavioural aspects of fish, mainly shoaling, feeding, dispersal and migration.

Three major technical developments in this century have made the fishery for pelagic, shoaling fish very efficient. These are: the use of sonar to detect shoals; the development of power blocks, making the use of large purse seines possible; and the development of large pelagic trawls (Fig. 4.15) with headline transducers towed by high powered trawlers of 1500–200 bhp.

Table 4.6 Some basic properties of the primary types of fishing gear and fishing methods.

Gear and/or method	Advantages	Disadvantages
Gill-netting	Rarely endangers stocks Highly species-selective Low energy consumption Modest vessel size Usually quality fish	Labour intensive Mostly seasonal Modest yield
Longlining	Rarely endangers stocks Acceptable catch at low abundance Low energy consumption Modest vessel size Species selective Quality fish	Labour intensive (mechanization possible) Modest yield Bait problems
Danish seining	Medium yield Modest energy consumption Works patches of ground Species variety modest Fish fairly good quality	Skilled crews required Cannot work rough grounds, nor deep water
Purse-seining	Usually high yield Species selective Modest vessel sizes Modest energy cons.	Mostly seasonal Often low priced species Can rapidly reduce stocks
Trawling	Steady yield High species variety Fair fish quality Usually satisfactory return from investment	Energy expensive Usually large vessels Capital intensive Can rapidly reduce stocks Little species-selective

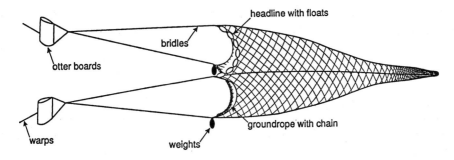

Fig. 4.15   Diagram of a single-boat pelagic trawl.

Parrish & Blaxter (1964) consider the following traits to be important in deter-mining fish responses to fishing gear; horizontal and vertical distributions, density of concentrations, response to stimuli, general shoal behaviour, and fish behaviour associated with light intensity, water temperature, currents and salinity.

The use of a particular gear can depend on economic factors, but also on tradition and on the vulnerability of a particular species to a given gear type. The availability, vulnerability, and catchability problems are discussed in Laevastu & Favorite (1988). Gear selectivity in respect of different sizes, ages and species is affected not only by the economics of fishing, but also to a large extent by the main objectives of fisheries management. These objectives include the conservation of resources to obtain the maximum and in most cases sustained yield, and to divide the available resources between those using different gear or fishing on the same stock in different areas and seasons. The ecosystem management aspects of the use of gear of different selectivity can and should make the greatest use of the differences in fish behaviour and their reactions to gear.

It is apparent from the literature on the world fisheries and catch statistics that the tendency of fish to shoal is the basis for most of the world's large-scale industrial fisheries. The bulk of the shoaling species of fish are caught by purse-seines. One of the greatest improvements in this fishery was caused by the introduction of the power block in the 1950s. The main problems in seining are tactical, such as manoeuvring during setting in order not to frighten the shoal. In some areas and seasons only about half of the attempted seinings are successfully completed. The location of shoals and identification of species and their age/size composition are often difficult. If a wrong shoal is seined, it is usually released and little or no mortality of fish occurs.

The trawl, especially the pelagic trawl, is the second type of gear used for catching shoals. Pelagic trawls allow some aiming with respect to depth and some sideways manoeuvring. The main problems encountered with pelagic trawls are avoidance of the trawl caused by ship and trawl noises, and visual avoidance of the trawl. Fish reaction to sound from vessels is variable in space and time.

Misund (1991) reported that 70% of the herring shoals avoided the vessel hor-izontally and that vertical avoidance was observed when the vessel passed over the shoal. Vessel avoidance was higher during pelagic trawling than during bottom trawling. Misund furthermore found that the average shoal speed was 0.8 m/s or about 2.5 BLs. Packing density and size of the shoals varied in space and time.

Most fishing on shoals occurs during spawning. The high density of spawning concentrations results in higher catches per unit effort, makes targeting of only one species possible and can minimize the capture of small fish. A conflict with management measures arises when these are designed to protect the spawning stock. There are, however, few indications that such measures have been effective. In case this measure is used, the fishery usually intensifies outside spawning areas and at other seasons.

The majority of fish species are dispersed or in small shoals. Large shoals of pelagic species also break into smaller shoals during the feeding season and these feeding shoals are usually distributed over a larger ocean area, the feeding ground. The main gears used for the capture of dispersed species are the moving gears, i.e. trawls and dredges. Of secondary importance are the luring gears, in which fish are lured to bait, longlines and baited traps. Trap nets are used for the capture of straddling species and on coastal migration routes of migrating species. In the tropics fish aggregation devices and artificial reefs have also become more frequently used.

Trawling is an expensive and energy-consuming fishing method. Therefore, this method is economical only where some degree of aggregation of fish, either higher density of dispersed feeding fish or a number of small shoals exist. During the trawling operation many factors affect gear performance and vulnerability of fish to gear. The latter factor is largely determined by the species-specific fish behaviour (see earlier chapters). Two of the determining factors are swimming speed and fish behaviour in front of the trawl.

Although burst speed might be important in the initial reaction to the trawl, a steady swimming behaviour in front of the trawl (Fig. 4.16) and corresponding endurance speed in relation to trawl speed are more important for the capture of different sizes of fish as well as for differentiating between different species, re the by-catch problem.

The observations by Engås & Ona (1990) indicated that the herding process during bottom trawling may be equally efficient by either day or night, and that hearing and tactile senses must play significant roles in trawl detection and avoidance by fish.

Knowledge of fish behaviour with respect to trawls has been used in the past to improve gear efficiency, by decreasing the resistance, increasing the sweep area and/or volume, increasing the speed while attempting to hold down the power requirements and fuel consumption, and above all decreasing trawl avoidance. Recently the problems of trawl selectivity in respect to size and species have again gained in importance. Sorting of the catch during trawling or dredging operations has recently been achieved in the shrimp fishery with considerably modified gear, and also in otter trawls to allow small fish to escape from the cod ends (Larsen and Isaksen 1993) (Fig. 4.17).

Longlining, of which there are several modifications, has been increasing recently in several regions, e.g. the North Atlantic and Japan. Longlining is especially suited to catching dispersed resources and has several other advantages, such as lower capital

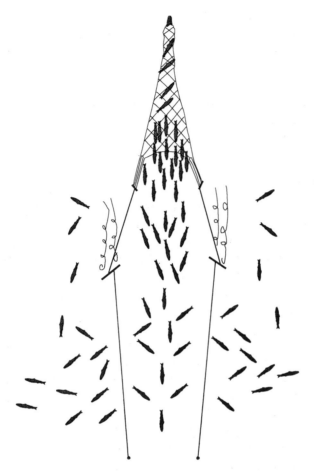

Fig. 4.16   Scheme showing the behaviour of fish in front of a trawl (modified, from Wardle 1993).

requirements and especially low capital investment per manpower engaged in fishery, lower energy requirements, greater selectivity for species and sizes, and usually higher prices for catches. Species selectivity of longline gear can be achieved by selection of hook and bait size and by the nature of the bait (Løkkeborg *et al.* 1989).

A variety of methods for catching dispersed resources are used in local coastal fishers, e.g. baited and unbaited traps and pots, pole and line fishing either in a fixed location or in trolling mode, jigging with vertical lines, and various set nets. Aggregation of some fish is achieved with lights at night. Furthermore, drift nets have been used extensively.

Capture by fixed gear relies on the active movement of fish, searching of food, involving in spawning migrations, or searching for protection. In gillnets and trammel or entangling nets, selectivity of species and size is achieved by mesh size, depth of net (surface and drifting nets and bottom nets), season and location. Trap nets, set nets and fyke nets are fixed or anchored gears, usually with a leader net leading fish to the

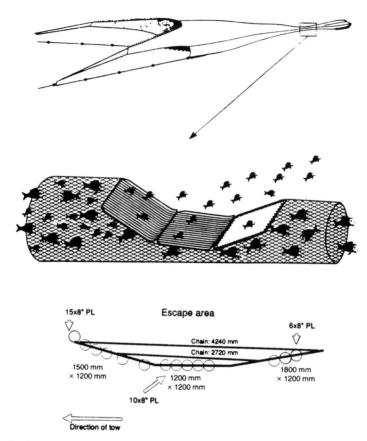

Fig. 4.17 Sorting grid in bottom trawl; (top) an indication of where the sorting grids are placed in the existing trawl; (middle) a sketch of the working principle of rigid sorting grids (sort-X model); and (bottom) details of the construction. The grid area covers $3.2\,m^2$ and the system is made of stainless steel, bouyed with 8 in (20 cm) plastic floats (Larsen & Isaksen 1993).

trap, which can be with or without fykes. They are mainly used in coastal fishers, and can be made species-selective in a number of ways.

Set nets and traps with leader nets are in use in selected locations near the coast to capture either local or seasonally migrating species. Traditional coastal (and artisinal) fisheries have existed for a long time in many areas, and therefore the local coastal ecosystems can be expected to be in balance with this fishery.

Use of different gear not only affects the economics of fishing directly, but also indirectly affects the ecosystem and the dynamics of the resources left in the sea for future catching. Different gear with different size and age selectivity have different effects on the stock left in the sea (Fig. 4.18). If a gear removes more older and larger specimens, the senescent mortality of a population is decreased. Recruitment to the fishable stock is also enhanced in many species.

In general, the degree of exploitation of a given marine ecosystem depends on its profitability. When the exploitation of a given resource becomes unprofitable with the

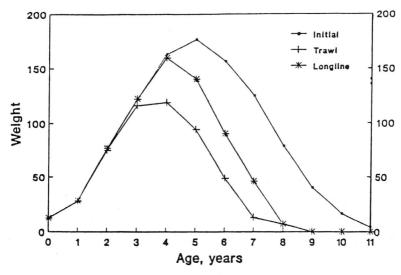

Fig. 4.18 Weight of fish of different age groups in the sea, initially and after 4 years of trawling or longlining ($F=0.2$).

use of one gear, the exploitation does not necessarily stop as it might still be profitable with other gears. Thus conflicts arise regarding the use of different gears as well as the socio-economic condition of the different exploiters (fishing communities) and the biological considerations of the level of desirable exploitation.

The more expensive gears in use, e.g. offshore trawlers, are the first to become unprofitable when the resource declines. Many desired resource management objectives, such as long-term survival of the stock, are at least partially achieved via the economics of fishing. Thus the economic efficiency of fishing gear is also a tool of fisheries management and also a means to minimize conflicts between user groups and to protect the livelihood of some semi-isolated fishing communities.

Chapter 5
Effects of Fishing on Stocks and on Marine Fish Ecosystems

5.1 Effects of fishing on major target species

The effects of fishing on stocks and the methods for computation of these effects have been described in the classical fisheries works by Beverton & Holt (1957) and Ricker (1975). However, modifications and simplifications have been made in these methods and there are several possible sources of errors, such as inadequacy and inaccuracy of data, which are not fully explained to fisheries managers (Parsons 1993). Laevastu & Favorite (1988) reviewed the causes of stock fluctuations and the difficulties of quantitative separation of these causes, as well as the sources of errors in stock assessment. In this chapter we review these effects from a fish ecosystem point of view. However, it is necessary to demonstrate first the effect of fishing on a stock of single, targeted species.

If recruitment and predation mortality were constant and the stock biomass did not change from year to year, then in a virgin (unfished) population the recruitment plus growth would equal mortalities. When fishing starts on a stock (i.e. the fishing mortality is introduced), the other side of the equation must also change or a new equilibrium must be established which means lowering of the exploitable stock in the sea. Usually several other stock or population changes occur with increased fishing, such as a decrease of 'old age' mortality, a decrease of average age of the population (rejuvenation) and consequent change in the mean growth rate of the population. It should also be noted that catch per vessel (or catch per unit effort, CPUE) decreases in some fisheries with increasing total catch. Using several simplified assumptions it has been postulated that there is a condition where net growth of the stock is at the maximum for a given catch, i.e. there is a maximum sustainable yield (MSY) for a given stock. However, owing to many interacting factors such an MSY is mostly an illusion (Larkin 1977).

Several models, for which mathematical procedures and formulae have been developed, have been in use for the computation of the effects of fishing. The simplest of these is the surplus production model, which does not distinguish between the effects of recruitment, growth, and various mortalities (including fishing mortality). The different analytical models, mainly developments of Beverton & Holt's (1957) work, divide the stock into age classes and follow their fate with them. Natural mortality as used in these models is usually only a rough guess, but is constant with age and does not distinguish between sources of mortality. The yield per recruit model, with constant recruitment, is an integral part of the analytical models. This sub-model is often used in management to estimate F_{max}, which varies from species to species and is not always significant for fisheries management (Parsons 1993).

The effects of different levels of fishing intensity on a given stock can be computed numerically when the age composition of the stock at a past fishing intensity is known. Furthermore, the age-specific predation mortality, senescent (or spawning stress) mortality, and apex predation (e.g. by mammals) must be known or estimated. The computations of the effects of fishing on a single-species stock assume no inter- and intra-specific interactions and not effect of environment on the stock.

The method of computation used here is very similar to cohort analysis of sequential computation (Ricker 1975), the essential differences being the use of a predetermined total mortality rate which changes with age (i.e. increases with age in exploited populations, being also higher in pre-fishery juveniles), and the assumption that the fishery removes an equal number-based percentage of fish from all fully-recruited age classes.

A long-term mean constant recruitment is usually assumed in these computations, although different recruitment assumptions are possible. The long-term mean age composition of a stock allows the computation of age-specific total mortality, from which senescent (or spawning stress) mortality, predation mortality and fishing mortality after full recruitment can be estimated.

Using this long-term mean age composition, the relative numerical strength of the exploitable stock can be computed with different fishing intensities (F), assuming constant fishing patterns and using previously-computed senescent and predation mortalities. Fishing mortality F is number based, and assumes that an equal fraction (or percentage) of fish from each fully-recruited year class is removed (i.e. catchability is constant with age). Furthermore, some fishing on two not-fully-recruited pre-fishery year classes is included in the computations.

The results of these computations that reveal the effects of fishing can be presented in different ways. Figure 5.1 shows the decrease of computed numerical strength of the exploitable part of the stock of Pacific cod with different fishing intensity, whereby the unfished stock is normalized with 100 specimens in the first fully-exploited age class (4 years old). The same information is given in Fig. 5.2 where the age class strength is presented as a percentage by number of the fully-exploited population left in the sea. This figure shows that with increasing fishing intensity the traction of first recruits (4-year-olds) increases, thus causing a rejuvenation of the exploitable population left in the sea.

These figures show also that with increased fishing intensity the amount of older, larger fish in an exploitable population and in catches decreases rapidly, causing the dominance of the first fully-recruited year class to increase, assuming constant recruitment. This condition is normally observed in the sea. For example, Nilsson *et al.* (1993) observed that in the Barents Sea the oldest cod declined with an increase of fishing, while the abundance of younger cod was generally stable. The corresponding effect of fishing on a herring stock in the North Sea is shown in Fig. 5.3.

From Fig. 5.2 we can deduce that it is possible to reverse the estimation, i.e. to estimate the long-term mean fishing intensity (fishing mortality) on a stock, if the age

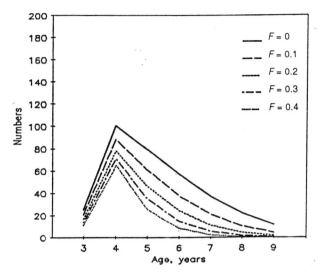

Fig. 5.1   Changes of relative numerical strength of different age groups of Pacific cod with different fishing mortality (number based). Initial number of fully-recruited age group (4-year-olds) in unexploited population = 100. *F* is number based and catchability is constant with age after full recruitment.

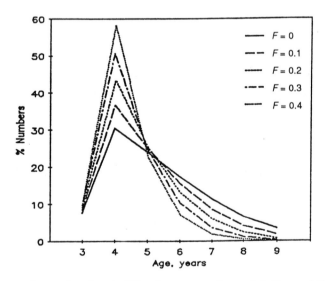

Fig. 5.2   Percentage change of numbers in different age groups of an exploitable population of Pacific cod with different fishing mortality (number based) and constant recruitment.

composition of catches and the predetermined senescent and predation mortalities are known, and the fishing patterns and gear mixture remained unchanged. This estimation can be based on the change of the slope of the normalized numerical age composition of catches. Using catch age composition, and senescent mortality data, we could roughly estimate the relative size of exploitable stock.

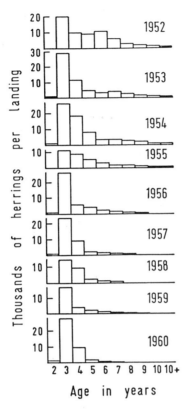

Fig. 5.3   The age distributions of the East Anglian herring fishery in stock density from 1952 to 1960. Note the severe loss of older fish during the second half of the 1950s (from Cushing 1968).

Figure 5.4 illustrates the compound percentage of new recruits in exploitable populations of three species, Pacific cod, walleye pollack and yellowfin sole, with a constant mean recruitment. Figure 5.5 shows the corresponding percentage by weight of annual recruits in the catches of the three species at different fishing mortalities.

Indices of recruitment strength can be estimated by observing the strength of pre-recruits in the catches and comparing their abundance with that of previous years, provided that the gear remains the same and that the fishery operates in the same areas as in previous years.

If we assume recruitment to be constant from year to year and the standing stock of pre-fishery juveniles to be constant, then the exploitable fraction of the total biomass decreases with increasing fishing intensity as shown in Fig. 5.6. However, a change of fishing intensity on the stock also affects predation mortality and cannibalism. When fishing mortality increases, senescent mortality decreases, as some fish which would otherwise die of natural causes will be caught. The decrease of senescent or spawning stress mortality (in weight) with increasing fishing mortality (number-based) is shown for the three species in Fig. 5.7.

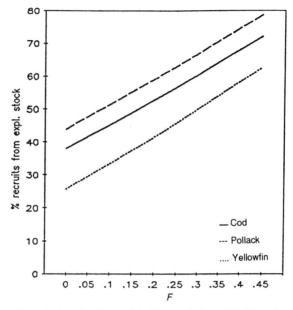

Fig. 5.4 Percentage of recruits (numbers) in exploitable population of Pacific cod, pollack, and yellowfin sole at different fishing mortalities.

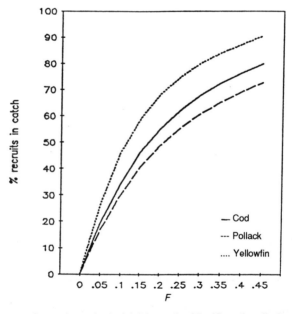

Fig. 5.5 Percentage of annual recruits (weight) in catch of Pacific cod, pollack, and yellowfin sole at different fishing mortalities.

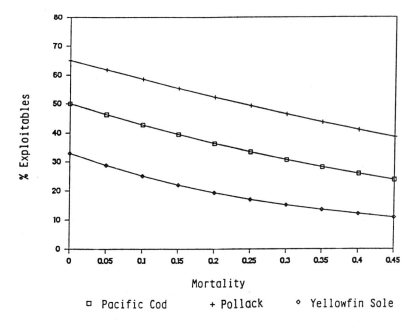

Fig. 5.6 Percentage of exploitable stock (long-term mean weight) at different fishing mortalities (numbers) of Pacific cod, pollack, and yellowfin sole from the eastern Bering Sea.

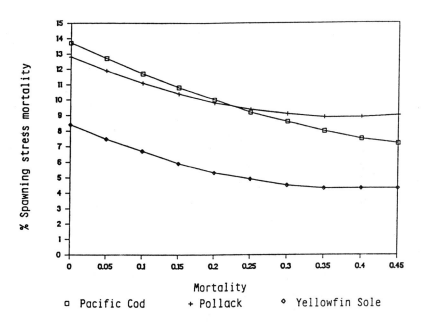

Fig. 5.7 Change in spawning stress mortality (senescent mortality as percentage of weight, long-term mean age composition) with increasing fishing mortality (number based) in Pacific cod, pollack, and yellowfin sole from eastern Bering Sea.

The growth rate (in weight) of recruited fishes decreases with age, first rapidly and later more slowly with increasing age. When the fraction of older fish in a given stock biomass decreases owing to fishing, the fraction of younger, faster-growing specimens increases if recruitment remains constant. This change in age composition results in an increase in the annual growth rate of the total stock biomass, which includes juveniles, with increasing fishing intensity, and is another manifestation of the rejuvenation effect of fishing on a stock.

The predation mortality of the total fish biomass (i.e. that being eaten) dominates all other mortality components. If in a given ecosystem the predation is both prey and predator density-dependent, and when a given prey density increases and its availability increases, predation on this prey item also increases as the fraction of this prey in the food of the predators is proportionally increased. The opposite happens when the density (availability) of a given prey decreases. When the amount of predators increases, they need more food and exercise a greater predation pressure on their food items. Again the opposite happens when the predator biomasses decrease.

The quantitative relationship between predation rate and prey density is, however, not well known. This quantitative relationship is most probably not linear, but close to the square root of prey density. The effect of assuming linear and square root density-dependence of predation on the change of stock size is shown in Fig. 5.8. The linear relation obviously has a more pronounced effect. Even with linear density-dependent predation, the stock stabilizes (reaches equilibrium) at a given lower level after about 5 years of fishing at a given intensity, provided that recruitment is not affected and remains constant. However, if recruitment is independent from spawning

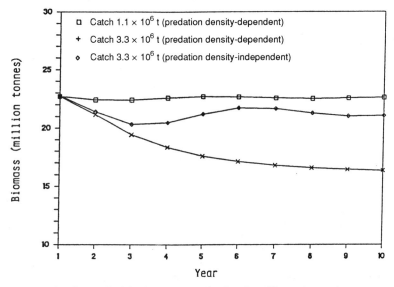

Fig. 5.8 Biomass of walleye pollack in the eastern Bering Sea (in millions of tonnes) as computed with ecosystem simulation: (a) with present catches and with density-dependent predation; (b) catches increased 3-fold and with density-dependent predation; and (c) catches increased 3-fold and with density-independent predation.

stock size and cannibalism is decreased by removing older more cannibalistic specimens, the fishing effects would be counterbalanced to a large degree by increased
recruitment.

There are three conditions which might affect fishing on pre-recruits and on the
remaining exploitable population: the growth rate of specimens in a given age group
varies, some growing faster, some slower; most fishes have a tendency to shoal by size;
and most fishing gear is size-selective.

Fish recruitment to most gears (e.g. the trawl) is related to fish size. However, gear
will also retain fish which are smaller than the fully-recruited size and 1 or even 2 years
younger than the fully recruited age group. Most of the younger fish caught are those
with higher growth rates. Faster-growing fish mature earlier and shoal and associate
with older fish (e.g. on spawning ground), and thus they will be subjected to more
intense fishing mortality than are slower-growing fish. Furthermore, maturation is
size-dependent and consequently faster-growing fish will mature earlier. If senescent
mortality is mainly spawning-stress related and increases by about 10% after each
spawning, then the survival of slower-growing fish is further enhanced by reduced
spawning-stress mortality, as they are subjected to this stress fewer times than faster
growing fish of the same age. The results of a size-selective fishing and the early
maturation of fast-growing fish is that the older age groups in an exploitable population would contain proportionally more slower-growing specimens.

Empirical evidence for the above hypothesis might be found in the analysis of size/
age composition of stocks and catches. Testing and verification of the hypothesis was
attempted on walleye pollack and Pacific cod from the eastern Bering Sea (Laevastu
1992). A sufficient number of age determinations of pollack was available for 1983
(7315 observations); with the sampling covering all the seasons and all fishing areas
and gear in use. For Pacific cod the number of age determinations for each year was
relatively small and so data from 1977, 1978 and 1979 were combined (total 3134
observations).

The pollack fishery was moderately heavy during 1983 and in earlier years; about
1.5 million t were caught each year from a total stock (including juveniles) of about 15
million t. The cod fishery was, however, relatively light in relation to total stock, only
about 50 000 being caught each year from a total stock of about 1 million t.

Figure 5.9 shows the mean and modal lengths of pollack at age in 1983. The
decrease of mean length below the asymptotic length at ages greater than 8 years, and
the decrease of modal length in relation to mean length, are illustrated in Fig. 5.9.
Both conditions are indicative of the presence of slower-growing specimens in older
age categories.

Mean and modal lengths at age and mean and modal ages of Pacific cod did not
show the same difference between the mean and modal lengths that were evident for
walleye pollack. The reasons may be that the fishing pressure on cod has been very
light in relation to the size of the stock, and cod has a much faster growth rate than
pollack, and thus the light fishing had no effect in changing the size/age composition
of cod stock.

The change of age or size composition of stocks with fishing has been observed

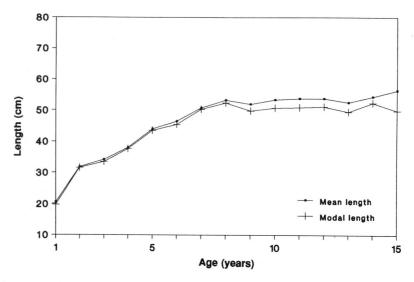

Fig. 5.9   Mean and modal lengths at age of walleye pollack from the eastern Bering Sea in 1983 (7315 observations).

elsewhere in the sea. Pope & Knights (1982) pointed out that the combined size frequency of all North Sea fish stocks shows that the mean size shifted towards smaller fish and size spectra now drop off more rapidly compared with less heavily-exploited areas.

Fishing also affects the relative population sizes of predator and prey. The quantitative evaluation of these effects is difficult because of 'prey switching'. Furthermore, the environmental fluctuations provide a noisy background.

The effects of predator change on the total fish stocks are in general small (Gulland 1987b). Gulland pointed out that the effects of fishing on the community structure is difficult to assess, because of the interaction of environment and processes internal to the ecosystems. One obvious example is the increase of prey species when predator species are reduced (see [3.5]). Other examples are the increase of one (e.g. the subdominant) pelagic species when the dominant pelagic species is reduced (Skud 1982). This change in dominance might also be an effect partly of competition and partly of predation on eggs and larvae.

Most fisheries data series are selective, with only a few species considered and the time series short. Therefore the changes in the total fish community structure in the sea are difficult to assess. The best data on fish ecosystem changes are from the North Sea. A classical example of fish community change there is the gadoid outburst in 1960s. The catches and stocks of cod and haddock rapidly increased at the beginning of this decade, whereas the stocks of mackerel and herring decreased owing to heavy industrial fishing. It is assumed that the decreased predation pressure on the gadoid larvae by reduced mackerel and herring stocks allowed gadoid recruitment to increase.

Many fish species are known to be cannibalistic. This occurs especially in the dominant species in an area, such as walleye pollack in the Bering Sea, and where the distributions of juveniles and adults overlap in space and time. Another condition conducive to cannibalism is that the adults become predominantly piscivorous at older ages. This occurs in the relatively fast-growing gadoids. In flatfishes cannibalism is relatively rare, partly because the larvae are pelagic whereas the adults are demersal and feed largely on benthos.

When the biomass of adult (exploitable) fish of a cannibalistic species is high, the cannibalistic predation on the juvenile biomass can be expected to be high. As predation is often the determinant of recruitment, recruitment could be reduced during high abundances of the older and cannibalistic part of the stock.

Fishing can be expected to interact with cannibalism in the following manner, which is beneficial to the stocks. When fishing removes part of the older, more cannibalistic population, the cannibalism of the juveniles decreases and recruitment to the exploitable population might increase.

Heavy cannibalism can also cause cyclic changes in biomass dynamics of a dominant species. When the adult biomass is high, the juvenile biomass is depressed through cannibalism, and so recruitment decreases. When the new adult biomass resulting from low recruitment becomes exploitable, its abundance will be below the long-term average, thus exercising a lower predation pressure and allowing a higher juvenile biomass and subsequent higher recruitment that will result in a high adult biomass, and the cycle is repeated.

Cannibalism varies from species to species and its effects depend on several factors, such as the degree of separation of juveniles and adults (e.g. the active migration of adults away from the juvenile nursery grounds of Arcto-Norwegian cod), and the ratio of adults to smaller juveniles in a given species is often a function of growth rate.

In considerations of the effects of fishing on stocks we encounter two terms which are often misunderstood and misused, namely 'fishing effort' and 'overfishing'. Fishing effort means different things to different people. Usually it means the amount of fishing done, and is at times correlated to a level of fishing mortality. If we intend to estimate the effect of fishing effort on the stocks or on the marine ecosystem, or use the limitation of fishing effort as a means of fisheries management, we must know and specify a number of other parameters, such as size and type of vessels and gear, area fished, hours fished, amount caught per days fishing, time trends of catch in fishing and the number of men employed. At times another undefined term could be used, 'fishing intensity', which could be further specified as amounts caught in unit time with specified gear, leading to another term, 'CPUE' (catch per unit effort).

The term overfishing has been widely used, often without quantitative definition, though several definitions in terms of fishing mortality rate exist, often without reference to stock abundance, which is also sometimes defined as 'spawning biomass threshold'. As seen in this chapter and as further discussed in Chapters 7 and 8, any quantitative definition of overfishing is subjective and very variable, with many conditions and objectives of fisheries ecosystem management. The definitions in terms of desired spawning biomass levels are not possible either because of the nonexistence of a spawner–recruitment relationship in most species.

Fishing reduces the size of the exploitable part of the stock left in the sea. The smaller stocks of fish that show little shoaling behaviour would lower the catch per unit effort, and thus the profitability for the fishermen. Thus it is of interest to review briefly the reaction of fishermen to the lower stocks. The following notes are extracted from a summary by Shepherd (1993).

When total fishing effort is low and stock size is high, there will be a tendency for fishermen to increase their effort and for new fishermen to join the fishery. This continues until the stock size and profitability fall and earnings from additional effort fail to cover the extra costs, at which time one might expect fishermen to reduce their effort. Fishermen, however, may respond to reduced profitability by increasing their effort to maintain their earnings.

The results of the unregulated response of fishermen are chronic unprofitability, low yields from the targeted species stocks, instability of catches and increased risk of poor recruitment. These changes take a few years to materialize fully. Indeed fishermen could experience a few years of short-term gains with increased activity and increased efficiency. These short-term gains are succeeded by long-term losses. The ideal bio-economic equilibrium is a mirage in the unregulated fishery, and the situation can be remedied with proper fisheries management (See Chapters 7 and 8).

Even mild fishing intensity/effort applied to a virgin stock lowers the average size and age – older, larger fish disappear. This is often called 'overfishing' by fishermen, but not by scientists or managers.

5.2   By-catch, mixed species fisheries, discards and their effects on the fish ecosystem

5.2.1   *Definition and classification of by-catch and discards*

The terms 'by-catch' and 'discards' have several different meanings. Thus, following the advice of Henry Pointcare, 'clarity in argument cannot be achieved before it is introduced into definition', we classify below the by-catch and discards into three categories and give some descriptive definitions. In a fishery which is targeting, or aiming to catch, a defined species, the catch of species other than the target species is usually called the by-catch. Discards are specimens which are thrown back into the sea for various reasons. The basic reason for by-catch is non-selectivity of gear. By-catch can also arise from fisheries regulations, either through specification of catch limits of defined species, or more often by exceeding the TAC (total allowable catch) for some species in a mixed species fishery where the catching of other species is not limited.

Alverson *et al.* (1994) have given some detailed definitions of by-catch and related terms, which are given in simplified form below.

'Econotechnical by-catch', which is mostly discarded, can arise for two economical or technical reasons: (1) the fish species or the size of specimens caught are not marketable; and (2) the economics of fishing or vessel operation make the species undesirable for stowing, either because there is no space on the vessel to stow the less desirable fish (e.g. on shrimp trawlers) or because retaining the fish would lower the bulk value of the catch and there is no time available on board for sorting it.

'Manreg by-catch' or regulation by-catch, which is also mostly discarded, arises from fisheries management and regulation measures: (1) there exists a minimum size limit for the species; or (2) a species is limited by catch quota and, if it is exceeded, the excess must be discarded, or (3) some valuable species have been declared prohibited species as of being owned by some interest groups who have some legal or economical power to do so.

Discards are fish which are caught but thrown back into the sea because they are below the legal size, too small and so unmarketable, deformed, and/or infested with visible parasites. Discards also include benthos of no commercial value. Most commercial fish species are gutted at sea and the offal is discarded. The offal might constitute food to birds, mammals and other fish.

Desirable by-catch in a mixed species fishery consists usually of marketable species, the catching of which is not restricted, but they form a minor part of the catch besides either targeted species.

It is apparent from the descriptions above that by-catch is seldom, if ever, fully quantitatively and universally definable. Furthermore, by-catch varies considerably from one type of fishery to another, from one region to another and even from one season to another.

By-catch can in some regions be an emotional and political issue used to rally political support and public opinion to serve the specific interests of some groups. Public awareness of by-catch problems was generated in the anti-gill-net campaign by US West Coast Pacific salmon interest groups and is now backfiring by focusing on other fisheries. The tide has indeed turned against commercial fisheries, partly because of by-catch.

In some regions, such as the north-east Pacific, the prevailing fishery regulations prohibit the retention of some highly-priced fish, such as salmon and halibut, which are protected by convention or international agreements through politically-influential fishery interest groups. The catch of reserved or protected species is called 'the incidental catch of prohibited species'. In some instances the fisheries are interested in only a part of the fish (e.g. roe), and discard the rest of the catch. In an industrial fishery, one that catches fish for fish meal and oil or for other chemical processing, nearly all the catch is retained. However, this fishery is often constrained by the by-catch of other commercial species or food fish, which have been limited or prohibited by national or regional regulations and quotas.

Mixed catches and by-catches arise because the gear, especially mobile gear such as the bottom trawl, catches nearly everything within range of its sweep. Different gears have, however, different selectivity in respect of species and size, and the species are not necessarily caught in the same proportions as they are present on or near the bottom. On the other hand, a fishery for pelagic species with purse-seines can be rather selective. The shoals are discovered with the use of sonar, and after species identification has been verified the seine is shot and retrieved. If a wrong species is seined, it is released unharmed. Longlines can be selective to some degree, depending on the hook size and bait used, whereas gill-nets are selective mainly with respect to mesh size. By-catch rates in trawls are usually much higher than in the outlawed offshore gill-nets.

The amount of by-catch depends also on the degree of targeting. Fishermen can be selective to a considerable degree if there is an incentive (usually economic) to do so. Table 5.1 gives the relationships between demersal fish catches and the corresponding biomasses present, to demonstrate the degree of targeting. In the pollack fishery with the pelagic trawls the by-catch is only 1%, whereas in the Greenland turbot fishery using otter trawls the by-catch is 36%. It could be noted that the turbot biomass is only 2% of the total fish biomass present.

Table 5.1   Relationships between demersal fish biomass and catch of target species (Alverson *et al.* 1994).

Target species	Target species percentage of total catch in each target fishery	Percentage of overall demersal species biomass
Pollock	99	42.0
Yellowfin sole	67	16.0
Cod	72	5.0
Rocksole	56	9.0
Pacific Ocean perch	16	0.9
Greenland turbot	64	2.0

While a given group of fishermen and/or fishing companies may be primarily interested in catching one species, they will obtain as by-catch other species which might be of primary interest to other fishermen. Targeting on a given species has arisen from such factors as the size of vessel, gear, markets and accessibility to home ports.

In the North Sea, vessels trawling for herring also catch small haddock, and vessels trawling for haddock often catch small herring. Prohibiting trawling for herring would also prohibit trawling for portion-size haddock. However, prohibiting the catching of herring became a necessity in the North Sea owing to extremely low stocks in the first half of the 1980s. Still, some herring were caught in other fisheries as by-catch, but often were not reported.

Sinclair's (1992) study indicated that age segregation of a cod stock would allow for year class targeting by the commercial fishery, and would thus minimize by-catch of smaller specimens. Age segregation of cod and many other demersal and semi-demersal species occurs by depth and temperature.

Regulations concerning by-catch or incidental catch can lead to conflicts between different groups of fishermen and can also inhibit the development of fisheries. To avoid this it is important that all fishermen understand and believe in the necessity and benefits of regulations, and that these regulations will promise better catches in the not-too-distant future. It is also important that each group of fishermen knows that the 'foreign' fishermen will also obey the regulations. To many fishermen a foreigner is anyone else.

The amount and nature of discards can be observed only on board the fishing vessels; in most regions of the world they are not reported as catch. Discards are a nuisance to fishermen and they would like to minimize them. The reasons for discarding part of the catch vary considerably by region and fishery:

(1) Part of the catch is unsuitable for processing, e.g. filleting, on board or in the processing plant; also the market price might not merit the retention of the species in the hold.
(2) A species is unwanted, unmarketable, or the size is unmarketable and would lower the value of the catch if retained.
(3) The quota for a given species was reached earlier in the year.
(4) The season for catching the species is not open.
(5) Catching the species is prohibited for a section of the fleet.

Of the manifold reasons for discards, some are due to fishery regulations which go back in history, especially those relating to the minimum size to be landed. Originally the minimum sizes and minimum mesh regulations were established on the premise that small fish should be left in the ocean to grow so that higher yield could be obtained later. However, any increase of mortality with age was not considered in this premise. The objective of preventing losses of smaller fish would be achieved better with mesh size regulation, which, however, although it diminishes the amounts of small fish caught, does not eliminate the problem entirely. Small fish are still caught in the coarse-mesh trawl even though some do escape through the cod end and wing meshes. Furthermore, when targeting different species the mesh sizes must change in order to retain smaller target species.

Another reason for discards is marketability. This could be avoided in some instances if part of the by-catch could be disposed of to the fish meal or other industries. However, space availability on the vessel might limit the retention of discards when all space is needed for the more valuable catch. This is especially so in small shrimp vessels.

A compelling reason for discarding fish is the elimination of diseased and parasite-infested fish. If these fish were retained the market value of the whole catch could be diminished. On the other hand, there is no sound reason for discarding valuable species, such as halibut and salmon, which are caught as incidental catch. Unfortunately regulations and fisheries management in the north-east Pacific requires this wasteful action. It would be more reasonable to sell the incidental catch of prohibited species with the condition that the fishermen would not profit but that the income from the sale would go into a benevolent or a research and development fund. Norway prohibits the discarding of commercial species, even though they might result in the total allowable catch (TAC) being exceeded. Instead, undersized fish are sold below the cost of catching, thus encouraging fishermen to be selective in targeting. In contrast, European Union (EU) vessels must discard all fish which are over the TAC.

Table 5.2 gives the mean landed and discarded weights and discard rates of groups of fish and other aquatic animals. Discards in the shrimp fishery are the highest and vary between 5 times the shrimp catch in higher latitudes and about 10 times the shrimp catch in the tropics. The present world catch of shrimp is in excess of 1.8 million t, and the total discards are about 9 million t annually. The discards from the shrimp fishery alone thus amount to about 35% of world fisheries discards or about 10% of the world's total catch of fish and other marine organisms.

Table 5.2   Global discards on the basis of the International Standard Statistical Classification of Aquatic Animals and Plants (ISSCAAP) species groups issued by FAO (the United Nations Food and Agriculture Organisation) (Alverson *et al.* 1994).

ISSCAAP Group	Mean discard weight (t)	Mean landed weight (t)	Mean discard rate
Shrimps, prawns	8 965 762	1 827 568	4.91
Jacks, mullets, sauries	4 865 463	9 349 055	0.52
Cods, hakes, and haddocks	2 943 930	12 808 658	0.23
Crabs	2 843 991	1 117 061	2.55
Redfishes, basses, and congers	1 256 807	5 739 743	0.22
Flounders, halibuts, and soles	1 067 611	1 257 858	0.85
Herrings, sardines, anchovies	1 018 239	23 792 608	0.04
Miscellaneous marine fishes	992 356	9 923 560	0.10
Tunas, bonitos, billfishes	220 936	4 177 653	0.05
Squids, cuttlefishes, octopuses	191 801	2 073 523	0.09
Mackerels, snoeks, cutlassfishes	160 081	3 722 818	0.04
Lobsters, spiny-rock lobsters	108 446	205 851	0.53
Salmons, trouts and smelts	38 323	766 462	0.05
Shads	22 755	227 549	0.10
Eels	8 359	9 975	0.84
Total	24 704 859	76 999 942	0.32

The amount of discards in other fisheries is very variable in space and time and from country to country, and depends on market conditions, type of gear and fishery in general, and especially on prevailing fishery regulations. In general, in major fisheries it varies between 10% and 30% of landings, but can be higher if fish are of poor quality or undesirable species, or are regulated by catch quotas.

Although discards are unavoidable, it is of some importance in resource assessment that discards be included in estimates of the total catch of a given species, and that their role with regard to resource assessment, future recruitment and food balance in the ecosystem be considered. At present, considerable effort is made in member countries of ICES (the International Council for the Exploration of the Sea) to account for the effects of discards in stock assessment by adjusting data on the reported landings and their age composition. However, in other parts of the world, such as in the north-east Pacific Ocean, the amounts of discards other than prohibited species are not estimated and consequently not accounted for in stock assessment.

Lost gear (e.g. snagged on the sea bed obstructions) might continue to fish minimally ('ghost fishing'), but the 'catches' are believed to be insignificant and most lost gear gets tangled fast.

5.2.2   *Utilization of fish behaviour to minimize by-catch and discards*

In two by-catch categories, econotechnical and manreg, there are a desire and a need to minimize or eliminate the by-catch. In econotechnical by-catch the fisheries usually find the proper and suitable measures to minimize by-catch. However, in the manreg category the minimization or elimination of by-catch is forced upon fisheries without the proper means and methods to achieve the prescribed goal. Thus severe conflicts between fisheries and management agencies arise within this by-catch category.

The following measures are used to minimize by-catch and discards:

(1)   Voluntary targeting and selective fishing by fishermen, caused by the economics of fishing and by regulatory measures affecting the economics of fishing and promoting selective fishing (see [5.2.1]).

(2)   Regulation of catching properties of the gear by its design. The most common regulation in this category is of mesh size. Other measures can include change of the size of bobbins in otter trawls, elimination of tickler chains, or introduction of separator panels and turtle excluders.

(3)   Regulation (e.g. limiting) of the use of different gear in selected areas and seasons in which by-catch of a species exceeds a predetermined level; and the TAC has been reached.

(4)   Closure of given areas in some seasons for all fisheries, preventing the capture of a given species which is especially abundant in the area or aggregates there.

(5)   Change of fishing tactics, such as fishing in certain depths during daylight or night hours only, moving into other areas if a high by-catch rate is encountered.

(6)   Use of gear which allows undesired by-catch to escape (e.g. sorting and escape devices).

A good knowledge of many aspects of fish behaviour is required in order that the voluntary by-catch and discard minimizing measures are properly selected and effective.

Minimizing gear-related by-catch and discards affects the size selection of the catch. This control is achieved by regulating the mesh size and shape in the whole gear or in parts of it. Mesh size regulation is, however, not totally effective in trawls as some fish which would normally pass through meshes are retained, especially when the cod end is filling. For longlines, size selection is partly achieved by selecting hook type and size. Some species selection is also possible for longlines by selecting different types of bait.

Trawls can be made slightly more selective of truly demersal, less motile species, such as crabs, by changing the tickler and footrope chains and bobbin diameters. A number of changes have been made to otter trawls to achieve partial species selectivity (Tomson 1993, see [4.4]). The most successful recent improvement has been the change in shrimp trawls, which allows sorting of the catch in the sea and allows all fish to escape. Experiments with similar sorting panels in fish trawls have also shown promise. Such sorting is not feasible in a mixed-species fishery.

Different types of gear are used in different fisheries. Each major type has different catching properties. Thus in some fisheries the by-catch regulation can be effected by specifying the gears to be used or to be forbidden. However, such regulations can seriously affect the economics of fishing.

Diurnal and feeding behaviour can be used only to a limited extent for by-catch minimization. The day and night differences in otter trawl catches have been well studied. The diurnal vertical migration and diurnal changes in shoaling affect catches, especially of pelagic species, but it is difficult to find species-specific differences in diurnal behaviour which could be used for minimization of undesirable by-catches.

There are large differences between day and night in the catchability of most demersal flatfishes, caused by their feeding behaviour. They are on the bottom during the day. Parrish & Blaxter (1964) reported that night-time trawl hauls contained more small fish than daytime hauls. However, this difference was not constant in all species or in all seasons, and might have been caused by differences in the reaction of fish to a trawl.

Diurnal differences in feeding affects the catchability for longlines, but these differences cannot easily be used to regulate bycatch.

Shoaling behaviour and shoal recognition offer the possibility of by-catch avoidance in purse-seine fisheries. However, it requires the ability to recognize the species, and possibly the size or age of the species, from acoustic (mainly sonar) images of the shoal, and its shape and depth. However, this ability is limited and depends largely on the past experiences of the skippers. Additional shoal recognition cues are the knowledge of the seasonal and spatial occurrence of the species and the depth and nature of the shoals of different species. In some fisheries the areal scouting of shoals (by day or mostly by night) recognizes the shoals of different species. Although some recognition of shoals of semi-demersal fishes is possible from shoal shape, little use can be made of shoal recognition of semi-demersal species in by-catch minimization.

The knowledge of seasonal migrations and aggregation of different species offers some possibilities of by-catch and discard minimization and their regulation. The basic principle in the application of this approach is that the fishery, or the use of a particular gear, would be limited by area and season if higher than normal aggregations of particular species occur, the catching of which should be limited for reasons of conservation, e.g. the TAC has been reached, or if by-catch is a serious problem. It could be noted that in many species the distribution of young, small fish is often different from that of older, larger and more mature fish.

A prerequisite to the application of this by-catch management principle is that the times and locations of aggregations, e.g. spawning areas, of the protected species are well known, and that these areas and times do not vary from year to year. If they do vary, the potential changes would be predictable to some extent. Furthermore, the major target species must be available in nearby locations. Unfortunately, these prerequisites are seldom fulfilled.

The aggregation of demersal and semi-demersal species is often specific to a particular depth and type of bottom. Therefore some by-catch minimization can be achieved by avoidance or selection of a given depth or given grounds for fishing with a given type of gear. However, the occurrence of the species mixture at a given location should be known, and it seldom is.

In summary, there are two basic aspects of fish behaviour which can be used to minimize undesirable by-catch and decrease the amounts of discards: (1) the behavioural differences of different species and age groups to different gear, and consequent selection of proper gear or its modification; and (2) utilization of the knowledge of seasonable behaviour, mainly migration and spawning, which cause either an overlap or separation of the species or age groups.

Most of the by-catch is unavoidable, and some minimization measures might be wasteful from the points of view of resource utilization and fishery economics. Minimization of econotechnical by-catch might limit the catch per unit effort of target species. Most of the manreg by-catch arises from conservation considerations, e.g. the exceeding of the TAC of given species. Careful consideration of the biological/population dynamics/ecosystem interactions is necessary to evaluate the effects of limiting the catch, which has been and still is a major fisheries regulation method.

Three regulatory measures are used to regulate by-catch, namely quotas, area and season closures, and gear specifications.

Annual, seasonal and spatial quotas have been established for some by-catch species. Sometimes these quotas have been designed to limit other fisheries, a practice which is not morally defensible. The TAC for one commercial species might become limiting as a by-catch problem in the fishery of other major target species. This is a problem to which a partial solution can be found, considering the seasonal behaviour of species and using area or seasonal closures or gear specifications. The by-catch quota system is unworkable in mixed-species fisheries. The question of disposal of limiting by-catch species has two possibilities, to discard them, or to land them and dispose of them below the cost of catching to discourage by-catch. The practice varies from one nation and area to another. Discarding is usually a wasteful practice, as the discarded by-catch of most species will not survive.

Area and seasonal closures have been used in catch and by-catch regulation in many regions. The catch limitation is achieved either by limiting the use of certain gear types or closing the area where high by-catch of limiting species occurs to all fishing for a given period of the year. The effectiveness and wise use of this method requires a good knowledge of seasonal peculiarities of the ecosystem and the reaction of fish to different gears. It should be noted that if an area is closed to fishing, fishing would be more intense outside the area. Thus area enclosures cannot be used to control TAC effectively.

The most common gear specification measure has been the regulation of minimum mesh size, which is most useful in reducing the amounts discarded. In addition, modifications to gear, such as the addition of a tickler chain, can be prohibited. Economically more limiting are the restriction of fishing to a particular gear only, e.g. gill-nets, or the prohibition of a given gear, e.g. the otter trawl.

Little can be done technically to decrease the amounts of discards. The selection of appropriate mesh size, alteration of the rigging of the trawl, and changing the area, ground and depth of fishing, are the main, but often ineffective, attempts to reduce undesirable by-catches. The regulatory limitation of by-catch of prohibited species has been designed to prevent foreign fishing or to shift it to other areas. The latter can work only to a very limited degree in specific locations. The shifting of vessels from an otherwise profitable fishing ground will nearly always result in considerably lower and, in most cases, unprofitable catches in other areas. Changes in the rigging of the gear have also been tried to decrease by-catch, but with little general success.

5.2.3   *Effect of discards on the fish ecosystem*

The amount of discards and discard rates in relation to target species varies considerably from one species to another (Table 5.3). The incidental by-catches of commercially-valuable species might affect recruitment to the respective exploitable stocks. Computations to investigate this possibility show that this effect can be very small compared with the change in predation conditions caused by more intensive fishing on target species. For example, intensive fishing on cod and pollack decreases predation on juvenile halibut and increases its recruitment to the fishery in the North Pacific (Laevastu & Marasco 1982).

Table 5.3   Catch and discards of all species in all Bering Sea/Aleutian Islands trawl fisheries during 1992 (Alverson *et al.* 1994).

Species	Retained catch (t)	Discard catch (t)	Discard/Retained Discard rate (%)
Atka mackerel	39 670	6 453	16
Arrowtooth flounder	539	6 738	1 250
Flounder	5 044	24 007	480
Turbot	317	223	70
Pacific cod	57 977	11 265	19
Pollack	1 115 730	75 734	7
Pacific Ocean perch	10 559	2 218	21
Rockfish	78	631	810
Rocksole	18 591	21 829	117
Sablefish	70	5	7
Yellowfin sole	90 995	31 252	31
Other	1 980	14 664	741
Halibut	0	5 136	
Herring	0	4 386	
Total groundfish	1 341 440	204 491	15

Discards by themselves will probably not appreciably alter the feeding and predation structure in the ecosystem. The return of discards to the sea, where they return nutrients after decomposition, cannot make any noticeable difference in the production cycle in the sea, as this source of nutrients is minuscule compared with the amount of nutrients present in the sea. In contrast, some concern has been raised that the decomposing discards from the flatfish fisheries in the German Bight and discards of the pollack roe fishery in the Bering Sea might cause low oxygen levels near the bottom. No proof of this is available. In the Bering Sea no bottom waters low in oxygen have been found, and in the German Bight the identified occurrences of low oxygen levels may well be caused by other natural factors. Furthermore, it is expected that much of the discards and offal will be consumed by other predatory fish before they decompose, thus serving as a good source for the ecosystem.

There are incidental mortalities in fishing that result from the operation of the gear, such as drop-outs of salmon from gill-nets, escape of hooked and injured halibut and salmon, predation on hooked salmon and halibut by seals and sea lions, etc. In addition to these incidental losses in the target fisheries, there are unknown losses that sometimes result from illegal and unreported catches. Fisheries by-catch also includes occasional catch of sea birds, turtles and mammals (dolphins and seals) in some fisheries. Knowledge about the extent of all these kinds of mortalities is currently highly speculative and judgemental.

The fraction of discard mortality from total fishing mortality varies greatly from species to species (Table 5.4). Alverson *et al.* (1994) commented on the consequences of discards given in this table:

'For pollack, cod and sablefish, discards constitute less than 10% of the observed fishing mortality, and because such losses are counted against the total allowable catch (TAC), discarding constitutes economic loss rather than a problem of unobserved or unaccounted-for mortality. Discard mortalities imposed on rock sole and unidentified flounders exceed the mortalities caused by the landed catch. However, the combined discard and reported fishing mortality for rock sole and flounders is sufficiently low to prevent overfishing, and hence in this instance discarding involves a wasteful fishing practice. For rockfish (*Sebastes*), a group historically suffering from excessive fishing, discards account for about one-half of the total fishing mortality, and discarding slows population recovery.'

Table 5.4 Annual discard mortalities by species and the percentage of total fishing mortality attributed to discards for fisheries in the North Pacific (Alverson *et al.* 1994).

	Annual discard mortality	Percent of total fishing mortality
Pollack	0.016	9.4
Pacific cod	0.013	6.8
Atka mackerel	0.008	15.1
Rockfish	0.004	50.0
Yellowfin sole	0.012	26.1
Sablefish	0.001	1.9
Rock sole	0.015	55.6
Flounder	0.02	83.3
Pacific Ocean perch	0.005	14.3
Halibut	0.043	14.2

For centuries, by-catches and discards have always occurred in the traditional, heavily-fished areas in the North Atlantic, e.g. the North Sea, Grand Banks and Barents Sea. So far there is no clear sign that total marine ecosystems have been greatly altered in these areas, except for the expected changes in fish stocks caused by fishing. It has been in fishermen's own interest to reduce by-catch, and they have been remarkably successful in this endeavour. An example of this can be observed in Table 5.5.

Table 5.5 Catches of species and groups of species in five targeting fisheries in the south-east Bering Sea in 1988 and catches of prohibited species (crabs – number/t of total catch; halibut – kg/t of total catch).

	Targeting fishery				
	Walleye pollack				
Species caught	Bottom trawl	Midwater trawl	Pacific cod	Yellowfin sole	Other flatfish
Walleye pollack	24 162	235 671	5 217	18 425	17 069
Pacific cod	5 103	2 508	55 965	9 985	11 314
Yellowfin sole	298	18	56	94 694	17 383
Other flatfish	1 655	114	3 753	30 262	49 931
Rockfishes	3	—	—	1	—
Crabs	3.5	—	1.2	4.2	3.1
Halibut	1.8	2.0	1.0	3.2	1.7

The discarding of juvenile or any undersized fish is a potential waste because of low survival rates. Survival varies from species to species with season, depth of trawling, time on deck, and other factors. Mortality depends also on sorting and handling the catch. Carr & Robinson (1992) placed discarded fish in large cages in the depth of tow for 24 h and found in two different cruises that the survival of cod was 12–51% and of plaice 44–66%. In contrast, the survival of fish escaping from the gear might be considerable. Soldal *et al.* (1993) found that the survival rate of gadoids which escape from the trawl through meshes can reach 90%.

For the improvement of stock assessment, estimating mortality rates, catchability and recruitment, and general evaluation of the effects of fishing on marine ecosystems, discarding must be taken into account.

5.3   Response of the total fish ecosystem to fishing and effects of marine mammals and birds

Most marine mammals and birds are apex predators in the marine ecosystem and compete with man in utilizing the living marine resources. In earlier years we might have considered marine mammals and sea birds as marine resources, but now many of them have been elevated into a sacrosanct positions through various national and international laws.

Consideration of the response of a marine ecosystem should include the effects of all apex predators on the system. The effects of mammals and birds is mainly via trophic interaction. Laevastu & Marasco (1982) reported a study of the trophic interactions of mammals and birds in the Bering Sea ecosystem, of which a summary is given below, followed by reports by other authors.

Besides preying on fish, mammals and birds also prey on zooplankton and benthos. Thus one might expect some competition between the fish ecosystem and apex predators. This competition would influence the fish ecosystem if the carrying capacity of the sea in respect of the fish ecosystem is materially affected.

The carrying capacity of any area is determined by the availability of basic food resources, zooplankton, squids, benthos and fish, and by the production and annual turnover rate of these biomasses, which are greatly affected by temperature and basic organic (phytoplankton) production. Over the shelf areas about approximately equal amounts of zooplankton, benthos and fish are consumed by the fish ecosystem, whereas in deep-water areas the main food resources are zooplankton, squids, and fish.

The standing stocks of zooplankton, benthos and squids are not accurately known; nor do we know how much of their standing stock and production can be consumed without affecting their recruitment and production. We know, however, that there are large seasonal variations and year-to-year fluctuations of the biomasses of these ecological groups.

The annual consumption of fish by birds and mammals in the Bering Sea is about 4.8% of the annual fish biomass production, of which about one-half consists of commercial species. Although the amount of fish eaten annually by marine mammals and birds is equal to the amount of the commercial catch (2 million t), only 1% of the annual mean standing stock taken by mammals is of commercial size. Thus the predation by marine mammals has little noticeable effect on the fish resources available to the fishery. The only noticeable effect of such predation might be on valuable salmon resource by some mammals (about 34 000 t annually). In contrast, the commercial fishery does not compete with marine mammals and birds for food, as the fishery removes larger specimens above the size taken by marine mammals. Furthermore, the fishery removes the larger, more piscivorous specimens, which would otherwise compete with marine mammals and birds.

Alverson (1992) has summarized the possible interactions of marine mammals with fishing, and so only a brief summary of these interactions by gear type is given here.

The use of gill-nets in the Bering Sea is at present minimal, being limited to subsistence fishing. Thus the tangling of mammals in nets in the Bering Sea is rare indeed. A few deep-diving birds might occasionally get caught in nets and drown, but these numbers are small in relation to the great number of birds present, and to the number of birds succumbing during heavy, lasting storms in critical periods. Some birds have always been caught in large-mesh gill-nets. The use of gill-nets was much more common in the past than it is now, but no noticeable decrease of bird populations due to entanglement has been recorded. At present most marine bird populations show some increase (see [3.4]).

Some gear, especially gill-nets, can get lost, e.g., in storms. In the 1960s it was postulated that the lost gear will continue to fish and affect fish stocks as well as birds and mammals. Several studies in this matter have indicated that the effects of such 'ghost fishing' are minimal indeed. Most lost gill-nets get tangled, fill with seaweeds, become fouled with other marine organisms and sink to the bottom.

Neither marine mammals nor birds get hooked in longlines during fishing in relatively deep water. However, during setting of longlines birds do considerable harm by stealing bait, and during hauling mammals can steal much of catch. Methods and repelling marine mammals, and sometimes birds, from the vicinity of fishing vessels are badly needed.

Trawling does not affect birds and mammals are very rarely caught in trawls during fishing operations in shallower water. However, during hauling of the trawl some mammals, which follow fishing vessels, attempt to get into the surfacing trawl to steal fish. Some of them occasionally get tangled in the wing meshes, but they rarely drown in the trawl. They can, however, cause considerable damage and endanger fishermen when hauled on deck. The process of returning a live, vicious 400 kg sea lion from the deck to the sea is a time-consuming and human life-endangering operation. Fishermen have reported that occasionally the numbers of marine mammals present are so high that the fishing vessels must abandon grounds with good catches and search for other areas where fewer mammals are present.

The incidental catches of marine mammals are very small, with the possible exception of porpoises in tuna seines. The most urgently-needed future action is the development of effective means of repelling mammals and/or methods which can keep mammals away from vessels during hauling operations of longlines and trawls.

Marine mammals off the Norwegian coast cause considerable damage to gear and catch, reaching into millions of NKr, for which compensation is paid by the Norwegian government (Bjørge *et al.* 1981). Great harm is caused to fisheries, nematodes, hosted by mammals and passed on to fish, making the fish unmarketable. The main nematodes are the cod worm *Phocanema decipiens* and the herring worms *Anisakis* sp. and *Contracaecum osculatum*. Up to 64% of all fish can be infested in some areas, with a mean of 8.5 cod worm larvae per fish (Bjørge *et al.* 1981).

In some locations sea birds can consume considerable numbers of fish larvae and small fish, such as salmon smolt leaving the rivers. Furness (1989) estimated that in the Shetland area sea birds consumed 27% of the sand eel production, and he argued that the predation mortality caused by sea birds should be included in multispecies models relevant to management.

The fish ecosystem fluctuates with time in any region, and these fluctuations are reflected in the food composition of marine mammals (Alverson 1992). For example, between 1960 and 1990 herring, Pacific Ocean perch, Atka mackerel and other rock fishes declined sharply in the Gulf of Alaska, whereas pollack, cod, salmon and several flatfish species increased in the late 1970s and 1980s. These changes in the fish ecosystem are reflected in the change of feeding patterns of fur seals and sea lions over this period.

In respect of the fish ecosystem biomass, the continental shelf can be considered as an area of decrease of biomass, whereas the open ocean is a source (growth) area. The reason for this is that the juvenile stages of many fish species are pelagic and feed on pelagic food further offshore. However, most of the species are bound to the continental shelf for spawning, thus returning from the oceanic regime to the neritic region. Furthermore, many older fish become partial benthos feeders on the shelf.

The return of fish biomass to the continental shelf is partly reflected in the fish catches in Table 5.6. This table shows that the Japanese catches in the 0–12 mile coastal zone are 18 times higher than offshore, and 42 times higher in the Seto Inland Sea.

The observation that many fish ecosystems in medium and high latitudes can withstand relatively large environmental perturbations and heavy fishing depends on the past adaptation of the systems to these perturbations (Ursin 1984). In contrast, tropical systems, e.g. in the Gulf of Thailand (Ursin 1984), are adapted to considerably smaller perturbations and thus cannot withstand heavy fishing pressure. Furthermore, higher latitude fish ecosystems have many flatfishes which are benthos feeders. Flatfishes are scarce in the tropics, mainly owing to the meagre benthos in these regions.

A further characteristic which affects the response of an ecosystem to fishing is that in higher latitudes most species have a relatively long life span, buffering the biomass against fluctuations of recruitment, whereas in the tropics the mean life span of most species is short.

Although the assemblages in middle-latitude ecosystems persist over long periods and their spatial configurations change mainly seasonally, changes in the total biomass and relative abundance can occur (Fig 5.10). The changes can occur in species which have great recruitment fluctuations and are heavily fished (e.g. haddock), and in species which are little fished, but can be competitors to declining species (e.g. skates and silver hake, Fig. 5.10).

The fish ecosystems in high latitudes contain relatively few species, some of which dominate quantitatively over others. If the basic prey species (e.g. capelin and herring in the Barents Sea) have recruitment failures, the predator species stocks (e.g. cod and haddock) might also suffer similarly (Jakobsson 1991). This happened in the late 1980s in the Barents Sea, where the whole fish ecosystem biomass became drastically reduced. The collapse of the capelin stock had a disastrous effect on the cod stock, its main predator. Even hungry Arctic seals migrated further south to the Norwegian coast, where they became a nuisance to local fisheries. The recovery of the Barents Sea fish ecosystem was, however, rapid, occurring within a few years.

Besides the direct effect of fishing and catches on the ecosystem one might think of other indirect side-effects, such as possible mortalities of fish escaped from the gear. Data on possible mortality of escapees is meagre. The few studies available indicate that the survival of escaping gadoids and flatfish is high, but that of herring and other seemingly sensitive pelagic fish is relatively low.

There is no proof of any genetic effect of fishing on stocks. However, there might be some minor effects of size-selective fishing (see [5.1]). The effects of trawling on sea bed is described in [6.1].

Table 5.6 Estimated catch per unit of sea area around Japan in 1974 (Sato 1977).

Zone miles from coast	Japan, USSR and others					Seto Inland Sea		
	Area 1000 km² (A)	Catch (1000 t)			Catch per unit of sea area g/m² (B/A)	Area 1000 km² (C)	Catch 1000 t (D)	Catch per unit of sea area g/m² (D/C)
		Japan	USSR & others	Total (B)				
0–12	390	3 100	500	3 600	9.2	19	400	21.0
12–200	4 120	2 200	—	2 200	0.5	—	—	—
Total	4 510	5 300	500	5 800	1.3	19	400	21.0

Note: The catch by USSR and other countries within the 200-mile Japanese waters may be more than the estimated 5 000 000 t shown in the table.

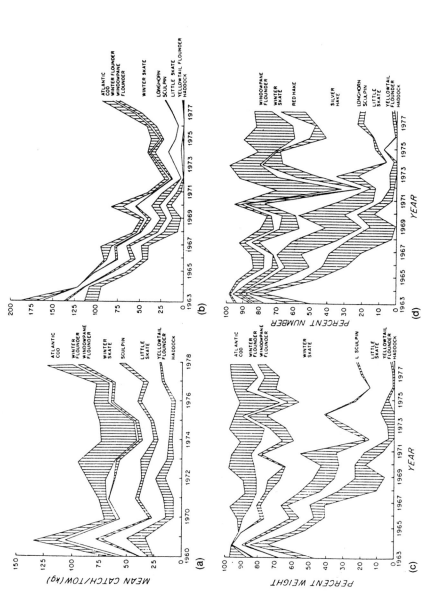

Fig. 5.10 Responses of species from the shallow water assemblage demersal fish community on Georges Bank over the period 1963–78. Panels (a) and (b) express cumulative absolute abundance, mean catch per tow (kg) for spring 1968–78 and autumn 1963–78 respectively. Panels (c) and (d) show cumulative percentage by weight and number respectively for autumn 1963–78 (Overholtz & Tyler 1985).

Recently the terms 'stock depletion' and 'stock rebuilding' have been used in the popular literature, both pertaining to the effects of fishing. These terms are, however, very subjective and difficult to define quantitatively, and several other considerations, such as stock distribution and its changes with time enter the picture. These concepts are further discussed in Chapter 8.

In the past the effect of fishing has been evaluated mainly in relation to the targeted commercial species. The effects of management measures (regulations), which all have various side-effects, have rarely been evaluated properly. Furthermore, we know little of the effects of selective fishing on ecosystems as contrasted to free, unregulated fishing. Nor have the various economic factors and their impact on fishing and on marine ecosystems been explored. The study of these effects has now been made easier with the use of holistic, dynamic marine ecosystem simulations, explored in the next section [5.4].

5.4   Numerical simulation of fish ecosystems and the effect of fishing

The world marine fish catch increased rapidly in the 1970s and levelled off in the beginning of the 1980s, indicating that the world marine resources might be close to full exploitation. Some pelagic resources collapsed in the 1970s, but many stocks recovered in the 1980s. Some heavily-fished stocks, such as gadoids in the North Sea, increased contrary to expectations from conventional concepts of the behaviour of fish stocks. It should be noted that the above-mentioned changes are not solely caused by the fishery, as most fish stocks fluctuate in abundance with time even in the absence of fishing (see Laevastu & Favorite 1988).

As the food resources of the world are limited, there is a moral requirement to utilize the marine resources fully, applying wise management measures for securing sustained yield. This requires accurate knowledge of the resources and their dynamics and response to fishing. Conventional marine resource evaluation methods, which are based on single-species concepts, have serious limitations especially for prognostication of the effects of fishing and other causes on fish stock and ecosystem fluctuations (Laevastu & Favorite 1988). A fishery that targets one species affects non-target species via interspecies interactions (mainly predator). What is needed, therefore, is a holistic ecosystem approach to resource evaluation, management and study of the dynamics of the marine resources and their response to environmental changes. A brief review of the holistic ecosystem approach for resource evaluation is given below.

All empirical sampling methods for the survey of marine resources are very expensive, especially of ships' time. They have shortcomings due to gear limitations and to the dispersed nature of the resources and their inaccessibility. The catchability of different species varies in space and time and with the conventional survey gear used, and it is nearly impossible to determine the catchability factor with the necessary accuracy. The distribution of the resources is patchy and this patchy distribution varies rapidly with time. Accurate resource surveys would require synoptic and intensive sampling. Despite these difficulties, we must continue with the resource surveys and should attempt to complement them with other means which become available.

Acoustic survey methods are relatively rapid, but suffer many shortcomings. They require quasi-simultaneous survey with several ships in order to achieve good coverage and to eliminate the effects of migration. Furthermore, high-speed sampling is necessary to determine the species and its size which is being recorded by the acoustic gear. There are also difficulties with the calibration of an acoustic signal, which varies from species to species and even with the cross-section aspect of the species being recorded. Shoaling fish are difficult to survey, and truly demersal species such as flatfish cannot be surveyed by acoustic methods. Nevertheless acoustic methods have a use in modern resource surveys.

Single-species population dynamics methods for resource evaluation, and other methods closely related to them such as multi-species virtual population analysis (MSVPA), also have serious shortcomings due to several simplifications. These methods use data obtained from commercial catches and their age compositions, both of which may contain considerable errors. These methods can be applied only as approximations on stocks that are under considerable exploitation and in areas where sampling of a given stock is sufficient, such as the North Sea. Each species is considered separately in these approaches without species interactions, except in MSVPA, where predation is taken into account. The natural mortality is not accurately known and is often applied as a non-age-specific constant. Furthermore, the spawning stock – recruitment relationships used in these models are highly variable in most species. The methods of assessment of fishery resources and their limitations are more fully described by Laevastu & Favorite (1988).

Since the early 1920s various methods of computing marine production, based on basic organic production, have been used, assuming some trophic levels and transfer coefficients of organic matter or energy between these levels. Although a great number of estimates of marine production have been made using these principles, the reliability of the results is very questionable. The main shortcomings of this approach are: (1) the basic organic production is variable in space and time and not known with the desired accuracy; (2) the use of this production by other ecological groups is also variable in space and time and is little-known; and (3) trophic levels cannot be explicitly defined in the marine ecosystem, because the composition of food varies with age or size of the species, and also in space and time for the same species.

As over-simplified approaches cannot be expected to yield realistic results in the evaluation of marine ecosystems, it is necessary to embark on holistic marine ecosystem evaluation using all pertinent available information on the processes in it.

One of the principles of a trophodynamic ecosystem evaluation is given in Fig. 5.11. The basic computation of this evaluation is to determine how much of the biomass of different species or ecological groups is consumed and dies per unit time in the given ecosystem. Given the biomass growth rates, the levels of the biomasses which can produce the amounts consumed can be computed with an iterative procedure. Mathematically it means to find a unique solution to a series of biomass balance equations. The simulation models have been described in detail by Laevastu &

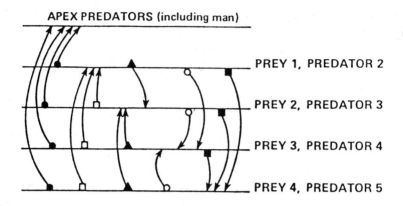

Principles: Determine who eats what and how much. Then determine how much of the prey must be there to produce the eaten amounts. (Food requirements and growth rates known, biomass can be determined.)

Advantages: Minimum values of the production and standing stocks of all prey can be computed.

Amounts of noncommercial (and nonsampled) species can be estimated.

Changes in one prey biomass are related to changes in other prey biomass via predation.

Fig. 5.11   Principles of trophodynamic ecosystem computations, based on food requirements.

Larkins (1981). A review of the available ecosystem models is given by Laevastu & Favorite (1981) and examples are found in the Appendix of this book.

The ecosystem simulation can be characterized as: numerical quantitative reproduction of a system by its structural parts, using deterministic formulations justified by empirical data, i.e. a Pythagorean dictum. The ecosystem simulation is used for:

● Synthesis of all pertinent information on the ecosystem available in a reviewable form;
● Determination of the magnitudes and present status of the living marine resources;
● Determination of the effects of exploitation and environmental anomalies on these resources, so as to ascertain the fluctuations in the abundance and distribution of the resources in space and time.

There are some basic requirements of ecosystem simulation, the fulfilment of which is essential for realistic simulation:

● The simulation must include all known components of the biota. This is necessary for realistic simulation of the trophodynamic process (feeding) and processes dependent on feeding, such as growth. The simulation must also include all known essential processes in the ecosystem, including environmental processes which affect the biota.

- Simulation must have a diagnostic phase, i.e. analysis of initial conditions, and a prognostic phase with proper time steps, monthly or seasonal.
- It must have a proper space and time resolution, which is defined by the region under consideration and by available computer resources.
- Mathematical formulae in the simulation must serve for quantitative reproduction of known distributions and processes; i.e. to simulate the known and proven. This means that the simulation must be deterministic and based on available data and tested formulae. Theoretical conceptualization, so common in modelling, should be avoided unless the theory has been tested and proved by empirical research.
- Explicit approaches, free from mathematical artefacts, must be preferred; i.e. the mathematical formulae used in the model must reproduce known processes, consistent with the data and functionally logical, rather than assume that a mathematical formula represents the behaviour of a system.

In addition, the following requirements apply to ecosystem simulations which emphasize the 'fisheries ecosystem':

- The simulation must be capable of solving the major part of the age-variable mortalities, especially predation, spawning stress and fishing mortalities.
- The system of equations should not be conditionally stable, except for unique solutions in defined conditions, as a marine ecosystem is not stable but fluctuates within varying limits.
- The simulation must include migrations with various causes as well as random movements. These migrations will affect the biomasses by spatial and temporal changes of predator–prey relationships.
- Total carrying capacity must reflect the space–time variable plankton and benthos production, the 'production buffers'.
- Major environmental factors, such as temperature, currents etc., must be included in the simulations and reflect the prevailing knowledge of their space–time variations.
- The simulation must be tailored to the availability of local data and local knowledge.

Verification of large ecosystem simulations is done by their components. This involves testing whether they reproduce empirically-known results and are otherwise correct according to recent knowledge. Validation of the results is carried out by comparing them with various independent survey results, knowing the limitation of these surveys. It is also possible to evaluate the probable errors of simulations by assigning plausible minimum and maximum values to uncertain input parameters as well as to quantitative formulations.

Fish stocks fluctuate considerably in abundance, in such a manner that the biomass of one species declines, the biomass of another species increases. The total biomass of all finfish tends to remain relatively constant, unless some basic food, e.g. benthos or abundant forage fish, changes materially. A summary of the fluctuations of fish biomasses, as determined with the PROBUB model is given in Table 3.7. The periods of fluctuations vary from 4 to 7 years. The magnitudes of fluctuations of fish biomasses vary from 35% to 80% of the mean equilibrium biomass over the fluctuation period, and can be as high as 120% in benthos, both epifauna and infauna.

Although a few holistic ecosystem simulations have been programmed and used and their utility has been well demonstrated, fisheries biologists have not used them further, complaining that they are too complex for them. Standardized, or 'canned', ecosystem simulations have little if any use. Separate and specific holistic ecosystem simulations must be programmed for each region. These simulations are tailored to the availability of data and to the region-specific conditions and problems. However, existing simulations can serve as examples, and some sub-routines can be borrowed from existing simulations with few necessary changes (see Appendix).

Simplified numerical models can be and have been used for special purpose, e.g. to obtain indications of the effects of different gear on the stock remaining in the sea. Bjørdal & Laevastu (1990) made a numerical study of the effects of trawling and longlining on code stocks. The results of this numerical study are given below.

The basic difference between the age composition of trawl and longline catches of cod is that the age/size of full recruitment to the exploitable stock is 1 year earlier in trawl catches than in longline catches. More pre-fishery juveniles are caught with trawls than with longlines, and consequently the amount of discards is higher from the trawl catch than from the longline catch. In the Bjørdal–Laevastu model the trawl was assumed to catch 26% of fish (by number) younger than the fully recruited age group (3 years old). The corresponding value for longlines was assumed to be 17% younger than 4 years old.

In the runs with a prescribed amount of catch both trawl and longline were assumed to catch equal, specified weights. However, if the catch is prescribed with a number-based fishing mortality coefficient F, the amount (in weight) caught by the same F is not necessarily equal owing to a higher catch of young fish by the trawl. The senescent (or natural) mortality remains higher than the fishing mortality.

If the recruitment to pre-fishery juveniles remains constant from one year to another, as was prescribed in the simulation runs, then with equal fishing mortality F a lower number of fish remain in the sea with trawl than with longline fishing (Fig. 5.12). This is mainly because the fishing mortality of trawl catches starts 1 year earlier than that of longline catches. The difference in weight of fish biomass remaining in the sea after 4 years of fishing with trawls versus longlines is even more noticeable than the difference in numbers (Fig. 4.18).

With increasing annual catches the number of fish left in the sea decreases. For the same amount (weight) of catch this decrease is considerably greater when the stock is exploited by trawl compared with longlines. Consequently the fish biomass in the sea decreases with increasing annual catch during the first 4–5 years. However, if the

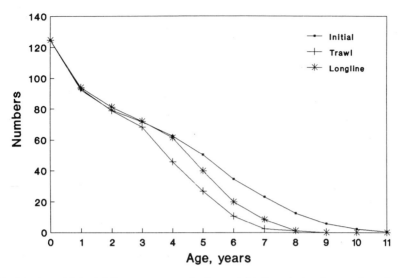

Fig. 5.12   Number of fish of different age groups in the sea, initially and after 4 years of trawling or longlining ($F=0.2$).

annual catch and recruitment remain constant, the biomass left in the sea reaches an equilibrium level that is dependent on the size of the annual catch. At the same catch level this equilibrium biomass is higher in longline than in trawl fishing (Fig. 5.13).

This numerical study demonstrates that the exploitation strategy may have a marked influence on the dynamics of a fish stock. In this case, it is predicted that if a given catch quota of cod is taken by longlines, a higher biomass will remain in the sea than if the same quota is fished with trawls. This effect is mainly caused by the different selective properties of the two gears, as the first fully-recruited year class in the trawl catches is 1 year younger than in the longline catches. Cod also become more piscivorous with increasing age. As the longline catches include more large fish, longlines remove more piscivorous and potentially cannibalistic individuals. If recruitment to the exploitable population is largely influenced by predation on juveniles, then longline fishing may also be more beneficial to recruitment.

The model predicts that, after sustained fishing, the biomass stabilizes around a certain equilibrium level determined by fishing method and exploitation level. With reference to Fig. 5.13, it is apparent that the choice of fishing strategy is relatively unimportant at low catch levels or in periods with good recruitment. However, with increasing exploitation rate, care should be taken as to the choice of fishing gear and strategy. The trends that are predicted in Fig. 5.13 also suggest that this simple simulation model can be used to determine the total allowable catch taken by different gears, if a biologically- or economically-determined minimum level of remaining biomass is desired or prescribed.

Because the ecosystem simulations can be made to reflect nature, they would be the preferred tool for management for all fisheries biologists and managers. In fact, for questions regarding multi-species management, e.g. how to optimally harvest the

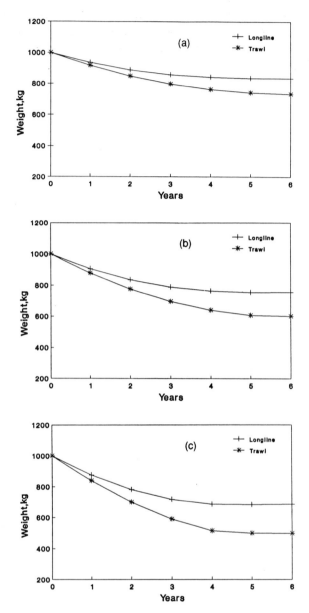

Fig. 5.13 Biomass reduction during 6 years of trawling or longlining with fishing mortalities of (a) 0.1, (b) 0.15, (c) 0.2 (initial biomass = 1000 kg).

components of a multi-species fishery, or to evaluate the possible impacts of an optimum exploitation rate of an abundant species on the production of less abundant species, there seems to be no recourse to methods other than ecosystem simulations or other similar multi-species models.

A brief description of a simple Skeleton Bulk Biomass model (SKEBUB) is given in the Appendix together with a gridded ecosystem simulation UUSDYNE. SKEBUB is a box model and is the simplest multi-species ecosystem model based on biomass, in contrast with conventional number-based models. The description also explains the essentials of fisheries ecosystem simulation.

Chapter 6
Effects of Man's Other Activities on Marine Ecosystems

6.1   **Effects of bottom trawling on benthos**

For the past half century there has been concern that bottom trawls may plough up the sea bottom and destroy fish, shellfish, and invertebrates that are prey for fish. As a consequence of these allegations numerous studies have been conducted (notably in European waters) to determine the impact of bottom trawling on the sea bottom and its biota. Real progress in these studies was made after World War II, when underwater observations by divers, underwater TV, sonar and submersibles became available.

The impact of gear dragged along the bottom depends on the type of gear, its rigging, and the type of bottom and its biota. Mollusc dredges, which cover a very small area, can cut a 20 cm deep trench into the sediment, whereas beam trawls with heavy tickler chains penetrate only 3 cm. Conventional otter trawls have the effects of harrowing, rather than ploughing.

The trawl doors on sand and silt bottom swirl up some sand, which settles quickly. On muddy bottoms the swirled-up mud settles in about 1 h, depending on the current speed and resulting turbulence near the bottom (Riemann & Hoffmann 1991).

Trawls have not been observed to kill many flatfishes. Few impacts have been observed on macrobenthic fauna, except for some slow-moving larger organisms such as starfish and sea urchins. Some bivalves can also get crushed. Otter boards dig out benthic organisms and crush some shells, e.g. *Artica islandica*, making these available as food for benthos-feeding fish such as cod (Rumohr & Krost 1991). Epibenthic vagile organisms regain their original density in disturbed areas within 24 h. Large starfish, e.g. *Asterias rubens*, which are large benthic predators (useless benthos) get damaged mostly by otter boards. This condition is also favourable to fish food production by the benthos, killing useless predators.

The damaged organisms, as well as the infauna that might have been harrowed up by the trawl, will be quickly preyed upon by fish and crabs. It has been reported that within 1 h of the passage of a dredge and/or trawl 3–30 times more fish and crabs were found in the path of the trawl's passage. The fish were apparently preying on organisms that would otherwise not have been available. Similar findings originate from a study of a hydraulic clam dredge in the south-eastern Bering Sea, where yellowfin sole quickly concentrated in the dredge's path, feeding on exposed organisms. This concentration lasted for up to 2 days.

The general consensus of the investigations on the effect of the trawl is that the overall impact of trawling on the sea bottom in beneficial, and it has been speculated that it might even result in the faster growth of benthos-feeding fish by making more food accessible.

Considerable literature on the effects of bottom trawling exists (Redant 1991) and results have been summarized by deGroot (1974). We have extracted a few of the results from a study by Eleftheriou & Robertson (1992), which is representative of such studies. These researchers used a scallop dredge, dredging 2, 4, 12 and 25 tows over the same track. They found that the infaunal community, consisting mainly of bivalve molluscs and crustaceans, did not show any significant changes in abundance of biomass (Figs 6.1, 6.2). Some sessile forms, however, showed some decrease in number. Some large epifaunal and infaunal organisms, such as *Asterias*, *Cancer* and some sand eels, were killed. Some small crustaceans and bivalves showed some

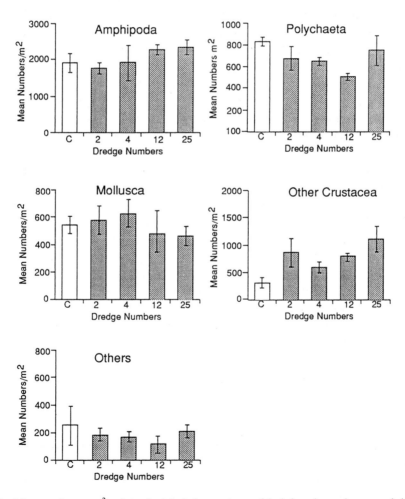

Fig. 6.1 Mean number per m² and standard deviation per taxon of the infaunal organisms recorded during a dredging experiment (Eleftheriou & Robertson 1992).

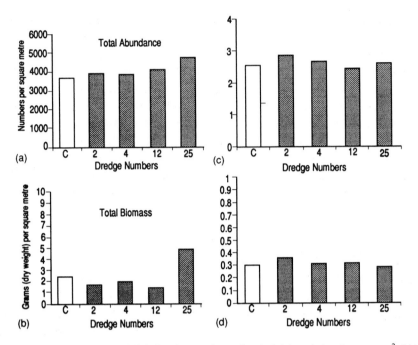

Fig. 6.2   Infaunal organisms recorded during the experimental period: (a) total abundance per m²; (b) total biomass grams dry weight per m². Indices of the benthic infauna during the dredging experiment: (c) diversity (Shannon); (d) evenness (Heip) (Eleftheriou & Robertson 1992).

increase at the end of the experiment (possibly recolonization), whereas highly-agile organisms such as *Carcinus* escaped the damage.

A moderate number of fish were seen to congregate within the dredged area to feed actively on exposed or dead organisms. Eleftheriou & Robertson (1992) stated that the epifaunal community of Firemore Bay, Scotland, has probably changed during the last 20 years, whereby *Chlamys* spp and *Pecten maximus* have been severely reduced owing to directed fishing on these species, and large individuals of *Asterias rubens* and *Cancer pagurus* have been severely depleted owing to beam trawling. However, the bottom community has maintained its original species composition. Furthermore, Eleftheriou & Robertson suggested that, because bottom trawling operations affect benthic life to some degree, and as the North Sea has been intensively trawled for at least the last six decades, the benthos communities, and especially macrobenthos, might already have adjusted to this condition and any additional long-term changes cannot be observed.

The International Council for the Exploration of the Sea (ICES) (1991) Report of the Study Group on Ecosystem Effects of Fishing Activities listed the past studies of the effects of trawling on the sea bed and summarized in general terms the penetration depths of different gear, but warned that the estimation of the potential patchy effects of these gears on the benthos is difficult to assess as the scales of the studies and the possible trawling effects do not match and are probably insignificant over large space and time scales. Furthermore, the effects depend on type of gear and the extent of its use.

The benthos is also affected by natural disturbances, especially wave action and currents, and ice in the higher latitudes. Very few comparative data are available on these natural disturbances of the sea bottom and benthos. The effects of natural disturbances of sediments by wave action and currents decrease with increasing depth, but can still reach considerable depths of, e.g., 40 m (van der Valk 1992). The most violent reworking of sediments by natural forces occurs during severe storms. According to van der Valk (1992) the depth of physical disturbance of the sediment by storms at approximately 12 m can be several times per year. The transition with depth is gradual, and at 20 m the disturbance due to natural causes equals the disturbance due to fishing and can also occur annually.

6.2   Marine pollution and eutrophication

Until about the 1960s the ocean was considered to be a dumping place for most of the wastes produced by man. This view is still defensible to some extent, provided the disposal is done properly and in selected places. In the last 25 years many studies have been conducted and a voluminous amount of papers and books written about marine pollution and its effects. Many international bodies have arisen to review and regulate ocean pollution. Only a general summary of marine pollution and its effects on marine ecosystems, i.e. a review of some aspects of ecotoxicology, is presented here. We are mainly interested in two aspects of marine pollution, the effect of the pollutants on the marine ecosystem, and the pathways and amounts of pollutants back to man.

The observation of the direct effect of pollutants on the marine environment and organisms is difficult and so most studies are carried out in the laboratory. For many reasons it is very difficult to extrapolate laboratory results to the real marine environment.

Sea water contains in solution all elements in various chemical forms. Marine organisms may take up these elements by different processes, absorption, adsorption, feeding etc., and concentrate them in various organs in to different degrees. The main pathway of these elements back to man is by consumption of fish and other marine products. Another aspect of marine pollution, not dealt with here, is the aesthetic effects of beach pollution.

The types of potential marine pollutants and their sources are diverse. First, there are the physical pollutants, plastic debris, dredge spoils and dumped solid wastes. Then there are chemical pollutants, various pesticide residues, including hydrocarbons (see [6.3], and metal compounds. Few of these chemicals are dumped directly, most originating from runoffs of polluted rivers. Sewage, containing nutrient salts (mainly nitrogen and phosphorus compounds) is also a pollutant, but is considered below under eutrophication. Sewage can also contain biological pollutants in the form of pathogens. Treated sewage and some industrial pollutants reached the sea from local coastal outfall and from river runoff; the latter has in the past been the major source of marine pollution in many areas. In the last decade considerable attention has been given to local nearshore and coastal pollution, resulting in improvements in techniques for waste disposal into the sea and in the pollutant loads in coastal waters.

How much liquid waste can be safely disposed into the coastal waters depends on the local mixing conditions and on the exchange between the offshore and coastal waters. The mixing is effected by tides (tidal flushing) and the transport is effected by coastal currents, which depend on wind conditions and on thermo-haline circulation, e.g. estuarine circulation. Many coastal waters are separated from offshore waters by recognizable coastal fronts that affect the distribution of pollutants from the coast. The offshore transport and diffusion through the front is dependent on tides and wind conditions. Offshore transport occurs in surface layers, and onshore in bottom layers (Blanton 1986). Most of the transport and diffusion of pollutants occurs along the coast, between the coastal fronts and coasts.

The effects of different pollutants vary greatly and depend on their nature as well as concentration. Only rarely are their effects physical, e.g. causing increased turbidity. Most effects of pollutants are via bio-concentration, bio-accumulation and bio-magnification in organisms, e.g. fish and shellfish, through which there is a possible pathway back to man.

The input of pollutants into the sea has been generalized as one-third coming from the atmosphere, one-third being land-based input arriving via rivers, and the remaining one-third being direct dumping. Heavy metals are found in solution in sea water and are also added as pollutants. They are removed by scavenging uptake by organisms and sedimentation. Halogenated hydrocarbons (DDT and PCB) are also taken up by organisms and might have some local effects on sea-birds. Chlorophenoids from pulp mills can cause occasional fin erosion of fish locally. Tributyltin oxide, an ingredient of antifouling marine paints, is known to cause shell deformities in mussels.

Low-level radioactive waste containing radioisotopes of strontium (Sr) and caesium (Cs) with a half-life of about 30 years has been dumped into some sea areas, e.g. the Irish Sea, for decades. Fish in these areas also have been monitored for decades with no noticeable ill effects to fish or to man. In general the total intake by man of 'produced' radionucleides about equals that of naturally-occurring radionucleides.

International bodies, including the United Nations Environmental Programme (UNEP), the International Union for Conservation of Nature (IUCN), produced assessments of the health of the oceans for the 1992 UN Conference on Environment and Development. However, owing to the great diversity of marine ecosystems, any general assessment of the oceans is futile, as the marine pollution problems must be considered as local and pollutant-specific problems.

It is acceptable to man to dispose of some wastes into the sea, either under conditions of rapid dispersal or in containers in the deep ocean. Some of these disposals are unavoidable. For example, the pumping of bilge water contributes as much oil to the oceans as accidental spills (Table 6.1).

Some local studies of marine pollution, e.g. pulp mill wastes and marine sewer

Table 6.1   Estimated inputs of petroleum hydrocarbons to the marine environment (10^3 t/year) (from: US National Academy of Sciences 1985, Hardgrave 1991).

Source	Best estimate	% of total
Atmosphere	0.3	9.2
Natural Sources		
Marine seeps	0.2 ⎫	—
Sediment erosion	0.05 ⎭	
Offshore production	0.05	1.5
Transportation		
Tanker operations	0.7	
Dry-docking	0.03	
Marine terminals	0.02	45.2
Bilge and fuel oils	0.3	
Tanker accidents	0.4	
Non-tanker accidents	0.02	
Municipal and industrial wastes and runoff		
Municipal wastes	0.7	
Refineries	0.1	
Non-refining wastes	0.2	36.3
Urban runoff	0.12	
River runoff	0.04	
Ocean dumping	0.02	
Total	3.25	

outlets, have been carried out since the 1920s. At the end of the 1960s the possible consequences of marine pollution to marine fish ecosystems were pointed out (Korringa 1968). Since then a flood of specific marine pollution studies and newspaper articles has appeared in the general literature as well as in newly-established special pollution bulletins. Reviewers who have been searching for large-scale effects of pollution in the seas have not found significant pollution. For example, Lee (1978) found that in the North Sea there is no evidence as to whether contamination of the waters by metals, pesticide residues, etc., has affected the well-being of the fish stocks. The levels of these contaminants in fish were not regarded as being hazardous to human health.

Many national and international bodies are involved in assessing the pollution of the ocean. McIntyre (1992) summarized the findings, stating that the open ocean is essentially unchanged, whereas local pollution occurs in spots along the coasts, especially in estuaries and near coastal cities, and some flotsam is visible on uncared-for swimming beaches.

Grey (1982) reviewed the possible effects of pollutants on marine ecosystems, based on controlled ecosystems experiments. His results cannot be interpreted directly in terms of real marine ecosystems and might apply only partially to specific coastal locations near outfalls. He concluded that continuous gross pollution leads to a dominance by small individuals that have life histories that allow rapid recolonization of disturbed habitats. Planktonic systems were little affected by moderate pollution, and little if any accumulation of organochlorines and other potential pollutants occurred above the natural level. Acute short-term pollution, such as oil accidents, recovered in about 3 years.

Toxic or noxious species exist within the phytoplankton community as a whole (e.g. *Chrysochromulina polylopis*). The so-called plankton blooms that attract public attention are usually not blooms at all, but increased local production of harmful species, and cannot in most cases be related to pollution or eutrophication (ICES 1991). The apparent increase in the frequency of plankton blooms is in most cases related to increased awareness and monitoring. However, blooms are usually associated with the presence of increased amounts of nutrients.

Studies of possible bioaccumulation of persistent contaminants in coastal areas are inconclusive and not normally feasible, because of the relatively small size of the waste fields and high motility of fish populations. There are considerable disposals of sewage sludges in the sea. Local benthos shows some response to these disposals, especially in increased biodiversity, and their effects are small, especially in relation to the effects of trawling on the sea bed. There is anecdotal evidence that some organic discharges may result in locally-enhanced fish populations. Maintenance of high diversity as one of the management goals is not incompatible with controlled waste disposal (ICES 1991).

Oxygen depletion of bottom waters can occur aperiodically during the summer in some areas. This condition is also affected by water exchange by currents. Increased hypoxia has been observed in some years in some areas, e.g. the German Bight and the New York Bight. Whether this is related to increased pollution and eutrophication is uncertain. This hypoxia has caused some temporary changes in local demersal fish ecosystems. It has also been assumed that increasing susceptibility to infectious diseases, such as lymphocystis and epidermal papilloma, appears to be associated with hypoxia in dabs in the German Bight and the Kattegat. Various diseases in fish have been observed since about 1870 (Waterman & Kranz 1992) and it is difficult to say whether these diseases have become more prevalent and are pollution related, or their apparent frequency has increased because of increased attention and monitoring.

A group of Dutch, English and German scientists recently surveyed the diseases in dab in the German Bight, attempting to define spatial variation in disease prevalence along the pollution gradient and to establish relationships with contaminant concentrations and other selected potential causal factors (Vethaak *et al.* 1992). The prevalence of epidermal hyperplasia/papilloma decreased with distance from the river Elbe estuary and with contaminant concentrations, whereas lymphocystis showed the opposite trend (Fig. 6.3). The spatial patterns of other disorders showed no clear trend (Fig 6.3) with the possible exception of skin ulceration which showed high prevalence at the outermost station. However, the aetiology of epidermal hyperplasia/papilloma, though probably infectious, remains uncertain (Vethaak (1992).

Plastic flotsam on the beaches has increased in the last few decades. Because of the high visibility of this flotsam, much concern has been voiced about the potential

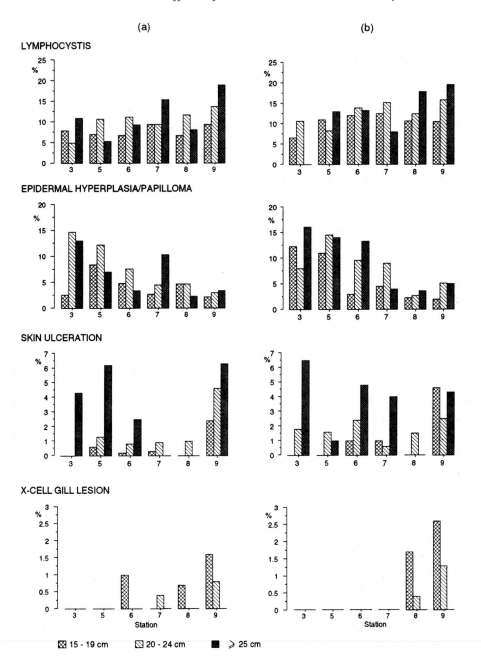

Fig. 6.3 Crude prevalence data of four externally visible diseases in dab *Limanda limanda*, plotted by length class for each station (pooled hauls): (a) 1st survey, (b) 2nd survey. Station numbers increase with increasing distance from the Elbe estuary (from Vethaak *et al.* 1992 a or b).

hazards of ocean pollution with plastics. Most plastic debris on beaches is of local origin and ocean disposal of plastic debris is now well regulated with national and international rules. No noticeable effects of plastics on marine ecosystems have been reported, and notes in some literature on this subject are purely hypothetical.

Eutrophication is a term from limnology which describes the process of adding nutrients, e.g. phosphorous and nitrogen compounds, to an oligotrophic water body, which would then increase the production of phytoplankton and sessile aquatic plants and can occasionally lead to oxygen depletion near the bottom. Although the addition of nutrients to limited water bodies can result in the above effect, this is not fully true in the marine environment, where the original amounts of nutrients in deep water are large, and where regeneration of nutrients from organic matter as well as mixing of deep nutrient-rich water with poorer surface water are of great importance.

Effects of eutrophication are, however, observable in some semi-closed seas and estuaries. They might also occur locally in some coastal ocean areas, but these are difficult to separate from naturally-occurring higher phytoplankton production in coastal waters. Several phenomena, such as phytoplankton blooms, ascribed to eutrophication are actually occurring naturally phenomena. Only occasionally do naturally-occurring algal blooms contain higher amounts of species of algae which produce neurotoxins and toxic domoic acid which can be concentrated by mussels, which feed on phytoplankton.

Harvesting of sea grass (*Zostera*), which was formerly used for house insulation, and kelp, as food, in the Orient and a source of potash and other chemicals (algin and agar), can temporarily lead to apparent eutrophication effects because the nutrients, formerly used by these sessile plants, are used by phytoplankton.

Where there is a continuous source of eutrophication, such as in estuaries, an unstable equilibrium will become established between addition and removal of nutrient by various processes, such as mixing and sedimentation. If and when there is eutrophication of coral reefs, benthic algal growth is favoured, potentially choking the system.

Although many articles in press describe the danger of pollution to coral reefs, there are also studies showing that deep ocean outfalls have no apparent negative environmental impact on coral reef ecosystems (Grigg 1994). Increases in fish abundance due to food subsidies (particulate organic matter) in the effluent have been observed on reefs around the outfalls.

Eutrophication can lead to enrichment of benthos, caused either by transport and sedimentation of organic matter or by sedimentation of the increased phytoplankton production. Josefson (1990) found that between 1972 and 1988 the benthos biomass in Skagerrak–Kattegat increased by a median factor of 1.7. The increase was positively related to water depth and paralleled freshwater runoff from western Sweden and the Danish inner waters, which increased the input of nutrients to the sea. Fish landings from the area have, however, not changed in any unusual way (Fig. 6.4).

Palacin *et al.* (1992) found that eutrophication in a small Mediterranean bay, brought about by continental runoff, caused an increase in the meiofauna biomass and some changes in the benthic nematode community.

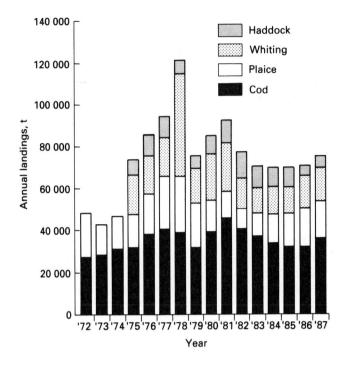

Fig. 6.4 Annual landings of four demersal fish species from the Kattegat and Skagerrak combined, over the period 1972 to 1987. Figures for whiting and haddock are lacking from 1972 to 1974 (from Josefson 1990).

Tiews (1983) believed that the decrease in abundance of some species in the German Bight could be related to pollution of anthropogenic origin. However, the knowledge of the effects of pollutants was too fragmentary to draw a definite conclusion.

Reconcentration of some chemical pollutants into biological systems limits the rate of introduction of these pollutants to the marine environment. The criterion for reconcentration is the potential damage to man if he consumes the biological object. The capacity of biological reconcentration also gives the opportunity to use selected biological materials as indicators of the spread or distribution of pollutants, which might otherwise be undetectable.

After the North Sea the Baltic Sea is the most studied sea in the world. The effects of man and especially the effects of eutrophication are evident in the Baltic Sea, which is a semi-closed sea with a dense human population along its shores. Hundreds of publications and many books are available on this subject. Here, we refer only to those studies summarized by Elmgren (1984, 1989) and Larson *et al.* (1985).

The primary effect of increased nutrient loads from rivers and municipalities on the coasts is a higher phytoplankton production, with blooms maintained during the whole vegetation period and an increased contribution from blue-green algae. The overall long-term trends in nutrient concentration is, however, not uniform, and there are great regional variations (Larson *et al.* 1985). Despite intensive research no clear trend in the open Baltic has been demonstrated.

The conditions in the deep water of the Baltic Sea are dependent on aperiodic inflow of salt water from the Kattegat, determined by specific meteorological conditions, and should not be considered in a search for eutrophication effects.

Increased benthic macrofauna biomass in the Baltic Sea is clearly identifiable as a result of organic enrichment. Appollov (1992) reported that there has been a considerable increase in benthos in the Baltic Sea, especially in the molluscs, which have increased 8–10 times. However, the deoxygenated area in the deeper part of the Baltic has also increased about 20%. Although fishery catches have increased from 450 000 in 1960 to 900 000 in the late 1970s, it is not possible to apportion this increase between that due to eutrophication and that due to more intensive fishing. The changes which have occurred in the Baltic Sea between 1900 and 1980 are summarized in Table 6.2.

Table 6.2 Baltic biological energy flows per year, expressed as carbon in organic matter, about 1900 and about 1980. Refers to the whole Baltic Sea inside the Danish straits (Elmgren 1989).

	About 1900 10^6 t C	$gC \cdot m^{-2}$	About 1980 10^6 t C
Primary production			
Phytoplankton	29–38	79–103	50
Benthos	2	5	2
Net import, including human wastes	3	8	4
Total available organic matter	34–43	92–116	56
Secondary production			
Macrobenthos	0.8	2	1.5
Meiobenthos	0.5	1.3	0.6
Zooplankton (meso-)	~4	8–11	~5
Fish food consumption	0.9	2.4	2
Fish production	0.17	0.4	1.36
Fish yield to man	⩽ 0.01	~0.02	0.085
Fish eaten by marine mammals	0.03	0.1	0.001

In general the increased primary production and increased catches in the Baltic and Adriatic Seas and in some coastal parts of the North Sea are the positive effects of eutrophication. This effect can be observed in reverse in the Mediterranean Sea off the mouth of the River Nile. When the flow of the Nile was decreased by the construction of the Aswan dam, the flow of nutrients to the eastern Mediterranean from the Nile delta decreased, resulting in a drastic decrease of local Sardinella stock and catches.

In the last 10 years the number of researchers investigating the impact of pollution on the marine environment has increased significantly. Many questionable techniques and criteria have been used, e.g. measuring the effects of stressors in biological systems, the induction of enzyme systems in response to given stressors', and a myriad of multivariate techniques for analysing the effects of contaminants on communities. Although some of the techniques might become useful in the determination of pollution, none has been fully successful. Furthermore, the time and space scale needed to observe the dynamic effects of pollution on the community/ecosystem are measured in decades and 10 km up to a global scale (Hardgrave 1991). Although it has been observed that local organic pollution causes increased biodiversity in intermediate levels of disturbance, the studies are hampered by high natural variability. Most of the local well-defined effects of pollution are also reversible. Apparently sludges can be disposed of with no noticeable effect if the disposal is done properly and high dispersion is achieved.

Obviously, marine and environmental pollution must in general be avoided and in many areas decreased. However, more objectivity in pollution research is also desirable. The disposal of wastes into estuaries and coastal waters is controlled by individual state and local regulations.

6.3 Effects of oil development and shipping

The sources of oil input into the marine environment are diverse. Most of it comes from tanker operations and accidents and from municipal wastes (Table 6.1). Oil pollution in the sea, wherever it occurs, is visible on the water's surface or on the beach. All oil pollution incidents must be reported to authorities in most countries (see example for Japan in Fig. 6.5) so that countermeasures can be taken where necessary.

Some past oil spills from grounded tankers have caused extensive temporary damage to beaches and have damaged local inter- and sub-tidal marine ecosystems. These coastal spills have received considerable attention in the news media and from the scientific community. However, no documentary evidence has been found of noticeable detrimental effects of past oil developments on fishery resources, excluding minor local impacts, despite many Draconian forecasts of the impacts of oil developments on marine fisheries and ecosystems. Many of these forecasts appear to have resulted from the incorrect extrapolation of selective laboratory observations on the effects of hydrocarbons on the physiology, genetics and mortality of fish. As a result of the misconceptions on the possible effects of oil developments on marine ecosystems, an antagonistic attitude between oil development and fisheries prevails in the United States, whereas in Europe and in eastern Canada a cooperative attitude exists which is based on multiple-use concepts of natural resources.

The main detrimental effect of local oil development on a fishery and its resources might result from an oil spill from a well blow-out or a pipeline rupture. On the other hand, an oil spill from a tanker accident may occur anywhere in the world where oil is transported.

Oil spills at sea spread on the surface, from where the greatest part of it evaporates, and the remainder dissipates through the water column by dissolution and emulsification. Weathered oil settles to the bottom, and if the accident happens near the coast some of the oil might be transported to the shore, where it adheres to the beach.

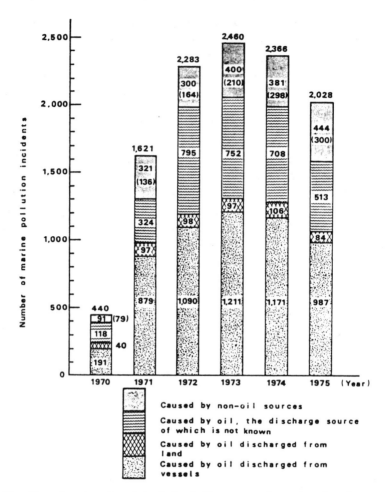

Fig. 6.5 Transition of number of marine pollution incidents reported to and confirmed by the Maritime Safety Agency of Japan. The numerals in parentheses denote the number of red tide occurrences. (Source: Maritime Safety Agency of Japan 1976; White paper on maritime safety 1976; Environment Agency 1977; Annual report on presesnt situations of pollution 1976).

If considerable concentrations of oil are found in the water column, e.g. dissolved or emulsified, it might have some lethal and sub-lethal effects on organisms, including fish, before mixing processes dilute the concentrations and the natural purification finally restores the environment to pre- spill conditions. Weathered oil on the bottom will, however, persist longer than in the water column, and may have some effect on benthic animals, including demersal fish, for a longer period.

It has been assumed in the past that some direct and immediate effects of an oil spill on fishing might be:

(1) Temporary loss of fishing area, due to presumption by the spill or clean-up activity.
(2) Possibility of fouling of vessels or gear (a rare and discounted possibility).

(3) Inability to sell the catch owing to tainting and consumer avoidance.
(4) Possible loss of catch, due to toxic mortality of exploitable stock, or of eggs and larvae affecting recruitment to the future exploitable stock.
(5) Acute but latent mortality to eggs, larvae, juveniles and adults.
(6) Effects on the habitat and alteration of the prey population and food chain.

Although possible genetic mutations are mentioned in some literature, no serious scientific evidence can be found to elaborate on this hypothesis.

A large amount of literature is available on the laboratory studies of the effects of oil on fish and other aquatic biota. The number of corresponding reports on field studies are limited, and are mostly qualitative, inconclusive descriptions of past accidents. Some summary works on the subject are also available, (e.g. Connell & Miller (1980a,b) and US National Academy of Sciences 1985). The objective quantitative evaluation of the laboratory studies and their applicability to the 'real world' is difficult indeed, and one has to agree with the conclusion of the US National Academy of Science report (1985):

'The single most significant gap existing to date is our difficulty in transferring the information obtained from laboratory studies to predicting and/or evaluating potential impact of petroleum on living marine resources in the field, especially in the case of spill impact on such commercially important stocks as fish and shellfish.'

The principal problems with the evaluation of the past oil pollution effect studies are:

(1) Most of the laboratory studies have been carried out with WSF (water soluble fraction) concentrations 2–4 orders of magnitude (100–10 000 times) higher than would occur in the ocean with the greatest feasible accident.
(2) Numerous different components of hydrocarbons have been used in these studies, with very different methods of exposure, on fish and other marine organisms.

Only rarely does a report state honestly the applicability of its results, as has been done by Duval & Fink (1981):

'Hydrocarbon levels in water following oil spills would rarely persist at the concentrations required to cause many of the physiological and behavioural effects observed during this investigation.'

Studies on the sub-lethal effects of petroleum hydrocarbons have also been summarized by Connell & Miller (1980b), US National Academy of Sciences (1985) and Malins *et al.* (1982).

The essential applicable conclusion from the numerous past studies is that WSF concentrations in excess of 100 ppb (parts ber billion) are lethal to fish eggs and larvae within a few days, and that adult fish tolerate concentrations in excess of 1 ppm (parts per million). The latter concentration can be taken as the lowest limit of WSF concentration which causes mortalities in fish within a few days.

The same concentration (1 ppm) can be taken as the lower limit which causes sub-lethal effects in adult fish. The latter effects are often ill-defined; pathological changes in the liver of flatfish, for example, occur both in oil-exposed and non-exposed fish (Malins *et al.* 1982).

Most marine animals, including fish, are capable of metabolizing hydrocarbons. Metabolic by-products are usually retained longer in the bodies than are parent hydrocarbons. Most of the hydrocarbons are taken up with food, especially benthic food. It was concluded from our literature review that fish can be considered tainted if the concentrations of hydrocarbons in the body are > 5 ppm. Hydrocarbons can be present in fish even when no tainting is detected (Grahl-Nielsen *et al.* 1976).

Frequent remarks on possible effects of oil spills on fish and fisheries can be found in the existing literature. These unquantified remarks are, however, unsubstantiated in the majority of cases. Only a few reports attempt to evaluate quantitatively the possible effects of oil developments and oil spills on fisheries. In addition, there exist few good local studies on the subject which cover and emphasize the socio-economic aspects of oil developments on local fishing communities, e.g. Canadian studies for Newfoundland and Nova Scotia.

An earlier study by Johnston (1977) concluded that losses reckoned as fish pro-duction or its approximate cash equivalent are very small even for a catastrophic oil spill. Another study by Norwegian scientists (Norges Offentlige Utredninger, NOU 1980:25) pointed out that the main effects of an oil spill on fish resources are through its effect on fish eggs and larvae. These effects would be delayed several years and are entirely masked by natural fluctuations of recruitment, and compensated for by the presence of several year classes of fish in the exploitable parts of the stocks.

Davenport (1982) reported that field studies have revealed no lasting damage by oil to the planktonic ecosystem, one of the food sources for fish. Conan (1982) described that in catastrophic oil spills reaching estuaries, e.g. the *Amoco Cadiz* spill, the estuarine benthos was affected by oil, whereas the resident fishes (flatfishes and mullets) were affected to a minor degree possible reduced growth and fecundity and some in rot).

A thorough examination of the oil pollution and fisheries by McIntyre (1982) concluded that no long-term adverse effects on fish stocks can be attributed to oil. There might, however, be some local impacts, such as in estuaries as reported by Conan (1982).

Laevastu & Marasco (1985) conducted an extensive study on the possible effects of oil developments in the Bering Sea on the fisheries and fish ecosystem. They assumed extensive well blow-outs and tanker accidents. They found that the presence and distribution of oil on the surface in offshore areas has no consequences for fish or fisheries. Obviously in some conditions oil on the surface could be beached, where it will be of local concern. Although some marine birds and mammals could be affected and even killed by surface oil, these kills are relatively small in offshore waters (most birds and mammals have avoidance reactions) compared with the great numbers of birds and mammals present in the Bering Sea.

After weathering in the water much of the residual oil precipitates to the bottom. Gearing & Gearing (1983) found that about 50% of the aromatic hydrocarbons with three or more rings and saturates with 10 or more carbon atoms were rapidly transported to the sediments where their half lives ranged from 33 to 80 days. In shallow water the concentration of oil in muddy bottoms might reach 100 ppm (Marchand & Caprais 1982).

Initially, the weathered and sedimentized oil accumulates in the near-bottom nepheloid layer. The existence and thickness of this layer is dependent on several environmental factors, such as water depth, nature of the bottom, and water movement over the bottom.

While weathered oil is no longer directly poisonous to organisms, it can be taken up by the benthos and via benthic food also by fish, causing tainting in fish. These tainting effects by sedimentized oil are considerably larger than the tainting from the WSF of oil. Tainting is a temporary condition, as most petroleum hydrocarbons are disseminated from the body by various processes. The main effects of tainting would be a necessary temporary area closure for fisheries.

Eggs and larvae of marine animals are most sensitive to dissolved and emulsified oil (WSF) in the water. The mortalities and serious sub-lethal effects start at concentration of about 100 ppb. The areas covered with WSF > 100 ppb are relatively small in a substantial blow-out lasting 15 days ($< 150\ km^2$). Even in case of such an unlikely event as a 200 000 t tanker accident with diesel fuel released almost instantaneously, the area covered by this concentration is $< 1200\ km^2$.

Most marine fish spawn over relatively large areas, and the pelagic eggs and larvae are distributed with currents and turbulence over very large areas. Furthermore, the spawning of most marine fish lasts 3–6 months, with peak spawning frequently lasting in excess of 3 weeks.

Of the species studied, the spawning of yellowfin sole and its eggs and larvae were found to be most affected by the simulated blow-out and tanker accidents in Bristol Bay. If all yellowfin sole in the Bering Sea spawned within a 2-week period and this spawning coincided with a large tanker accident, only 1.2% of the eggs and larvae would be killed. However, in reality the yellowfin sole's spawning period is about five times longer than that used for the simulated accident – thus less than 0.3% of yellowfin sole eggs and larvae would be affected. The proportion of eggs and larvae of other fish species that would be killed is less than this.

The lethal effects of the WSF of oil on fish commences at the 1–10 ppm range. In the studies reported here the lower value (1 ppm) was used to achieve MEC (Maximum Effect Condition). In evaluating the lethal and serious sub-lethal effects it was also assumed that concentrations of weathered oil on the bottom (tars) in excess of 5 ppm affect juvenile and adult fish.

Figure 6.6 gives the computed areas covered resulting from the assumed large oil spill by different concentrations of water soluble fraction of oil (WSF) and weathered oil on the bottom (tars) in 0–50 days.

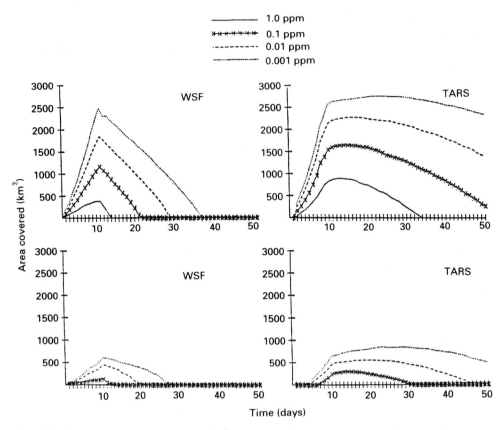

Fig. 6.6 Time series of total area covered (km²) by WSF and by tars at concentrations greater than 1.0 ppm (parts per million); 0.1 ppm, 0.01 ppm, and 0.001 ppm for the accident (upper) and blow-out (lower).

Of the species considered in this study, yellowfin sole and king crab were found to be most affected by the hypothetical oil spill. It was found that an extensive well blow-out would kill or seriously affect only 0.03% of the yellowfin sole and king crab populations in the eastern Bering Sea, which is nearly three orders of magnitude less than the accuracy of resource estimates.

An unusually large tanker accident as envisaged in our study might kill or otherwise seriously affect 0.15% of the adult yellowfin sole population. This amount is about two orders of magnitude less than the accuracy of resource estimate, and at present less than 2% of the catch, i.e. about an order of magnitude less than the error in the estimation of catch. However, a 0.15% fluctuation of the resource would have no effect whatsoever on the catch.

As discussed above, fish can be temporarily tainted with petroleum hydrocarbons by direct exposure as well as by food uptake of contaminated food, mainly benthos. The uptake of petroleum hydrocarbons and their dissemination with time were computed in detail with numerical models. The percentage of some species biomasses in the computation area with an internal contamination > 5 ppm (lower level of tainting) were determined, and the areas covered by these contaminated fish were found significant (Fig. 6.7) in the case of the blow-out and/or accident when they should be temporarily closed for fishing to prevent tainted fish from being caught and marketed.

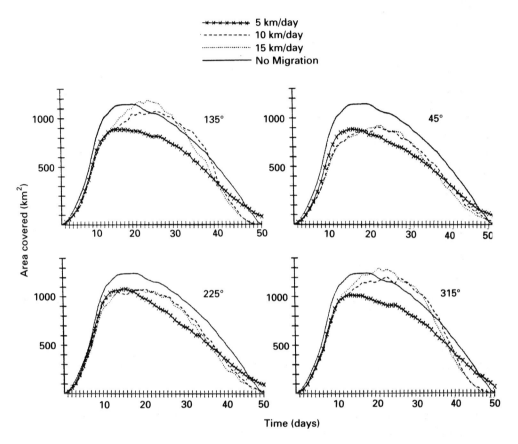

Fig. 6.7 Area covered by tainting (contamination > 5 parts per million) of a pelagic fish species from a model run with no migrations (solid line) and with migrations of 5, 10, and 15 km/day. Migration directions are shown.

Offshore oil development might cause some interference with fisheries by offshore structures such as platforms, wellheads and pipelines. The effects of these structures on marine ecosystems is minimal and local. More significant local effects may be caused by geophysical prospecting using explosives. Løkkeborg & Soldal (1993) found that seismic explorations with an air gun reduced catches in the cod trawl fishery over a considerable area by more than 50%. The catch mixture was also changed. Fishing returned to normal when the seismic work ended. There are also a few anecdotal data on long-range effects of submarine explosions on salmon in pens in northern Norway.

A few decades ago it was assumed that oil and gas pipelines on the bottom of the sea could hinder fisheries and could easily be damaged by trawling. This fear was largely unfounded, and at present pipeline burial is no longer required in the North Sea. The pipes are allowed to sink naturally into the sediment, as also are submarine cables. Most of the well-heads are now below the sediment surface, and the small number of 'Christmas trees' above the bottom are a small fraction of the hazards for trawling compared with over 20 000 known ship wrecks in the North Sea. These wrecks provide shelter for a variety of smaller and juvenile fish. There are also large quantities of explosives from World War II on the bottom of European seas.

The local extraction of sand and gravel from the sea bed is also insignificant relative to the large size of most fishing areas, e.g. the total North Sea. There are obviously local and limited effects of sand and gravel extraction, including destruction of the bottom and benthos in extraction area, attraction of some fish to feed on crushed benthos, and increased turbidity affecting feeding of some fish. Recovery of the benthos varies from a few months to a few years. In some locations the amount of benthos increases. Dredging is well regulated in most countries.

Artificial reefs are used in some areas, mainly to promote local sports fishing. However, they prevent trawl fishing in the area (FAO 1986).

The greatest nuisance, as well as damage to coastal ecosystems, are caused by large tanker accidents, such as the well-publicized *Amoco Cadiz* accident on the coast of France. Follow-up studies of the recovery (Glemarec & Hussenot 1982) have shown that benthic communities are first taken over by opportunistic, pollution-tolerant species and gradually change towards more normal communities, the characteristics of which depend on the type of location and bottom.

Nature itself has cleaned up the *Amoco Cadiz* spill and restored the abers and beaches. It was, however, a gradual restoration of the benthic ecosystem over 5 years, whereas local coastal and estuarine fisheries returned to normal within 2–3 years.

The effects of shipping on marine ecosystems have not been extensively evaluated, as they were expected to be small and unavoidable. One of the effects might be the influence of ships' noises, especially low frequency propeller noises, on fish shoals, affecting their migrations and/or the break-up of shoals. Another effect is the introduction of 'foreign' species, either on the hulls of ships or in the ballast water. An example of the latter is the introduction of the comb jelly *Mnemiopsis leidyi* into the Black Sea (Caddy 1993), where this species multiplied rapidly and caused considerable harm to the pelagic ecosystem, including pelagic fish.

Anderson (1993) reported the occurrence of the destructive starfish *Asterias amurensis* in Australian waters, transported there with ballast water from the North Pacific. It now poses a considerable threat to the Australian shellfish industry.

6.4   Comparison of the effects of fishing and other actions by man with natural processes affecting marine ecosystems

Man has altered many of the earth's rivers during the last two centuries by constructing dams for hydro-electric power, diverted some water for irrigation, and built canals and large harbours in the estuaries where most large coastal cities are located.

He has also polluted the rivers with agricultural, domestic and industrial wastes. As a result of these man-made changes in rivers many local anadromous fish stocks, such as salmon from the River Rhine, sturgeon *Acipenser sturio* from most rivers in Europe and the shad *Alosa alosa* from the Rhine disappeared in the early years of the twentieth century. The disappearance of these stocks has been a very small, but unavoidable, price to pay for technological development and human population increases.

Other fish stocks disappeared as a result of land reclamation in the Zuider Zee (Korringa 1967) during the early 1930s. The first species to disappear was the Zuider Zee herring (earlier catches around 13 500 annually) and this was followed by the anchovy. The enclosure of the Zuider Zee crippled the traditional fishery in the estuary, but made possible eel culture in the resulting freshwater lake.

The effects of fishing on stocks was described in Chapter 5. Fishing and stock fluctuations were reviewed by Laevastu & Favorite (1988) and the possible effects of climatic fluctuations on fish stocks and fisheries were summarized by Laevastu (1993). A summary of the main changes in a fish ecosystem brought about by fishing follows:

(1) Moderate to heavy fishing would decrease the stock size left in the sea. Recruitment might be seriously affected when the spawning stock size is very low. Many processes and factors affect recruitment (Anderson 1988) and it is not possible to quantify the separate effects of these factors in any given case. Stock size is also changed by several natural processes, such as predation and emigration.

(2) For many species, fishing removes larger, older and more piscivorous specimens, thus relieving predation pressure on prey species and on small juvenile fish, which might increase as a result. In cases of highly cannibalistic, dominant species, predation pressure on juveniles is also decreased, increasing recruitment.

(3) When fishing mortality increases, natural mortality (senescent and spawning stress mortality) decreases.

(4) Heavy fishing on several dominant and sub-dominant species in a given region might lower the total finfish biomass. If the total finfish biomass (carrying capacity) has been limited by food availability, then the fish left might exhibit increased growth rates and mature earlier.

The effects of fishing on marine fish ecosystems cannot, however, be generalized. For example, no all gear has the same effect on the stock left in the sea. A comparison of the effects of trawling and longlining is given in Chapter 5 and by Bjørdal & Laevastu 1990). Furthermore, the amount of by-catch and discards varies from one gear to another. Bjørdal (1989) showed that trawl and seine net catches contained 19% small cod while the corresponding values for longline and gill-net were 6% and 2%, respectively. He also compared the conservation aspects of trawls and longlines and, although data are scarce on several conservation topics such as discards, survival after escape and environmental effects, existing knowledge clearly indicates the conservational superiority of longlines over trawl.

In some areas, such as the Baltic Sea, there are several stocks of the same species, e.g. herring. These stocks fluctuate independently of each other (see Fig. 6.8). Daan (1980) believed that an increase of the exploitation rate has triggered large-scale changes in fish ecosystems in many areas. He also suggested that the changes in the North Sea may be compensatory in nature, whereas in Californian and South African waters a species replacement (sardine–anchovy) has occurred. Daan concluded that there is no general ecosystem response to exploitation and that for each particular situation a unique response may be expected.

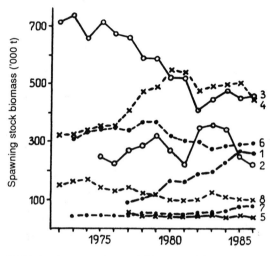

Fig. 6.8 Dynamics of Baltic herring spawning stock biomasses (SSB) as assessed by the ICES Working Group. The curves refer to ICES statistical areas: 1—Sd 22–24 + Sd IIIa; 2–Sd 25–27 (coastal); 3—Sd 25–27 (open sea); 4—Sd 28–29S; 5—G. of Riga; 6–Sd 30E + 29NE; 7—Sd 31E; 8—Sd 32 (Ojaveer 1989).

El Niño is often cited as the cause of a multitude of changes in marine ecosystems (Bragg 1991). However, many changes attributed to El Niño are caused by local surface wind anomalies and related surface current anomalies.

One of the best and longest monitoring studies of a marine ecosystem has been done off Plymouth, UK (Southward & Boalch 1986). Many changes in this ecosystem have occurred over the last 60 years (Figs. 6.9–6.12), the causes of which are difficult to find. The effects of man cannot be linked to any of these changes and the hypothesis of possible climatic change as a cause is possible but likewise unverifiable.

An ICES Study Group on Ecosystem Effects of Fishing Activities (1991) concluded that 'even if there are habitat-mediated effects of fishing practices on fish populations, these will probably be local effects, with negligible consequences of the population level.'

Unfortunately some environmentalists complain that scientists have found it difficult to draw definitive conclusions about the changes in a marine ecosystem, and advocate the closures of many important fisheries of the world (Hey 1992), despite the fact that it cannot be proved that fishing alone has been the cause of ecosystem changes.

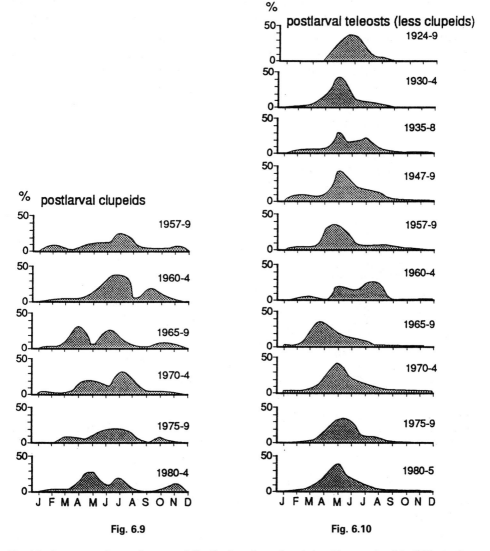

Fig. 6.9

Fig. 6.10

Fig. 6.9 Long-term changes in seasonal distribution of post-larval clupeids at station L5 off Plymouth. Average monthly means for 5-year periods shown as percentages of the annual sums (Southward & Boalch 1986).

Fig. 6.10 Long-term changes in seasonal distribution of all post-larval teleosts, other than clupeids, at station L5 off Plymouth. Average monthly means for 5-year periods shown as percentages of the annual sums (Southward & Boalch 1986).

Although size/age-selective harvesting affects the fish population structure and may affect growth rate and age/size at maturity (Laevastu 1992), there is no evidence that fishing has any direct genetic effect on the populations, although there is a theoretical possibility that genetic diversity might be affected in very heavily-fished populations.

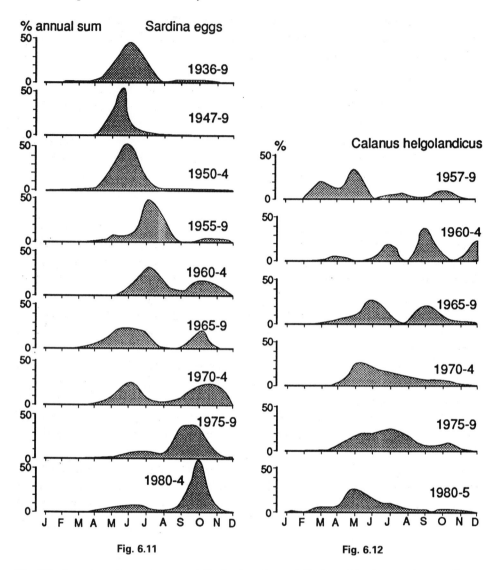

Fig. 6.11

Fig. 6.12

Fig. 6.11 Long-term changes in seasonal distribution of eggs of *Sardina pilchardus* at station L5 off Plymouth. Average monthly means for 5-year periods shown as percentages of the annual sums (Southward & Boalch 1986).

Fig. 6.12 Long-term changes in seasonal distribution of the copepod *Calanus helgolandicus* at station L5 off Plymouth. Average monthly means for 5-year periods shown as percentages of the annual sums (Southward & Bolach 1986).

Fears have been expressed that the entanglement of marine animals in discarded netting present a danger for the populations of these animals. Objective consideration of this subject brings out the following facts:

(1)   The amount of discarded netting is so minuscule, when compared with the area of the bottom of the traditional fishing grounds, e.g. the Bering Sea shelf, that it cannot rationally be considered a hazard.
(2)   Discarded netting is usually bundled, and either sinks to the bottom immediately or, if its artificial fibres are lighter than water, floats on the surface. There is no reason for the mammals to seek out these bundles and wilfully entangle themselves in them.

The use of gill-nets has in most areas, especially in the North Pacific, drastically decreased since the 1960s and has been, for example, almost absent in the Bering Sea since 1979. At present there is no scientific evidence directly linking the possible fluctuations of marine mammal populations, of which little is known, with any fishing activity.

As discussed above, the discharge of domestic and agricultural wastes causes local coastal eutrophication. Because long-term objective quantitative data on algal blooms are scarce, it is not possible to ascertain whether these cases of local eutrophication cause a higher frequency of occurrence of red tides. The growing attention to these phenomena and more frequent reporting in the press, accompanied by various alarmist views, might be responsible for the apparent higher frequency of red tides. After all, eutrophication of previously oligotrophic areas, such as the Baltic Sea, might be partly responsible for higher fish production and catches (Fig. 6.13, Kalejs & Ojaveer 1989).

Increased attention has been given to naturally-occurring diseases in molluscs and crustaceans (Rosenfeld 1976) and fish (Dethlefsen 1984). In the absence of earlier extensive investigations and long-term data and monitoring, it cannot be ascertained

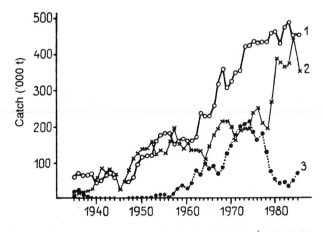

Fig. 6.13   Herring (○), sprat (●), and cod (×) catches in the Baltic Sea (Kalejs & Ojaveer 1989).

whether there is an increased frequency of diseased organisms or that the possible increased frequency can be ascribed to increased pollution or increased ocean dumping of wastes. Conversely, we cannot carelessly dump wastes into the ocean assuming no ill effects.

Marine ecosystems might have been influenced locally by the introduction and acclimatization of 'foreign' marine organisms by man. This introduction might have been accidental, e.g. carried by ships and their ballast water, or planned. A few examples of shipborne organisms are: the 'wollhand' crab *Eriocheir sinensis* from the Orient to Europe; two molluscs, *Crepidula fornicata* (a competitor for oysters, from North America to Europe); and the oyster drill *Ocenebra japonica* from Japan to North America and Europe. With the opening of the Suez Canal many species could also pass between the Indian Ocean (Red Sea) and the Mediterranean (see Table 6.3).

Table 6.3 Auto-introduction from the Suez Canal of various hydrobionts (according to Walford & Wickland 1973).

Group	Number of species	
	Passing from the Red Sea to the Mediterranean	From the Mediterranean to the Red Sea
Algae	1	2
Sponges	2	3
Coelenterata	29	—
Worms (Polychaeta, Sipunculoida)	10	2
Crustacea	14	1
Pantopoda	1	—
Mollusca	20	—
Echinodermata	3	—
Ascidia	6	—
Fish	25	7
Total	111	16

There have been many successful transplantations and acclimatizations of organisms such as salmonids, several species of oysters, the anadromous *Alosa sapidissima*, and the rock fish *Roccus saxatilis*, from the Atlantic to the Pacific coast of the United States. Polychaetes have also been transplanted from the Azov Sea to the Caspian Sea, leading to increased production of benthos in the latter. Table 6.4 gives examples of purposeful and successful transplantations of fish to the Hawaiian Islands.

In some cases the accidental transplantation has caused some damage to some native species, e.g. the oyster drill, or some nuisance to fisheries, the 'wollhand' crab. However, no profound changes in the ecosystems have occurred with the exception of the accidental introduction of the ctenophore *Mnemiopsis leidyi* into the Black Sea.

Another example of the introduction of a species with neutral or positive effects is that of the exotic sea-grass *Zostera japonica* to the North Pacific Coast of North

Table 6.4 Successful acclimatization of marine commercial species of fish and invertebrates in the waters of the Hawaiian Islands (Walford & Wickland 1973).

Species	Origin	Date of transplantation	Results
Bivalve molluscs			
Tapes philippinarum	Japan	1920	Numerous, utilized.
Citherea meretrix	Japan	1926, 1939	Well acclimatized.
Crassostrea virginica	Atlantic coast of USA	1871 and 1883, unsuccessful 1893 and 1895, successful	Common on the rocky littoral of Kaneohe Bay.
Crustaceans			
Scylla serrata	Samoa	1926–35	Common on all islands, valuable food product.
Fish			
Marquesas sardine *Herclotsichtys vittatus*	Marquesas	1955–7	Well acclimatized.
Cephalopholis argus	Society Islands	1956	Acclimatized and fished.
Lutianus kasmira	Marquesas	1958	Rather common, commercial
L. vaigiensis	Marquesas Society Islands	1955–8	since 1962. Common, fished since 1969.
Mollienesia latipinna	Texas	1950	Brought in to destroy mosquito
Tilapia mossambica	Singapore	1951	larvae. Adapted well to salty water. Well acclimatized to brackish and fresh water.

America in the first half of this century along with oyster shipments from Japan (Baldwin & Louvorn 1994).

Commercial aquaculture of various species, such as salmon and shrimp, has increased rapidly in the past 10 years. In 1990, aquaculture produced 12 million t, more than 10% of the total world catch; 55% of this production was finfish and much of it was from fresh water. In marine finfish aquaculture, the most important species are salmon (see Fig. 6.14). The 1989 salmon production would require about 1 million t of feed, mainly cheap marine fish such as sardines and blue whiting.

Although several suggestions have been put forward regarding the possible harmful effects of salmon aquaculture on marine ecosystems (e.g., local pollution, increases in diseases and genetic effects on wild stocks via escapers) none has been proved.

Another fish culture practice that interacts with marine ecosystems is ocean ranching of salmon. Smolt or juvenile salmon are released into the sea or into estuaries, and returning adult salmon are caught by coastal fisherman in the estuaries. This practice is common in the Baltic Sea, Japan and Iceland.

The separation of the effects of fishing on stocks from the effects of other factors affecting stock fluctuations, e.g. environmental changes and changes of predation conditions, within an ecosystem is quite difficult and often uncertain (Laevastu & Favorite 1988). Sharp & McLain (1993) described these difficulties in the Chilean fishery:

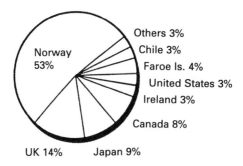

1989 Total: 217,000 t

Fig. 6.14   Estimated farmed salmon production by the major producing nations, 1989 (Anon 1991).

'Interpretation of the effects from human activities is complicated. In the northern Chilean fishery, older age classes of sardines have been quite heavily exploited since 1985, and until about 1987 fish < 3 tears were rarely caught. By 1987 the adults (fish older than 3 years) had begun to move away from the coastal areas that are most accessible to the fishery. The information needed to understand this situation is complicated by the distribution of fishing activities. The spotter-planes, and therefore also the sardine fishery, operate exclusively at night. The fisheries distribution are limited by the willingness of spotter plane pilots to fly distances off-shore. The pilot's decisions about whether to proceed offshore, explore, and direct fishing vessels are tempered by their concerns about safety. Their willingness declines with distance offshore.'

Fishing causes changes in the parameters of the target species stock and in the interspecific processes. Some of these changes in the hake stock on the Argentinean continental shelf have been described by Prenski & Angelescu (1993) (Fig. 6.15). Cannibalism in hake is high. The biomass removed by cannibalism is equal to or greater than the commercial catch. The increase in commercial catches causes a decrease in cannibalism because the fishery removed the older, more cannibalistic specimens and also lowered the hake stock. This resulted in a decrease in the natural mortality rate (M). The biomasses of the other species which served as prey for hake increased.

The evaluation of the success or failure of past management actions and attempts to rebuild the stocks is difficult. Man has been attempting to influence the fish stocks with a reverse approach to fishing, i.e. to allow the stocks to recover by limiting the fishery on given species or imposing a total moratorium on fishing. This rebuilding has had various other forms in the past, such as limiting or forbidding fishing on spawning grounds or during spawning times, or regulating the use of the gear.

Before judging the success or failure of the stock rebuilding/recovery efforts it is necessary to recall some basic properties of the marine ecosystems and factors affecting changes in the ecosystem (Table 6.5):

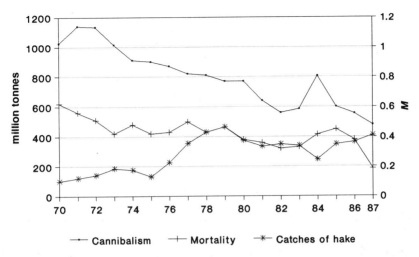

Fig. 6.15 Cannibalism, natural mortality (*M*) and commercial catches of hake on the Argentinian continental shelf from 1970 to 1987 (Prenski & Angelescu 1993).

Table 6.5 Consequences of changes in biomass, stabilization mechanisms and limiting factors.

Cause for biomass change	Consequences and stabilizing effects	Limiting factors
Increase of fishery	Decrease of adult biomass (might lead to decreased recruitment) Decrease of spawning stress and natural mortality (stabilization) Relative increase of pre-fishery juvenile biomass, thus increase of growth coefficient (stabilization) Decrease of cannibalistic predation (stabilization)	Decreased availability to fishery Decreased recruitment (recruitment overfishing)
Increase of growth rate (temperature anomaly effects)	Increase of biomass Increase of predation (higher food requirements) (stabilization)	Availability of food Biological limit to growth
Increase of predation (including environmental effects affecting predator-prey relations)	Decrease of recruitment to adult biomass Decrease of biomass growth coefficient (aging of population) (destabilization) Increase of individual growth is possible if food was limiting growth (stabilization)	Decreasing food density (availability)
Increase of recruitment, caused by:		
(a) decreased predation, resulting in increased larval survival (also environmental effects possible)	Increase of biomass Increase of biomass growth coefficient (rejuvenation) Decrease of individual growth is possible (food availability) (stabilization)	Increasing prey density Increasing food requirement
(b) increased spawner biomass	Increased spawning stress mortality (stabilization) Increased juveniles (increased relative predation) (stabilization) Decreased food availability (stabilization)	Increased predation on juveniles (higher prey density; increased cannibalism)

(1)   The various marine ecosystems and their components are interconnected mainly through prediction, but also through competition of various kinds.

(2)   The marine ecosystems are not stable, but fluctuate in space and time in respect of the quantitative composition and dominance of the species without the effects of fishing.

(3)   We seldom know with any degree of accuracy the quantitative composition of the fish ecosystems in any given region. However, changes in abundance of important commercial species can be noticed in fishery and various resource assessments. Thus we do not have any good knowledge of the mean (or normal) state of the fish ecosystem when we want to rebuild the stock.

(4)   The fisheries management failures, including rebuilding failures, are seldom reported and less investigated as to their causes. Thus we can list only a few general examples of rebuilding successes or failures.

(5)   The recovery or rebuilding of a stock depends on successful recruitment. In most cases our quantitative knowledge of the factors and processes controlling recruitment are incomplete. There is often a desire to rebuild a stock to a given high level. This desire can seldom be fulfilled, or, if fulfilled, the costs are high for the fishery at large.

Some depleted stocks recover rapidly and others recover very slowly. For example, the Pacific Ocean perch has not recovered in 15 years, despite intensive management measures to this end. The Pacific herring stocks have also not recovered during the last 15 years to the level seen in the 1970s, despite the absence of offshore fishing in the Bering Sea on this species.

Some of the recent recoveries have been summarized by Jakobssen (1991). The rapid recoveries of some herring stocks (e.g. Icelandic summer spawners, Fig. 6.16,

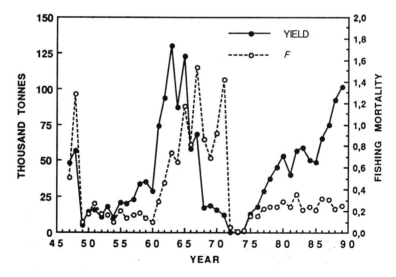

Fig. 6.16 Yield and fishing mortality in Icelandic summer spawning herring, 1947–89 (Jakobsson 1991).

and North Sea herring) contrast sharply with other herring stocks that have not recovered (e.g. Atlanto-Scandian and Icelandic spring spawners). The drastic change of the feeding migrations of Atlanto-Scandian herring during and after the stock collapse is also interesting. The collapse of the North Sea and Atlanto-Scandian herring stock had several possible causes.

Another example of rapid collapse and recovery in the mid1980s is that of the Icelandic capelin. As a result of the collapse of the capelin stock, cod growth decreased severely, but increased after the recovery of the capelin stock. The collapse of the Barents Sea capelin stock in 1986 had a disastrous effect on the whole Barents Sea ecosystem. The rapidly-growing cod stock (1983 year class) grazed down all other available food, including its own progeny, and suffered starvation; thereafter average weight at age was reduced by about 50%. Starving Arctic seals were forced to migrate southwards to the Norwegian coast in search of food. It is still debatable whether the breakdown of the Barents Sea ecosystem in the second half of the 1980s was due to natural causes or to fishing. Most probably both contributed to the disaster.

Finally, we must recognize that marine ecosystems can change and adapt to man's actions, such as fishing and eutrophication. If these actions remained quasi-constant in time, the adapted ecosystems and processes in them would remain in balance with the changes caused by man. This adaptation and balancing has most probably already occurred in the North and Baltic Seas. Consequently, we might not be able to rebuild stocks as they were in a quasi-virgin state and must accept large-scale changes as inevitable.

Chapter 7
Fisheries and Marine Ecosystem Management and its Objectives

Ancient chronicles are devoid of any mention of marine fisheries management. However, localized fish culture in ponds, rivers and estuaries flourished in the Roman Empire and in China. The focus of earlier cultures was clearly on securing provisions, with little understanding of the processes and factors that influence supply. The accepted wisdom, until almost the present century, was that man's ability to impact oceanic resources was infinitesimal. Oceans were large and the factors affecting fish availability were mysterious.

Marine fishery resources were thought to be inexhaustible until almost the beginning of the twentieth century. The issue of inexhaustibility first came into focus at the International Fisheries Exhibition held in London in 1883 (Larkin 1980). The close of the nineteenth century was marked by a recommendation by a British Select Committee for the curtailment of fishing activities because of the concerns over resource depletion. The changing state of fish stocks stimulated interest in determining the optimum level of harvest. This interest led to the emergence of fishery science and the eventual formulation of the idea that fish stocks should be harvested to provide a maximum sustainable yield (MSY).

Yield maximization dominated interest during the first half of the twentieth century. It was recognized that fisheries filled an important need and it was suggested that fisheries should be managed to maximize net economic return. More recently the focus has broadened. It is now generally recognized that biological, economic, social and political factors required consideration in the development of management objectives.

7.1   Need and objectives of fisheries and marine ecosystem management

The fisheries and their resources and objectives of fisheries management vary greatly from one geographic region to another! Maintaining the economically profitable level of commercially important stocks in a given region has been the most frequent objective of fisheries management, and most management-orientated fisheries biological research has been directed to that end.

A frequent management objective of local fisheries has been to retain the viability of a traditional fishery or an existing fleet, i.e. to preserve the way of life of coastal communities and local fishermen by restricting development of competitive fisheries or excluding fishing by other fleets in the given protected area. The traditional fishery could also be retained by protecting its markets by using various means.

In the last decade it has been fully recognized that the fluctuations in fish stocks can also be caused by factors other than fishing. This understanding has led to the consideration of multi-species and ecosystems management concepts, which have not, however, been fully implemented as yet (see Chapter 8).

In July 1992, a declaration of the United Nations Conference on Environment and Development (UNCED), adopted by a majority of the world's coastal nations, recommended that nations of the globe:

(1) prevent, reduce, and control degradation of the marine environment so as to maintain and improve its life support and productive capacities;
(2) develop and increase the potential of marine living resources to meet human nutritional needs, as well as social, economic and development goals; and
(3) promote the integrated management and sustainable development of coastal areas and the marine environment.

The term 'overfishing' is often found in the fisheries literature in connection with fisheries management, implying that the objective of fisheries management is to prevent overfishing. The term overfishing is, however, subjective and has no exact, universally accepted definition. Four other related terms have been coined, which have better definitions, but are also subjective; recruitment overfishing, growth overfishing, ecosystem overfishing and economic overfishing. Recruitment overfishing is difficult to define quantitatively because no simple relationship exists between spawning stock size and recruitment, except possibly at a very low spawning stock. Growth overfishing would depend on recruitment to capture by particular fishing gears in use. Economic overfishing would depend on the various fishing methods and vessels in use and on market conditions. In ecosystem considerations the problems of overfishing take the form of an evaluation of the effects of different fishing intensities on all species in the ecosystems as a whole, and are dependent on region-specific management objectives.

Although it would be difficult to agree upon any general and universally applicable set of marine ecosystem management objectives, the following three could be listed as the basic objectives:

(1) To sustain the fisheries in the long run (an old desire, the search for the elusive MSY of the past). This objective depends on the intensity of fishing, recruitment variation, economics of fishing, properties of gear in use, and several social considerations. It has been mainly a biologically-derived constraint in the past. Any fishing reduces the stock in the area. Thus one has to consider the future biological and ecological consequences of the reduction of the stock abundance, the efficiency of fishing and expected future benefits. The difficulty of predicting recruitment is one of the uncertain components of this objective.
(2) To optimize the economic benefits for the fishery, nation, and world community. This requires the consideration of the often complex economic characteristics of the regional fisheries, methods of harvesting, and variations of the

costs and harvest. Future consequences and benefits are also important considerations.

(3)   The principle of *Mare liberum*. This requires the consideration of fair and
equitable distribution from production to distribution of the product, the concepts of rights, power base, coastal states and acquired privileges. Furthermore,
considerations of different needs are linked to this objective, e.g. the need for
animal proteins, jobs and investment profits. The principle of *Mare liberum*, the
freedom of the seas, was initiated in the sixteenth and seventeenth centuries.
Freedom to fish and to navigate were the important arguments in this concept.

These three objectives are summarized as follows: how many fish can be harvested
and what will happen to the harvest in the future; who may fish, and what catching
methods can be employed; and how can the marine ecosystem be maintained in a state
which is acceptable to the public at large. Jakobsson (1991) stated that sound management policy for the fish resources has to be based on thorough knowledge of all the
characteristics of each ecosystem which is being managed.

The determination of explicit management objectives is essential to each local or
regional marine ecosystem and in some cases even to each species affected in the
fishery in this system, because such specific objectives would provide criteria for
selecting the management measures and exploitation strategies. Each region has its
own economic setting, social problems, characteristics of fishery and different marine
ecosystem.

In most fisheries, management implies that the catch of some species will be limited.
The multi-species nature of the catch is especially problematical for management, in
that catch limits have to be based on the principal components of each catch. The
result is often that the total allowable catch (TAC) of one species might be reached,
whereas a fishery on other species in a multi-species fishery could otherwise continue.
Each fisheries management action is subject to oversight and enforcement. Both are
expensive and difficult because cheating occurs at all levels.

Earlier fish population dynamics models still in use show the increase of yield with
the increase of catches to a given maximum over a longer time period, after which the
increased effort results in future lower yields. The more-recently-introduced bio-
economic models indicate that the biological MSY is not a satisfactory management
principle, and define the overexploitation in a free-entry fishery in economic terms,
without seriously affecting any biological criterion. When fishing intensifies, catch per
vessel in most fisheries declines as the exploitable part of the stocks declines. The
smaller exploitable stocks do not necessarily mean that the stocks are depleted.

Parsons (1993) believed that the MSY and maximum economic yield (MEY)
concepts, though containing some useful arguments, cannot for various reasons be
fully implemented in fisheries management. Furthermore, optimum yield (OY) and
other similar concepts are elusive. One of the main difficulties in fisheries management is that many and endless semantic arguments about its objectives exist.

Ideally, fisheries management must be a collective and co-operative process, which
steers not only the fishing levels, but also fishing patterns and quantitative harvest

strategies, and considers the fair and equitable allocation of resources and catches and their taxation.

Assessment of the state of the ecosystem and the abundance of the stocks within it must be made regularly and the future stock behaviour and allowable harvest levels should be forecast by different methods with different favourable and unfavourable assumptions. Thus the status of the system is being continually updated.

Fisheries management objectives also change with time as more knowledge and experiences are gained. Daan (1987) pointed out the following:

'More important, however, are the implications with regard to the selection of management objectives, which have until now been primarily based on some optimum yield per recruit concept. The yield per recruit appears to be not only a function of the level of exploitation on a particular stock, but it must also be viewed as a multi-dimensional plane, in which all stocks represent different axes. Because there is not just one optimum solution in a biological sense for exploiting the fish community, some common value must be given to the catches of different species. It would follow that within the multi-species approach, economic aspects cannot be avoided. This problem needs further attention, but for the time being it would seem reasonable to relate any biological management advice to stabilization of the fisheries at their present level, and not to propose dramatic reductions in fishing effort on the basis of unrealistic single-species yield per recruit concepts.'

Arnason (1993) grouped the fisheries management measures into two broad classes, biological fisheries management and economic fisheries management. The latter was further divided into two additional groups of measures, direct restrictions and direct economic management.

In this book we have dealt mostly with fisheries management problems from the scientific-technical and economic points of view. Most of these views have not fully prevailed in fisheries management politics and practices in the past (Underdal 1980). Different professional groups, e.g. biologists, economists and administrators, have been involved in the management process, with each group having its specific interests. Therefore fisheries management must be a co-operative undertaking where checks and balances must be present.

Marine ecosystem management would also include management of marine waste disposal and of other uses of the sea, such as oil development, aspects of navigation and coastal engineering and social or recreational uses. A myriad of local management measures cover these activities and are not considered in this book.

7.2   Methods and means available for fisheries management

In most cases of fishing on a stock there will be a decrease in the abundance of the target species, except for highly cannibalistic dominant species, and the prey species for the particular given species might increase as a result of decreased predation (see Chapter 5). Fisheries management has been using the 'stock' as the basic biological unit to be managed. The difficulties associated with defining the stock concept have been discussed by Dickie (1979a). In most cases the stock is an abstract term that provides the rationale for aggregation of catch data. Many biological factors hinder our definitions of a unit stock, e.g. lifetime migration of the species, uncertainty of spawning areas and origin of the fish from these areas, as well as changes in fishing patterns, often caused by changing migration of the assumed unit stock.

There are also many problems in obtaining and collecting the accurate data underlying the stock assessment as well as collecting catch data. The catch per unit effort (CPUE) value used earlier as an index of stock abundance is unreliable, especially for pelagic and semi-pelagic species. These species exhibit shoaling behaviour and the fishery is conducted on shoals which are detected by sonar and echo-sounders. It is said that to catch the first or the last shoal requires the same effort.

The various methods of stock assessment and their shortcomings have been reviewed by Laevastu & Favorite (1988) and Gunderson (1993). Sinclair *et al.* (1991) found that the current abundance surveys have a coefficient of variation in excess of 30%, and that there are numerous causes for the lack of precision and accuracy. Thus the determination of the initial state of the stock upon which the management measures are based is already difficult and somewhat uncertain.

A variety of methods have been in use in biological fisheries management and for the regulation of fisheries. The most used method is to set an upper limit to catch (TAC) and to divide it into quotas. In the past, the biological guidance for fisheries regulations was derived mainly from single-species models, which do not take into account biological–ecological effects and the economics of harvesting of predator/prey systems, e.g. the cod/capelin system in the Barents Sea. Past regulatory techniques are summarized in Table 7.1. Figure 7.1 depicts a general fishery management system, which was common in the past. Present practices of fisheries management are described in [7.3].

Several ill-defined or undefinable terms used in fisheries management, such as 'long-term and current potential yield', under-, fully- and overutilized stocks', are sparingly used in this book, as acceptable definitions for them are not available and local usage is variable.

Table 7.1 A typology of regulatory techniques (Underdal 1980).

Specific purpose	Subject of restriction	Kind of restriction
Protect young (immature) fish	Fishing effort	Minimum mesh size, or other measures affecting gear selectivity. Closed seasons/closed areas.
	Catch	Minimal landing size.
Limit overall fishing intensity	Fishing effort	Overall effort limits/quotas. Closed seasons/closed areas.
	Catch	Total allowable catch/quotas.
Redistribute access to or control over resources	Fishing effort	Exclusive national fishing zones.

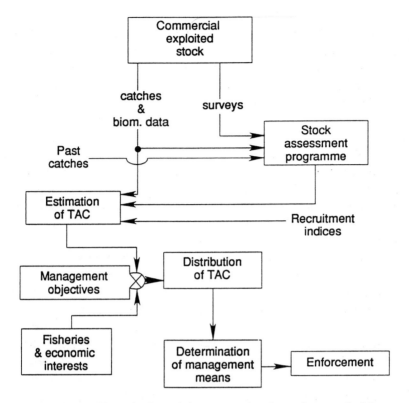

Fig. 7.1 Schematic diagram illustrating interrelations among the primary elements of a fishery management system. Arrows indicate directions in which information flows.

7.2.1   *Control of catch and total allowable catch (TAC)*

In the past, most fisheries have been managed on the basis of MSY. However, it has been demonstrated that MSY is universally undefinable or might not exist in reality (Larkin 1977). The fluctuations of the stocks also preclude the existence of a long-term sustainable yield. If MSY or a long-term allowable catch is set when stock abundance is low, resources would be wasted if the allowable catch is not increased when abundance increases. If the allowable catch is set when the abundance level is high, overexploitation could result or stock recovery might be inhibited if the allowable catch remained fixed.

Serious doubts have been expressed about the relevance of the so-called biological reference points used in single-species equilibrium models to estimate a biologically-advisable TAC (Mesnil 1986). Many fisheries are mixed-species fisheries. The TAC system is unable to limit fishing mortality in a mixed-species fishery. Landings exceeding quotas are forbidden, but not the discards. Furthermore, species mixtures depend on areas, gears and seasons. For a discussion of the control of by-catch and discards see [5.2.2].

Management of a fluctuating stock on an annual basis requires an accurate annual evaluation of the resource and the state of its fluctuations. The effect of present and predictable future environmental anomalies on the stocks should also be considered in the assessment. Furthermore, the response of the stock to an existing or a proposed fishery must be determined in the context of an ecosystem where species interactions and stock fluctuations are taken into account (see Chapter 8).

In most developed fisheries there are more than the minimum number of vessels required to harvest the allowable catch (over-capitalization). This situation arises as a consequence of market demands and the common property nature of fisheries resources. Thus it is thought that limiting the size and number of boats and gear would prevent over-harvesting of the resource and, above all, control the prices on the markets. If fishery resources were privately owned, fishing effort would be adjusted to minimize costs and maximize long-term yield and profits.

Under these circumstances, the need for regulations and the bureaucratic infrastructure required for their enforcement would greatly diminish. Generally, however, fishery resources are publicly owned, and therefore there is seldom a specific statutory authority to exclude anyone with the desire and capital to participate in their harvesting. With every increase in the number of entrants into a fishery having a fixed quantity of allowable catch, there must be a corresponding decrease in catch per boat, return per dollar of fixed and variable operating cost, and length of fishing season.

Hannesson (1993) reviewed the various methods of catch control and their applicability and shortcomings. The TAC alone will intensify competition between types of fishery and between boats. Limiting the fishing season has the same effect, as well as many undesirable economic side-effects. Quotas by vessels are considered by Hannesson to be one of the better means to control the catch in regions where it can be applied, e.g. in a given Exclusive Economic Zone (EEZ).

Fishing effort has been controlled by allocation of days at sea or hours of fishing and by the capacity of the vessels. Control of fishing effort by control of capacity usually requires long- and short-term vessel licenses, whereby long-term licenses control the aggregate fleet capacity and short-term licenses control the catch quotas. This method cannot be all-inclusive, because fishing vessels vary in size and fishing capacity and the license criterion should also include a cost-minimization input.

Regulations other than direct curtailment can be effective in influencing the number of vessels participating in a fishery. Catch quotas and allocations have the greatest impact on vessel participation. The economic realities of diminished catch per vessel following drastic cuts in allocation, the increased cost of developing gear and strategies for avoiding prohibited and by-catch species, and the increasing risks and costs of being caught in violation of catch regulations, have all contributed to the reduction of, for example, the Japanese fishing fleet. Such factors as user fees may also become an important consideration with operators having marginal profitability.

7.2.2   *Quotas or shares*

Quotas or shares can be used when a given stock is fished by only one state. Quotas are usually sold by authorities each year to limit the TAC. The price of quotas must vary with the size of the stock. Quotas can be transferrable, provided that the initial allocation is fair. If quotas are sold for longer periods a taxation system must usually accompany quotas. It should be noted that quotas promote misreporting of catches.

Quotas tend to eliminate the basic common property consideration of fisheries. Furthermore, transferable quotas tend to revert to the most efficient fishing firms, a result that is often contrary to the socio-economic considerations in some fisheries. In some instances a uniform implementation of a quota system is not advisable because quotas cannot always be freely interchanged. Consequently, a more restrictive procedure must be implemented to give property rights for some fish stocks to long-established coastal fishing communities.

The individual transferable quotas have been introduced in New Zealand. Annala *et al.* (1989) described their advantages and shortcomings, such as difficulties with widely-fluctuating stocks and in by-catch management.

Boat licenses or fishing licenses are somewhat similar to quotas, but these can be used in multi-national fisheries to control catches. Boat licenses and related licenses give the holders the right to participate in the fishery. The fishery still remains a common property, and the licenses must be linked to quotas commensurate with the annually-determined TAC. Boat licenses are easier to monitor than quotas. Furthermore, in combination with a tax, the fishing effort can be controlled; i.e. with a low stock abundance the cost of fishing plus a tax can make fishing unprofitable. The effect of a general tax is to make fishing less profitable and to squeeze out the less economically-efficient fishermen or capital-intensive fisheries. Socio-economic aspects can be included in the fisheries taxation systems.

Shepherd (1993) believes that catch quotas limit fishermen's earnings even more directly than restrictions on fishing effort. Furthermore, effort restrictions are easier to enforce effectively than catch quotas.

7.2.3 *Area and seasonal closures and gear specifications*

Area and season closures and gear specifications are used to control by-catch (see [5.2]) and can also be used to attempt to achieve specific management options, e.g. protection of spawning stocks if this protection is necessary.

Area and seasonal closures for catch and by-catch regulation have been in use in many regions. The by-catch limitation is achieved either by limiting the use of given gear type or closing the area to all fishing for a given period of the year. The effectiveness and wide use of this method requires good knowledge of the seasonal behaviour of the species, the seasonal peculiarities of the ecosystem, and the reaction of fish to different gears.

Area closures cannot be used to control the TAC, mainly because fishing intensity usually increases outside the closed areas, and fish may migrate in and out from the closed area. The seasonal fisheries closures to protect a spawning stock mostly have little effect on securing good future recruitment (Shepherd 1993).

Area closures are often used with gear specifications. For example, in many countries trawling is prohibited near the coast, with the objectives of protecting resources for coastal artisan fisheries and sports fishing or to protect nursery areas.

Gear specifications are made to select gears that minimize retention of juveniles and other by-catch, optimize economic returns (e.g. by decreasing fuel consumption and capital requirements), and minimize ghost fishing by lost gear. For example the use of longlines or traps (pots) will probably provide the best-known method for the selective harvesting of demersal fish resources without affecting other species or the bottom habitat (Wyman 1994, Gorman 1994). Traps have the ability to keep the catch alive until they are retrieved, thus increasing the market value, and have minimal ghost fishing potential if rigged properly with an escape door or panel. Traps can also be size-selective through the use of escape rings or web panels of a certain size to allow the release of juvenile fish.

The most common gear specification measure has been the regulation of minimum mesh size. In addition, use of different modifications to gear can be prohibited, e.g. the tickler chain. Economically more limiting for some sectors of fisheries are the compulsory use of a given gear only, e.g. gill-nets or the prohibition of a given gear, e.g. the otter trawl. The use of effective gear, such as the otter trawl, is prohibited in some coastal areas to protect coastal and sport fisheries. The effect of mesh size regulation can also be achieved by changing the gear mixture (Fig. 7.2). It should be noted that increased mesh size also decreases, at least initially, the catch per unit effort, and decreases the amount of discards.

All fishery regulations have some side-effects and the application of some regulations can be of questionable value. The species mixture in multi-species fisheries, which can contain species for which the TAC has been exceeded, is difficult to control with either spatial or temporal regulations (closed areas and seasons) or with gear specifications.

Mesh size regulations can have only limited success because a mixture of species with different growth rates are taken, fishing effort usually increases, and maintenance of the spawning stock is questionable with mesh size regulation. Minimum size regulation usually causes the decrease of discards.

Fisheries regulation through the use of closed areas usually causes an intensification of fishing effort outside these areas, and closed seasons can intensify the fishery during the open season. The increased cost of fishing caused by regulations is different for different types of fisheries. There are also a myriad of local side-effects, especially socio-economic, caused by various fisheries regulations, and the achieving of desired biological and ecosystem effects also varies from place to place.

7.3   Past single-species assessment and management and their limitations

The bulk of the world catch comes from relatively few, mainly pelagic and semi-pelagic species (Table 7.2). The open access philosophy to fisheries (*Mare liberum*) beyond the narrow territorial seas prevailed until the end of the 1960s. This common property approach led to heavy exploitation in some areas and to the suppression of some widely-distributed stock by a few irresponsible countries, such as the USSR. As a countermeasure, the Exclusive Economic Zone (EEZ) regime was established world-wide in the mid 1970s.

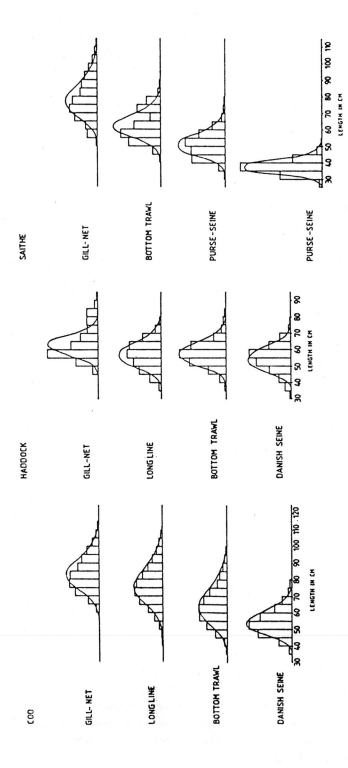

Fig. 7.2 Size composition of catches of three gadoids with different gear. The columns show the length distributions (average 1979–83) for cod, haddock and saithe from some of the units best covered by sampling. The curves show the LOG-normal distribution used in the model (Jakobsen & Nedreaas 1985).

Table 7.2 World fisheries: top 20 species.

Species	Catch in metric tons		
	1988	1989	1990
Alaska pollack	6 658 607	6 320 902	5 792 813
Japanese pilchard	5 428 922	5 142 930	4 734 922
Southern American pilchard	5 382 681	4 530 393	4 253 718
Chilean jack mackeral	3 245 699	3 654 628	3 828 452
Anchoveta	3 613 107	5 407 527	3 771 577
European pilchard	1 366 325	1 558 305	1 539 708
Atlantic herring	1 685 904	1 631 298	1 538 418
Atlantic cod	1 955 675	1 775 567	1 499 153
Silver carp	1 508 371	1 359 695	1 423 381
Chub mackerel	1 825 770	1 685 485	1 391 250
Skipjack tuna	1 281 625	1 221 486	1 238 972
Common carp	1 204 943	1 090 225	1 229 434
Grass carp	608 832	960 922	1 042 429
Yellowfin tuna	896 072	936 763	986 529
Capelin	1 142 325	897 657	982 434
Largehead hairtail	617 616	682 818	752 711
Bighead carp	716 365	653 566	677 687
Atlantic mackeral	708 673	591 099	657 385
Pacific cupped oyster	707 671	649 327	654 706
Blue whiting	671 636	662 655	576 673

Most fisheries of the world are still being managed on a single-species basis only, despite the fact that complex interactions in the fish ecosystem as well as in fishing practices has been well demonstrated. Despite the past theory of fishing having assumed steady-state equilibrium conditions which do not exist in reality, and most assessment methods having been based on this assumption, many of the management approaches worked to some extent in heavily-fished areas and guarded against recruitment overfishing, but with some exceptions. The different assessment methods at hand have recently been compared and evaluated by the ICES Working Group on Methods of Fish Stock Assessment (ICES 1993).

Two basic sources of information commonly used to estimate the abundance of exploitable marine resources are commercial catch data and directed scientific surveys. Both sources of information have their biases, and both suffer from a lack of precision. An increasing number of studies indicate that these two sources of error are both serious and extensive.

Bax *et al.* (1989) noted that there is no lack of theory or models in fisheries science to estimate the abundance and population dynamics of exploited species and that some of the models would perform well given adequate and accurate data. However, it is the adequacy of the data that is unfortunately lacking in most instances. This lack of data was compounded by the tendency of fishery scientists to concentrate on only one species at a time, as though the species existed in a vacuum. In doing so, additional data and insights that could be brought to bear on the analysis were lost.

Fisheries assessment which precedes management measures is essentially an analysis of a series of relative abundance indices upon which the harvesting strategies are based. An example of regional assessment and management methods is given for South Africa in Table 7.3.

Table 7.3 Summarized categorization of the present bases used for assessment of a number of South Africa's major marine resources, together with the harvesting strategies applied to the results of those assessments to provide scientific recommendations for TACs. The harvesting strategy notation is explained in the text. Seals are not harvested at present in South Africa, but the strategy indicated would apply if they were (Butterworth *et al.* 1992). For assessment methods see further Table 7.4.

Resource	Assessment method	Harvesting strategy
Anchovy *Engraulis capensis*	Bayesian cohort analysis with research survey data	Constant proportion of 20%
Pilchard *Sardinops ocellatus*	VPA with research survey data – as 'ADAPT'	$F_{\text{status quo}}$
Hake *Merluccius* spp.	Dynamic production model with CPUE, research survey data	$f_{0.2}$
Kingklip *Genypterus capensis*	Age-structured production model with CPUE data	No directed fishery
Horse mackerel *Trachurus trachurus capensis*	Dynamic production model with CPUE and research survey data	RY
Agulhas sole *Austroglossus pectoralis*	VPA tuned with effort data	$F_{\text{status quo}}$
South African fur seal *Arctocephalus pusillus pusillus*	Age-structured production model with aerial pup counts	(RY)

A danger in management exists in reading too much into noisy fisheries data and our inability to measure and estimate the basic parameters needed for assessment. Furthermore, some models underlying the assessment methods are too simplistic and/ or flawed.

The single-species population dynamics models for fisheries evolved in the 1950s, culminating in Beverton & Holt's (1957) development of a comprehensive treatise on the subject. Numerous minor changes, improvements and additions were suggested later, some of which have found application as auxiliary methods, such as cohort or virtual population analysis (VPA). As computers were not available to fisheries scientists in the 1950s, Beverton & Holt had to weigh the importance of the factors to be considered in the formulations of the dynamics and had to make decisions on what to include and what to leave out from these formulations. The detailed consideration of predation and the effect of predation of one species upon another were included in one parameter, M, the natural mortality, partly because a detailed calculation of predation is a formidable task for manual computation and partly because the spatio-temporal variable food composition of the species was not then well known.

One of the greatest shortcomings in the application of the single-species population dynamics is the lack of data for the determination of M. As we know now, M is largely a function of age and size of the fish; the greatest part of M in juveniles is predation mortality, whilst in adult fish it also includes spawning stress and 'old age and disease' mortality. Another difficulty in VPA is in the estimation of terminal fishing mortality F, the fishing mortality for the last year.

Various information is used for assessment of fishery resources for management purposes (Fig. 7.3). The simplest approach is to use catch data to generate indices of abundance (vertically up the right side in Fig. 7.3) or combining these data with trawl

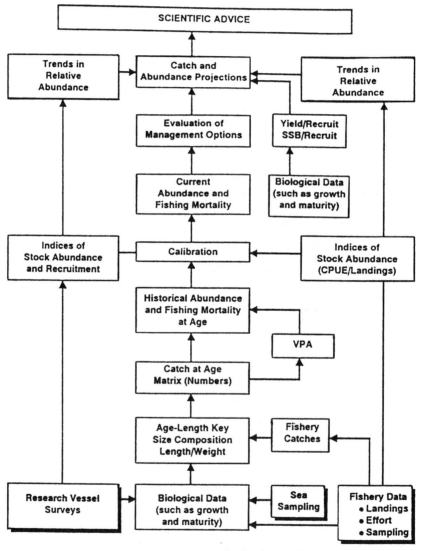

Fig. 7.3 Diagram of possible ways in which fishery-generated data and research vessel data (lower right and left boxes, respectively) are combined to provide scientific advice on the status of the stocks (Anon 1993).

survey data (vertically up the left side in Fig 7.3). In addition a variety of other biological information is used having different levels of sophistication and reliability (centre of Fig. 7.3).

Various other procedures and structures are set up for setting the TAC. An example of these processes from South Africa is shown in Fig. 7.4. Fisheries management in Japan has a long tradition (Asada *et al.* 1983). The coastal fishery in Japanese waters is considered to be the fishermen's bona fide property and is largely regulated by licensing which contains restrictions and prohibitions. Thus the Japanese use a community approach to fisheries management which uses the following means:

(1) Fisheries with high productive capability, e.g. trawlers and purse seiners, can be limited by area closures in nearshore areas or areas of operations may be restricted in respect of the given gear in use.

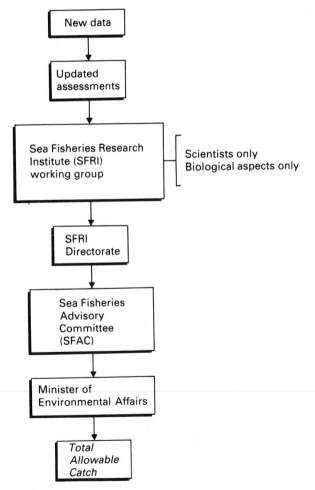

Fig. 7.4 A schematic illustration of the current process by which TACs are set for South Africa fisheries (Butterworth *et al.* 1992).

(2)   Most fisheries are subject to closure of the fishing season.
(3)   Restrictions can be placed on the horsepower of trawlers, mesh size of gear, and intensity of fish-collecting lights in night-light purse-seining.
(4)   The catch can be limited by species and body size and the ports of landing.

Abundant pelagic species, such as sardines, anchovy, mackerel, saury and squids, are usually not regulated in Japan, because the fluctuation of their stocks is not caused by fishing, but has some other, natural causes. In other words, normal reproduction is given preference over yield per recruit considerations. Furthermore, diversification in the capture of different pelagic species is encouraged in pelagic fisheries.

The institution of the EEZ regime is forcing some fisheries to become small-scale, locally-based fisheries and fisheries management to become more regional than before. A regional fisheries management council system has been tried in the USA. However, it has not been fully successful and has lost some credibility (Bevan *et al.* 1992). There are several reasons for this:

(1)   conflicts of interests of council members, to further their own financial interests;
(2)   major problems are not being attacked and objective scientific advice is ignored if it does not suit vested interests;
(3)   the process is expensive and time consuming; and
(4)   the politicians and bureaucrats make the ultimate decisions.

During the last half century, fisheries management and the basis for it (i.e. the scientific/technical management advice) has been worked out by ICES. The following is a summary of current ICES approaches to fish stock assessment and management advice, summarized by Ulltang (1989) and reproduced below from his report by permission.

The Northeast Atlantic Fisheries Council (NEAFC) area, for which ICES gives the management advice for about 80 stocks, is divided into three regions. Region 1 is the northernmost and is dominated by single-species fisheries (gadoids, capelin and herring). Region 2, which includes the North Sea, has a more mixed trawl fishery where fishing for food fishes, mainly gadoids, and a small-mesh industrial fishery dominates. Region 3, the southernmost area, has a mixed, relatively small-mesh trawl fishery predominantly for sardine and mackerel. There is an adequate database collected by a number of countries, the quality and quantity of which decreases from north to south.

The basic stock assessments are carried out annually by about 20 multinational stock assessment working groups, and their reports go to the Advisory Committee on Fishery Management (ACFM) of ICES. This committee then carefully checks the assessments and formulates the management advice on behalf of ICES. The members of an assessment working group are appointed at the national level and include scientists directly involved in the assessment work on the stocks in question at their research institutions.

The advantage of this system has been that the full utilization of specific local expertise has been guaranteed, while at the same time the later scrutiny of the working group reports by ACFM has secured a certain uniformity in quality and approach in all areas. The risk of contaminating the basic description with political considerations has also been reduced.

The introduction of annual catch quota regulations has demanded an accurate yearly prediction of catches at different levels of fishing mortality (F). To make an accurate prediction of next year's catch, taking into account that usually some data are not available for the year of assessment, the following information is needed:

(1) Estimated stock size and its age composition at the end of last year.
(2) Expected catch level and exploitation pattern, i.e. distribution of fishing mortality by age, for the present year (year of assessment).
(3) Estimated recruitment of the fishable stock for the present year and the next year.

Each Working Group establishes its own database, which is updated yearly and stored at the ICES Headquarters. The data include total catch and catch in number by age, by country, and possible catch per unit effort data and fishery-independent data such as trawl surveys, acoustic surveys and tagging data. Apart from catch data, the database is rather variable.

The assessment techniques applied depend on the database. Where catch and age composition data are available, these data are aggregated for a combined estimate of total catch in numbers by age for each year, and a VPA (Pope 1972, Ulltang 1977) is carried out.

However, the VPA alone does not give a reliable stock size estimate for the last year. Assuming that the catch data are correct, the VPA estimates of F and stock size coverage towards true values as one back-calculates year by year, but the estimates for the last year and the most recent years depend on the input to the analysis of F or stock size by age for the last year. The problem is then how to decide on these input values, and the solution of this problem varies widely depending on the database.

One can distinguish between three cases:

(1) Absolute stock size estimates from acoustic surveys or tagging data are available for the last year and possibly earlier years.
(2) Relative abundance indices from, for example, catch per unit of effort data are available. For using such data one needs a time series, including the last year.
(3) No quantifiable data on relative or absolute stock abundance are available.

In the first case, one can as a starting point use the absolute stock size estimates, corrected for mortality back to the beginning of the year or forwards to the end of the year, as input to the VPA. However, taking into account that the stock size estimates always have some variance, often unquantifiable, and also may have a bias, the results of this VPA are carefully checked to see whether they seem reasonable or give some strange, inexplicable features, e.g. VPA estimates for earlier years that are not comparable to independent stock size estimates for those years.

In the second case, where only relative abundance indices are available, a common procedure is first to run a VPA based on, for example, the same input F values as used by the Working Group last year. The next step is to determine whether the VPA estimates of stock sizes correlate with the relative abundance indices over the time series available. If there is no correlation even for years backwards, when the VPA should have converged, one is left with the third case, if no other abundance indices are available.

If there is a correlation, different approaches are used to determine input values of F for the last year. One approach is to estimate the regression between VPA and abundance indices for the years when the VPA is expected to have converged more or less, and then use the estimated regression line to predict stock size in last year from the abundance index. Another approach is to run several VPAs, trying to find the VPA which gives the best correspondence with the abundance indices over all years, including the most recent. In any case, the final VPA should not show inexplicable features as, for example, change in exploitation pattern or exploitation rate that is inconsistent with available independent information.

In the third case, the question whether it is at all possible to decide on a VPA that has some relation to reality for the last and most recent years depends on the availability of qualitative information that can be used to tune the VPA. For example, if one has an indication that there has not been any significant change in the overall level of fishing effort or the fishing pattern over the most recent years, and one is able to produce a VPA that shows a stable situation, then this VPA may be accepted. In other cases one simply has to realize that the available data do not allow an analytical assessment to be done.

In all three cases one often experiences difficulties with the exploitation pattern. If one assumes that the exploitation pattern has been constant over some given number of recent years, the so called Separable VPA (Pope & Shepherd 1982) is an efficient method to automate the procedure of generating an internally-consistent VPA in the sense that the exploitation pattern is more or less constant. The fundamental difficulty with this method is that there is seldom a guarantee of the validity of its principal assumption.

There are several examples of an assessment that has gone completely wrong because drastic changes in the exploitation pattern were not observed or accounted for.

One difficulty with the VPA technique is the inherent uncertainty about values of natural mortality. A discussion of this problem is found in Ulltang (1977).

In many cases it is difficult to determine whether an assessment based on the techniques described above can be accepted as an analytical assessment. Because of unavailability of data to calibrate the VPA, or because the available data show low correlation with the VPA, ACFM has on a number of occasions rejected the assessment done by a Working Group.

Having agreed on a VPA, one is then left with the two problems already mentioned before a prediction can be made, expected catches in the year of assessment and recruitment estimates.

Even where catch quotas exist, there are often large uncertainties in the expected catches because of lack of enforcement of the regulations in some areas. In a number of instances two possible assumptions can be made, firstly that the quota will be taken, and secondly that fishing mortalities will be the same as in the preceding year.

The situation concerning recruitment estimates varies between stocks. Sometimes annually-conducted young fish surveys give indices of abundance which correlate with VPA estimates of abundance, and regression lines are established to predict absolute numbers of recruits from the index. In other cases an estimate of the absolute number of recruits from acoustic surveys are available. However, often no measurements of the number of recruits exist and one simply has to make assumptions, e.g. that the recruitment will be on the level of a long-term average.

The problem of recruitment estimates is especially important in two situations, when dealing firstly with short-lived species and secondly with stocks where the fishing mortality is high.

In both situations, any prediction will be heavily dependent on the strength of the incoming year classes, and without accurate estimates of their strength it is impossible to predict catches corresponding to different levels of fishing mortality.

The Advisory Committee on Fishery Management (ACFM) starts by carefully examining the assessment carried out by the Working Group for the stock in question. This process may result in one of the following three decisions:

(1)   The assessment is accepted.
(2)   ACFM makes a reassessment.
(3)   The available data are not adequate for making an assessment.

In most cases ACFM will accept the assessment.

Once an assessment has been agreed upon, the next step is to translate it into management advice.

In the early years, management advice was usually given in the form of a single catch figure which was estimated to meet some biological objective. This objective could, for example, be a reduction of F to the level maximizing equilibrium yield from a 'per recruit' consideration (F_{max}), a gradual decrease towards this level or maintaining F since the present value of F was thought to be near the optimum level.

However, it was realized that where the stock was not considered to be in an endangered state, it was difficult to recommend a single figure on purely biological grounds. Instead it was decided to give the managers some options 'within safe biological limits'.

The procedure now adopted is made clear in the following introduction to the 1983 ACFM reports (ICES 1984).

'The stocks are grouped for the purpose of providing management advice into:

(1)   Stocks which are rapidly depleted and suffering from recruitment failure. In these cases, ACFM shall not calculate options but shall recommend a single figure.

(2)   Stocks which are fished at levels largely in excess of the levels indicated by biological reference points. In these cases, ACFM shall give options inside safe biological limits, and shall recommend one of these options, according to the general principles of aiming at more stable levels of stock and catch.

(3)   Stocks which are fished at levels not very different from the biological reference points. In these cases, ACFM shall give options inside safe biological limits, but shall not recommend any particular one of these. It shall only indicate a preference, which is in line with the general principles mentioned above.

(4)   Stocks where at present it is not possible to carry out any analytical assessment with an acceptable reliability. In these cases, ACFM shall indicate precautionary TACs to reduce the danger of excessive effort being exerted on these stocks.

(5)   In cases where fisheries on a stock are not subject to TAC regulation, there may be a danger of catches taken from stocks of the same species in adjacent areas being misreported as having been taken in areas of unregulated fisheries. To reduce the risk of this happening, ACFM, on occasion at the request of management bodies, has advised an implementation of TACs, and their levels on this basis. Since in the majority of cases the data on these stocks are inadequate for analytical assessment, they too will generally be recommended as precautionary TAC based on historic catch levels.'

In order to allow more flexibility to the management authorities, the type of recommendation given for a Category 2 stock is that fishing mortality should be reduced to one of the biological reference points $F_{0.1}$ or F_{max} as quickly as possible, or (in some cases) towards one of these points.

One difficulty with this procedure for providing management advice is the definition of safe biological limits. If a stock is exploited to an extent which endangers recruitment, because of low spawning stock size, one is certainly outside the safe biological limits. The problem is, however, that for most of the stocks one has not been able to show any clear relationship between spawning stock and resulting recruitment for the range of spawning stock sizes observed. An exception is some herring stocks (Ulltang 1980).

The solution has in many cases been to define a lower limit below which the spawning stock should not be allowed to fall. This lower limit has been arrived at in various ways, e.g. taken from a scatter diagram of recruitment against spawning stock, or set as the lowest level observed in the historic series.

In addition to giving advice on TACs, ACFM has made recommendations on the improvement in the exploitation pattern, e.g. to increase the minimum trawl mesh size, to introduce minimum landing size, and to close seasons or areas dominated by young fish. The policy of giving management options has partly changed from underlined recommendations to more general recommendations stressing the need for improvements and explaining how these can be achieved.

So far, in only a very few cases have interactions been taken into account when giving management advice. The reason for this is simply that multi-species analysis is still new because of the lack of basic data on interactions, for example predation data. However, work on this is in progress in ICES.

A bothersome fisheries management problem is the existence of an over-capacity of harvesting in most traditional fishing areas in the Northern Hemisphere. Reduction of over-capacity is not strongly recommended by management bodies because it would not affect the immediate problem of controlling fishing effort and stocks, but it is assumed that the existing over-capacity will reduce itself on economic grounds when the TAC on most commercial species is limiting fishing and catches.

Small-scale coastal fisheries have been managed with a myriad of local regulations adapted to local conditions and desires. In many areas the small-scale, including subsistence, fisheries provide employment to relatively large numbers of families and can be a considerable source of food. This type of fishery usually provides low income and lacks capital and adequate fishery resources for expansion. Small-scale fisheries are endangered by the intrusion of large-scale vessels in their traditional fishing areas.

7.4   Problems hindering fisheries management

The problems hindering fisheries management could be classified into the following categories:

(1) Inability to make correct resource assessment.
(2) Inability to predict the natural behaviour of the resources and their interactions, especially predation (natural mortality), recruitment and immigration/emigration.
(3) Inability to prognosticate correctly the effects of management measures.
(4) Inability to determine the effects of fishing, and of the biological reference points.
(5) Inability to control the fishing.

The implications of the knowledge of fisheries and applications to management have been reviewed by Dickie (1979a), who notes that:

'A concern with a more balanced approach to fisheries regulations takes us beyond the simplistic mathematics that, disguised as biology, has dominated management technology for 20 years.

The fact that the regulations did not work suggests that at least part of the blame for the failure of management must be borne by management technology.'

Parsons (1993) has pointed out that the progress in fisheries science has been very slow for the main issues, stock size estimate, determination of recruitment, and quantitative determination of the interactions in the marine ecosystem. Parsons also found it somewhat deplorable that fisheries biologists and population dynamicists, although often aware of the limitations of data, methods and conclusions, have seldom communicated these limitations to fisheries managers or to the fishing industry.

The assessment method most commonly used is virtual population analysis (VPA). It relies on parameters measured on catch and estimates of unknown parameters, such as natural mortality. Furthermore, the terminal fishing mortality is difficult to estimate and must be tuned with other independent estimates, which can also be in error. Total sampling of the catch is impossible. There are changes throughout the year such as fishing intensity, stock abundance and availability, and gear selectivity. However, VPA emphasizes the aggregate catch, which could be achieved if many unknowns could be accounted for. Most natural mortality rates are not known. The rate of computed fishing mortality is a combination of the changes by season, area, gear competition and environmental interactions, and in terms of sampling can be variable in space and time and according to method of catching (Fig. 7.2).

The results obtained by different resource assessment methods can be very different (Fig. 7.5). A summary of more common stock assessment methods is given in Table 7.4.

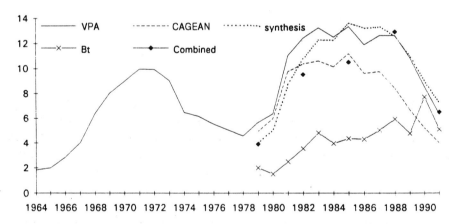

Fig. 7.5 Eastern Bering Sea biomass trends as estimated by different models: virtual population analysis (VPA), catch-age analysis (CAGEAN), synthetic analysis, bottom trawl survey (Bt) and combined bottom trawl and hydro-acoustic mid-water trawl survey (Wespestad 1993).

The limitations of different stock assessment methods have also been reviewed and summarized by Laevastu & Favorite (1988). Weber (1989) investigated the reliability of the different input data for cod stock assessment in the Baltic Sea. He concluded that the initial estimate of F was too high and that recruitment was underestimated. With time, however, improved research data led to estimates of bigger stock sizes.

Because our quantitative knowledge on the processes affecting recruitment is incomplete and our efforts to define a quantitative spawner/recruitment relation have

Table 7.4 Some commonly used stock assessment methods.

Method	Data requirement	Advantages	Limitations
Trawl survey	Catch per unit area; size/age composition	Direct measurement, comparable to fishing	Catching efficiency variable, expensive operation
Acoustic survey	Back-scattering strength (target strength) required	Useful for pelagic species, concurrent abundance index	Many assumptions involved, expensive multi-ship operation
VPA and cohort analyses	Catch data, age composition of catch, (mortality data) required	Provides age structure of stock, and past history	Natural mortality not known, affected by change of fishing patterns and immigration/emigration
Multi-species virtual population analysis (MSVPA)	As VPA, and in addition good food composition data	As VPA, natural mortality input improved	As VPA, food composition variable in space and time
CPUE (index)	Catch per unit effort	Gives relative abundance change	Relative abundance change only, affected by changes in fishing patterns
Ecosystem simulations	Great variety of available data, especially food composition data (time dependent model required)	Provides estimates of all stocks, possible to experiment with fishing	Complex for others to understand, time-consuming to set up

not been entirely successful, it has been assumed that it is more meaningful to present the effects of fishing on stocks in relative terms, e.g. yield per recruit. Also there has been the search for the elusive MSY, which cannot be defined in any reasonable terms and which would be affected by changes in many processes in an ecosystem or in the fishing technology. Species interactions in a fish ecosystem will also affect the fluctuations of any stock, the main interaction being predation. There is also an uneven balance between predator and prey.

Fisheries management advice is given by different specialists, such as biologists and economists, but management is effected mostly by bureaucrats and administrators, who have a different understanding of fisheries problems.

In general, fisheries scientists have failed to emphasize to managers the limitations of the models used, especially the fact that the interactions between species have been neglected, that stock estimates are often independent of catches, and that catchability can be independent of population size. Economists have used in their models simple biological assumptions and biologists have used simple economic and technical assumptions. The complexities of local ecosystem conditions and local fisheries social problems have seldom been considered, whereas environmentalists' views have recently been brought to the forefront.

Public acceptance of management is required. However, conflicts will arise owing to differing priority values and lack of communication. Management decisions should be clear so that their benefits and reasons can easily be understood. In some countries, e.g. Canada, fisheries management is based on fishing rights with very limited transferability and divisibility of quotas (Crowley & Pálsson 1992). This system can cause conflicts between efficiency and the maximization of employment. However, this system is necessary for social reasons, especially for maintaining traditional small fishing communities.

It has been pointed out that the common property nature of oceanic resources forces fishermen to overexploit the fish stocks against their own better judgment, as each fisherman wants to grasp as large a share of the potential yield as possible (Arnason 1993). As long as there are economic rents in a fishery, there will always be an incentive to bypass the management regulations and restrictions. In a competitive fishery an equilibrium will be reached when the stock size is down to the point where total fishing costs equal the value of the harvest.

Free access to fish stocks and competition result is increased production, but lead to economic losses in large-scale ocean fisheries. Restriction of fishing effort is designed to realize the potential economic benefits of the fisheries or to restrict investment in fishing capital. In most cases these restrictions aim to keep the stocks of the main target species on an economically profitable level.

Thus the economists consider open access to major fishery resources to be wasteful. However, rights of exclusive use, such as quota systems, also have shortcomings. Furthermore, attempts to control fishing with price regulations and taxation have achieved only limited success in special cases.

Limitation of access to a fishery can cause a multitude of economic and social problems, e.g. idling of capital and unemployment. Fisheries management requires the consideration of complex interactions between the biological, economic, technical and sociological aspects. Though the access to fish might be free, the fishing is not. It costs manpower, fuel, boats and gear. Fishing can be considered also as a benefit, providing jobs and income.

The greatly variable economic conditions and conflicts complicate fisheries management more than the uncertainties in biological assessment and the TAC settings. Although total revenue could be increased with well-managed fisheries, the distribution of fishery resources and fishing rights might not fit the social needs and maintenance of employment levels. Furthermore the total cost and total revenue are not comparable in subsidized fisheries.

There are a myriad of general and local problems that affect fisheries and marine ecosystem management. For example, not enough attention has been given in the past to the utilization of large by-catch (e.g. the shrimp fishery), or to the possible impact of this large by-catch on the corresponding ecosystem from which it has been taken.

The amount of energy used for catching unit weight of fish is increasing with the reduction of the stock abundance. Furthermore, with the increase of fuel prices, the profitability of fishing decreases. These changes affect fishing patterns and fisheries management, but have not been directly considered in management in the past.

The free fishery outside the 200 mile EEZ has been increasing in many areas, affecting the resources and their management within the nearby EEZ as well. Recent (1993) UN conferences have attempted to find solutions to these international problems, as well as how to manage widely-migrating stocks.

Some high seas fisheries, e.g. for squids, have recently been banned rather than regulated, mainly owing to concern by environmentalists about the well-being of marine birds and mammals. In addition, many fisheries management approaches have been narrowly focused on some valuable species, e.g. salmon, or on marine mammal protection, or on commercially important species, without consideration of other affected species and their interactions as well as effects of variance in the dominance of species.

The management in most regions has been 'top down', allowing little or no say by fishermen. Nor has enough effort been spent on educating fishermen on the needs and effects of fisheries management. Although the final fisheries management decisions are made by bureaucrats, the management advice has come mainly from fisheries biologists. Parsons (1993) pointed out that fisheries science has unfortunately become synonymous with fisheries biology. He called for a change in this attitude, asking that fisheries economics, sociology, physical oceanography and general marine ecology be included in fisheries science, and that fisheries management move towards an integrated bio-socio-economic approach. However, this change will not occur overnight.

In many places, fisheries management has also been plagued by differing and biased advice provided by consultants, hired by fisheries, government or environmental interests, who provide information from their sponsors' point of view. These different views, not based on scientific facts about the fish ecosystems but on one-sided economic considerations, are especially bothersome in international fisheries (shared stocks). Therefore, the arguments that more impartial international organizations, such as the UN and FAO, should step in to solve the international fisheries conflicts, are becoming more frequent. One result was the UN Conference on Straddling Stocks and Migratory Species in 1993.

Chapter 8
Management Revived as Ecosystem Management

The 1982 UN Convention of the Law of the Sea gave the coastal states the right to manage and distribute the living marine resources in the 200 mile zone off their coasts, the Exclusive Economic Zone (EEZ). Some coastal states have gained economically and some nations with distant-water fishing fleets have incurred losses. Overall the fishing effort in the high seas has increased and there are numerous claims that the fish stocks in many coastal zones have been overexploited and have seriously declined. Thus there have been shortcomings in past fisheries management and corrective actions are called for.

8.1   New trends in fisheries management

The current trends in fisheries management could, in some ways, be considered off-shoots from previous and sometimes not fully successful fisheries management concepts. In this chapter we present some potential approaches to fisheries management using fish and other marine ecosystems as the central unit of consideration.

Marine and fisheries ecology has followed the path of other sciences and has attempted to study the dependence on the effects of single factors of the environment one at a time. However, strict experimental solutions are seldom possible in marine ecosystem studies, and consequently scientists have been forced to deduce logical dependence and must draw intuitively deterministic conclusions, i.e. postulate that the events at different times are connected with some laws which are often only partially known in such a way that some prediction is possible.

Fisheries and marine ecosystem management has historically involved two separate aspects. The first aspect is the management for social, aesthetic and economic value. Fisheries management is concerned with the determination of what and how much can be taken annually from a given fish ecosystem or from a given stock under certain biological, aesthetic and economic constraints. This aspect is usually called biological management advice. In an ecosystem context there is a need to know the potential effects of the removal of a quantity of a given species on the future functioning of the ecosystem, and what the future yields could be. The ecosystem management context, which has recently received increased attention, pertains to the 'health' of the various ecosystems and to the possibility of harvesting the optimum amounts of desirable living marine resources from the specific ecosystems without materially diminishing the output of desirable fish production and other economic values, while maintaining the species and genetic biodiversity of the system. In addition, certain ethical and human relationships to marine life might require the special protection of some species or species groups, e.g. marine mammals.

The second aspect of management is the operational element of allocation and regulation of the harvesting methods, and includes the executive and enforcement parts of management. Licensing, effort control, taxation and quotas are examples of methods of management. In the past, various economic considerations have taken a dominant place in making decisions about how to manage the ecosystem.

In reality, the structure of fisheries management can be complex. It can involve multinational cooperation, such as ICES and NEAFC, or national entities such as those in South Africa (Fig. 8.1). In other national settings, such as in the USA, management is effected by government-appointed councils, representing various fisheries interests. Such councils can become quite partial, where members further

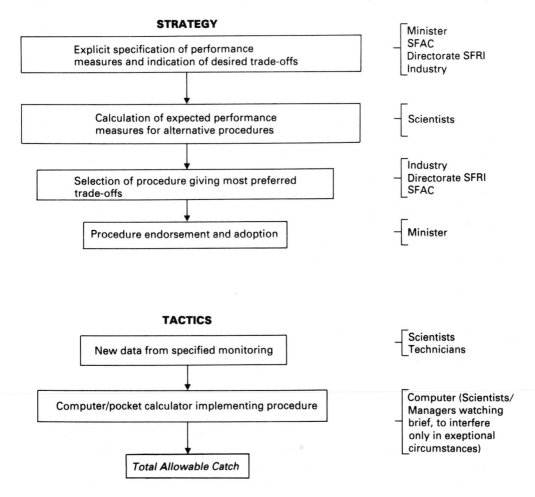

STRATEGY

Explicit specification of performance measures and indication of desired trade-offs
— Minister
  SFAC
  Directorate SFRI
  Industry

Calculation of expected performance measures for alternative procedures
— Scientists

Selection of procedure giving most preferred trade-offs
— Industry
  Directorate SFRI
  SFAC

Procedure endorsement and adoption
— Minister

TACTICS

New data from specified monitoring
— Scientists
  Technicians

Computer/pocket calculator implementing procedure
— Computer (Scientists/ Managers watching brief, to interfere only in exeptional circumstances)

Total Allowable Catch

Fig. 8.1   A suggested structure for the adoption and implementation of management procedures for South African fisheries (Butterworth *et al.* 1992).

their own financial interests (Bevan *et al.* 1992) rather than follow biological and national economic advice.

Parsons (1993) reviewed the relationship between the scientific advice given to managers and the managers' need for information to make quick and correct decisions. He found that the advice and information provided is often inadequate and uncertain. Fisheries scientists have also realized that their estimates of stock abundance and other vital parameters, e.g. mortalities and recruitment, have large margins of uncertainty. In recent years, the fishing industry, environmental groups and fisheries managers have become critical of the validity of the scientific methods and this credibility gap is increasing. Nearly everything about management advice pertains to the future and is in essence a prediction and by its nature uncertain. Therefore managers have been encouraged to select *status quo* TACs as a conservative management approach. A similar situation exists between management and economic advice. The complex bio mathematical models developed by academic economists are mostly irrelevant to fisheries in that they often require voluminous and specific data and expertise that go beyond the legal authority of management ethics and political regimes.

Owing to the variable nature of fisheries management in many regions and many past management failures, new concepts and terms have been introduced into possible management procedures, such as 'risk assessment' and 'management under uncertainty'. These terms are difficult to interpret and more difficult to apply, since the character of the uncertainty is difficult to define. It would be more rational indeed to explain the different uncertainties in quantitative fisheries assessment and management advice and be conservative when establishing yield potential. This approach would be better than attempting to define uncertainty and devising uncertain statistical parameters to account for it.

Uncertainty in fisheries management decision-making can arise from a multitude of diverse causes. Examples of these causes are:

- difficulty in defining a lower critical level to which a stock could be allowed to fall because of heavy exploitation;
- observation and reporting errors in catch and in resource assessment;
- lack of adequate knowledge in functional relationships in the ecosystem;
- the stochastic nature of the dynamics of the resources;
- the effects of exogenous variables, e.g. environmental anomalies, on the resources;
- the difficulty of establishing life history parameters precisely.

Some of the uncertainty can be partly removed by using either more and different information and different methods or a multidisciplinary approach. Unfortunately, there will always be a certain amount of objective uncertainty, especially in forecasting events and projecting results of actions into the future.

Related to uncertainty is the new concept of 'precautionary principles' which does not have a universally accepted definition as yet but is in general considered to mandate that conservation measures be in place before the available scientific data supporting specific action are at hand, or before fishing may begin or be continued.

Another new term, which is equally difficult to define, is 'risk assessment' (Fogarty *et al.* 1992). In this concept we cannot define any threshold of the population size below which the probability of poor recruitment increases appreciably. Furthermore, owing to the natural fluctuations of most stocks, it is difficult to define any critical level of population collapse. The natural fluctuations of stocks, in respect of recruitment, availability and mortalities, and assessment errors complicate the formulation of management criteria as well as the possibilities of evaluating management performance.

The risk assessment concept is, in essence, the old problem of spawner/recruitment relation at a low level of spawners (low spawning stock). Fish stocks fluctuate because of variations in recruitment, which in most stocks cannot be accurately predicted but can be foreshadowed to some extent in species which are fully recruited to the fishery when a few years old and where indices of abundance of pre-recruits can be obtained.

Stock fluctuations have substantial economic consequences for the fishing industry. However, methods are available to reduce the unfavourable effects of fluctuations to the fishing industry (Doubleday 1980), such as catch insurance, controlling the use of different fishing gears, limiting of fishing mortality rate and influencing the composition of the fishing fleet.

The concept of fish stock and unit stock has been an important parameter in fisheries resource assessment as well as in management. The latest idea for stock separation is to recognize stocks in terms of genetically homogeneous units (Gauldie 1991). This genetic stock identification can be achieved by comparing red blood cell antigen frequencies and protein allele frequencies. The results might provide information on whether stocks are migration-linked and exhibit heterogeneous growth rates (Gauldie 1991).

The considerations stated above pertaining to uncertainty and risk in fisheries management lead to the conclusion that effective marine ecosystem management will require the ability to predict, or at least foreshadow, what will happen in the ecosystem if and when a component of it is either affected by man's actions (fishing) or influenced by anomalous environmental forces. Furthermore, it is imperative that the effects of management actions can be reasonably predicted. At times, this is possible using reliable numerical ecosystems simulations (see [5.4] and Appendix).

The concept of marine fish ecosystem management started to develop with the development of methods for its evaluation, i.e. numerical simulation models. The more substantial of these were the Andersen & Ursin (1977) model, which in its simplified form gave rise to MSVPA (multi-species virtual population analysis), and DYNUMES and PROBUB models (Laevastu & Favorite 1978a,b, Laevastu & Larkins 1981).

Ecosystem simulation emphasizes trophodynamics and predation mortality as the important recruitment-controlling processes. The Andersen–Ursin model is number-based and has no spatial resolution (i.e. it is a 'box' model). The DYNUMES model described in [8.2.3] and [8.3] is a biomass-based model with spatial resolution (i.e. it is a 'gridded' model).

The multi-species assessment with MSVPA, which is an offshoot of the Anderson–Ursin model, has been explored by the ICES Multispecies Assessment Working Group (ICES 1989), to define its limitations and the advantages of MSVPA over the single species approach. Daan (1987) pointed out that the development of MSVPA, which assesses interspecific and intraspecific predation through an analysis of stomach contents, has verified the hypothesis that predation among exploited fish species contributes significantly to their natural mortality and that predation, and thus natural mortality, are inherently variable from year to year. Furthermore, MSVPA has considerable implications for fisheries management. For instance, MSVPA suggests that the effects of mesh size and discarding need re-evaluation and that year class strength may not be as fixed as previously assumed.

In more recent developments, the properties and effects of fleets have been included in the management models. For example, O'Boyle *et al.* (1989) programmed a bio-economic age-structured model for groundfish resources where the resource was exploited by a multi-gear fleet. With this model they demonstrated that the yield and employment picture was superior for longliners compared with trawlers. The effort of the joint fishery could be controlled by regulating only trawler activity, leaving longliners unregulated.

Sunnanå *et al.* (1986) used a biomass-based ecosystem simulation, NORFISK (similar to DYNUMES), to assess the fisheries resources off Norway and in the Barents Sea. This simulation model has incorporated both spatial resolution and seasonal migrations. The schematic representation of the NORFISK model is given in Fig. 8.2.

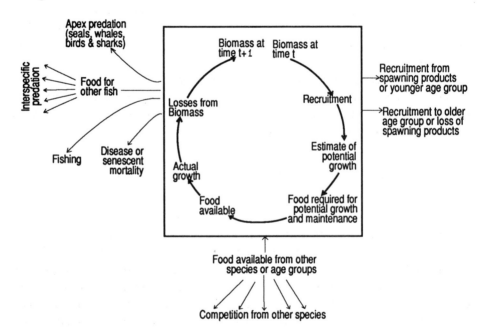

Fig. 8.2 Schematic representation of processes occurring for one age group in one time-step (2 weeks) using the simulation model NORFISK (Sunnanå *et al.* 1986).

Sparre & Willmann (1992) developed a bio-economic analytical model (BEAM) which was meant to be used as a tool for the rational management of fisheries and which integrates fish stock assessment and economics. The BEAM model is also an example of the successful collaboration between biologists and economists. The economic part of the model accounts for the costs and earnings of each fleet and processing plant. It calculates a suite of measures for the performance of the harvesting and the processing sectors. Examples of measures for economic performance are profit, profitability, gross, net and national net value added, employment, and resource rent. Migration of animals between fishing grounds and the spatial distribution of fishing fleets are also taken into account in this model. Figure 8.3 gives the flow chart of the BEAM model and Fig. 8.4 that of the economic sub-model.

Fisheries assessment and estimation of the effects of fishing on stocks and on marine ecosystems are difficult, complex and full of misunderstandings. Extensive application of statistical theories that do not pertain directly to the understanding of the causes of uncertainties in the assessments, such as risk analysis, could contribute to deterioration of the progress we have made using available methods and understanding of the functioning of marine ecosystems.

The methods used to manage and control fisheries and the exploitation of fishery resources are not new, but their use changes in space and time. A fisheries management technique that is being used more frequently is the concept of individual transferable quotas (ITQ), which might solve some allocation problems in national fisheries and provide more effective control of effort when tailored to local conditions and characteristics of local fisheries. They are more difficult to use in long-term stock allocation or utilization in international fisheries. The latter case may require establishment of national quotas that can be subsequently structured into an ITQ system. Implementation of an ITQs system will limit the traditional free access to fishery resources.

Limited entry is both an economic and a biological tool to limit fishing, in both national and international fisheries. It can usually be effected via vessel licenses and by limiting new capital-intensive large units into the fishery. Small-scale coastal fisheries can be favoured with this method.

8.2   Marine ecosystem management concept and practicability

Specific to each ecosystem is its quantitative manner of functioning mainly as a material and energy transfer predator/prey system. Thus a given system is defined by its components, the interactions between them, and the interaction of components with environmental physical and chemical properties. Ecosystems are hierarchical in content and are space- and time-dependent. The term 'ecosystem integrity' as currently used seems to imply intrinsic soundness and completeness. However, considering the dynamic character of many ecosystems, the term ecosystem integrity may not have a scientific basis, but could be defined in terms of desirable human values, both ethical and socio-economic.

Marine ecosystems management is in essence the controlled utilization of the

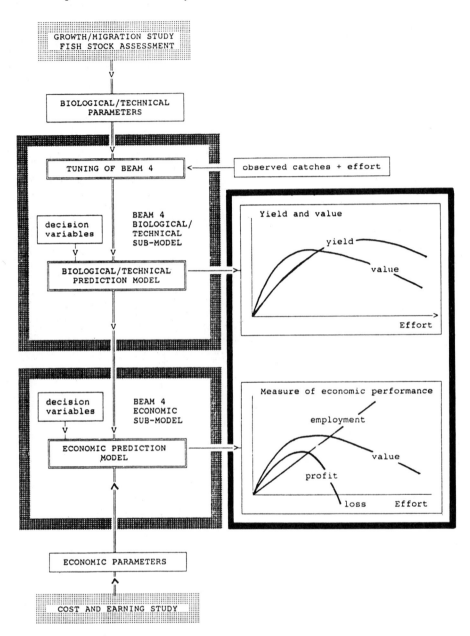

Fig. 8.3 Flow chart for the BEAM 4 methodology (Sparre & Willmann 1992).

components of an ecosystem, mostly living marine resources, without unduly harming its potential of biological production, and maintaining its living components and abiotic features in a state which is acceptable to society; i.e. a sustaining utilization of the marine ecosystem and guarding its health. Society makes the decisions about what should be the state of health of an ecosystem. The tasks of scientists would be to describe the ecosystem and its functioning in quantitative terms. They should

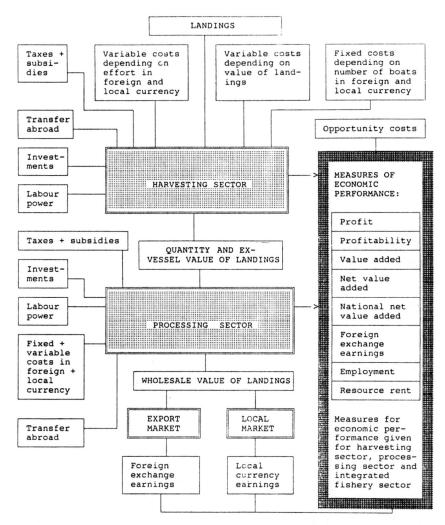

Fig. 8.4 Flow chart of the economic sub-model of BEAM 4 (Sparre & Willmann 1992).

also determine and describe what will be the effects of different intensities of man's exploitation on the ecosystem. The information provided by scientists should allow the concerned and informed public to decide on the state in which they would like to keep the ecosystem under consideration regarding present and future economic and social needs and the conservation of nature, and which patterns of use are acceptable. Having these constraints in hand, the managers can then decide about the exploitation and controlling measures for the proper implementation of which they are responsible to the society.

Most marine fish ecosystems are affected by four processes, migration, predation, fishing and environmental change and anomalies. The environmental processes can affect recruitment and effect the advection of pelagic populations. Data on ecosystem processes are often meagre and their quantity and quality are variable from one region to another.

Bax *et al.* (1989) used a multi-species analysis model, with a spreadsheet, and analysing the predation problem using fisheries survey data and catch data from the Barents Sea. Output from the spreadsheet highlighted discrepancies between the biomasses estimated from survey and commercial fishery statistics on the one hand, and losses in biomass estimated via food habits and catch data on the other. Such discrepancies indicate where data are inadequate or biased, and will help to focus subsequent research. In Fig. 8.5 the causes of the biomass loss are shown, for the five analysis models used in the multi-species analysis model. The major loss of the shrimp and the prey is to fish, whereas the major loss of polar cod and redfish is to mammals.

This spreadsheet model demonstrated that it is possible to simplify the analysis of major multi-species interactions and ecosystem functions and to foreshadow possible changes. More importantly, such models emphasize errors and deficiencies in the data, the knowledge needed for a multi-species approach to ecosystem management, and the sources of uncertainties.

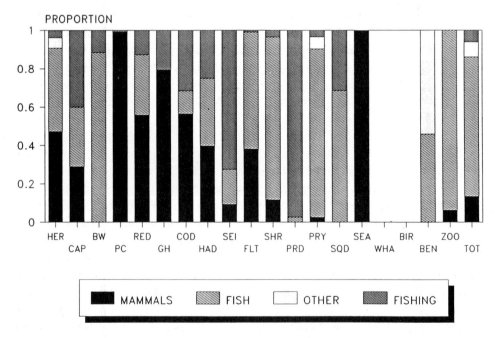

Fig. 8.5   Proportion of the total biomass lost from May 1984 to April 1985 to mammals, fish, other species and fishing, in the Barents Sea (Bax *et al.* 1989). HER, herring; CAP, capelin; BW, blue whiting; PC, polar cod; RED, redfish; GH, Greenland halibut; COD, cod; HAD, haddock; SEI, saithe; FLT, flatfish; SHR, shrimp; PRD, other predators; PRV; other prey; SQD, squids; SEA, seals; WHA, whales; BIR, birds; BEN, benthos; ZOO, zooplankton; TOT, total.

8.2.1   *Marine ecosystem management concept*

The concepts and objectives of marine ecosystem management must be defined in conformity with the nature and behaviour of marine ecosystems, and in conformity with man's utilization of the living resources from these ecosystems. The basic principles in qualitative terms of marine ecosystem management are:

(1) Regulation of the removal of species biomass by a fishery so that the repro-duction potential (recruitment) is not lowered below a level which would not produce satisfactory recruitment in future years, assuming a static environment. Consideration must also be given to the effects of natural recruitment fluctua-tions and interspecies interactions via predation.

(2) Reduction of the potential damage to the non-harvestable part (e.g. undersized and undesirable from market standpoint) of the ecosystems below a level aes-thetically, economically and biologically acceptable in respect of maintenance of biologically safe biomass levels and composition of ecosystem (i.e. conservation of the nature of the ecosystem at a given state of its exploitation without interruption of its normal functioning).

(3) Maintenance of biodiversity and reversibility of the ecosystem in an acceptable level consistent with the ecosystem utilization. The conditions do not present a contradiction in marine ecosystem management.

(4) Regulation of the species biomass removal (fishery) must consider the demands and economics of a fishery for different species and the balancing of the different fisheries within the biological processes in the ecosystem. Thus two different reference points, biological and economical, of minimum biomass level of commercial species should be determined for each identified regional ecosystem. Figure 8.6 illustrates some relationships between exploitation, benefits and efforts on the ecosystem.

Detailed ecosystem management goals cannot be universally defined as they depend on many local national and regional factors. However, some of the goals are similar in general terms in all regions, such as sustainability of the desired outputs and conservation objectives pertaining to minimization of risks of extinction of targeted and non-targeted species. These management goals are often summarized in the term 'sustainable development', a term that is in some sense an oxymoron.

Marine ecosystem management should not run contrary to existing fisheries management, except that all species and sub-systems should be considered together and balanced to objectives. The means of managing the harvest would remain the same, catch limits (TAC) and quotas, or effort restriction if necessary, gear specifi-cation (e.g. mesh size), and restriction of fishing in space and time (closed areas and seasons). In addition, the efficiency and economics of different gear and their effects on the various components on the ecosystem, i.e. the by-catch problems, species interactions and habitat maintenance become more significant in ecosystem man-agement.

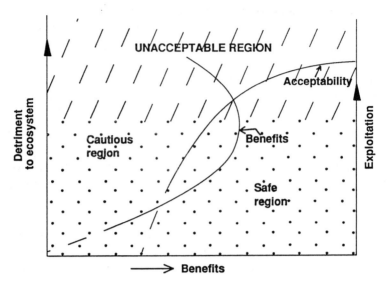

Fig. 8.6 Scheme of relationships between the exploitation of a marine ecosystem, its benefits to man and acceptability of possible changes in the system caused by man. Benefits increase with increased exploitation, reach a maximum and then decrease. The acceptability of ecosystem changes decreases with increasing exploitation, whereas detriment to the ecosystem increases with increasing exploitation from safe via acceptable to unacceptable.

It is desirable to manage the fisheries ecosystem of a given region by 'steering' the utilization of its components and sharing the different resources in this system to compensate for the natural fluctuations of these resources in this region.

Knowing the behaviour of an ecosystem and the factors affecting it allows societies to manipulate the ecosystems to some extent. Thus steering of a marine ecosystem refers to man's active intervention in the marine ecosystem to change the abundance of certain species to achieve defined social goals. A morally ideal objective of fish ecosystem steering might be to obtain sustained maximum amounts of human food from a given regional ecosystem, while maintaining the species and genetic biodiversity of the ecosystem components. This objective also suggests the maximum utilization of fishery resources, i.e. elimination of under-utilization by adjusting fishing mortality to obtain the maximum possible yield from the whole ecosystem. This could lead to a decrease in senescent mortality and apex predation to a balanced utilization of possibly all fish and shellfish species, i.e. including those species at present considered discards and scrap fish.

The steering of a marine fish ecosystem can be achieved by four groups of methods: species selective fishing; mariculture; transplantation; and productivity changes (e.g. eutrophication, large-scale mixing, mariculture and transplantations. Species- and size-selective fishing is expected to be the most commonly used method in the future. Several different approaches are possible:

(1) Economic fishing under strict management, whereby some species and some sizes are allowed to be fished heavily, above the sustained production but not

threatened with extinction (e.g. to remove the predatory part of the biomass), whereas other species are periodically fished only slightly, either to sustain food bases for other species or to build up a suitable population age composition to obtain economic production.

(2) Subsidized fishing has been little used in the past. One of the main objectives with this type of fishing is to remove predators and scrap fish and to increase recruitment of other, more useful fish.

(3) Mariculture is gaining ground, not only for salmon but also for other species such as cod and halibut, where juvenile phases are cultured and released.

(4) Transplantation of species has been tried in the past, but only a few attempts have been successful. Transplantation requires international cooperation in a form quite different from that practised at present.

(5) The change of productivity of limited areas or semi-closed seas has proceeded without directed actions by man. An example is the Baltic Sea, where eutrophication has increased pelagic fish production. Large-scale mixing of nutrient-rich deep water with surface water is a future possibility in some areas but might not be economical owing to high energy requirements.

All steering methods can be tested to a considerable extent with numerical ecosystem simulations. All ecosystem steering also requires close international collaboration and investment.

One of the main objectives of ecosystem and fisheries resource management is also a basic moral principle, equitable sharing and full utilization of the available, renewable resources. Full utilization implies the reduction of senescent mortality and increased exploitation of the older fish, which would otherwise die if not caught, and which are in most species more piscivorous than the younger biomass. This can be explained also in two other terms, full utilization of surplus production and decrease of natural mortality, mostly senescent and predation mortality, and at times also emigration.

In order to evaluate the fish ecosystem and quantify the processes in it, numerical multi-species and/or ecosystem simulations are necessary. Multi-species ecosystem simulations rectify some of the shortcomings of single-species approaches for determination of what and how much can be removed by man from the ecosystem, and what are the consequences of this removal. Furthermore, the numerical ecosystem simulations can complement resources surveys. A comparison of single-species versus ecosystem approach for fisheries resource evaluation is given in Table 8.1.

It cannot be expected that a marine ecosystem will remain unaffected, retaining its natural state, when harvesting of its constituents starts. Unfortunately we frequently do not know and cannot define this initial or natural state, mainly because large spatio-temporal natural variations in different space and time scales have occurred and will occur in these systems.

Bax & Laevastu (1989) have stressed the importance of predation interactions in the marine ecosystem. These interactions can be evaluated from stomach content analyses. Little attention has been paid to early life stages, where death through

Table 8.1 Information for fisheries management and its availability from single-species population dynamics and ecosystem approaches.

Type of information needed	Availability from	
	Single-species approach	Ecosystem approach
Size of the resource (stock)	Estimation with cohort analyses if data available	Equilibrium biomass computed; less stringent data requirements
Natural fluctuations	Not available	Computed, including the effects of environmental anomalies
Response to fishery	Computed for target species, no interspecies interactions included	Computed for all species, fishing and natural mortality interactions and effects on non-target species included
Interactions between species	Not included	Included in the computations via predation, competition, by-catch
Possible optimum yield	Computed without interspecies interactions	Computed with consideration of the whole ecosystem
Rate of change of biomass and recovery rates	Only rate of change due to fishery computed	Computed as caused by all factors within the ecosystem
Recruitment to fishery	Can only be estimated	Computed, function of predation, env. anomalies and other factors
Spatial distribution and vulnerability to gear	Not possible to compute	Computed in models with spatial resolution

processes other than predation may also be as significant. This is not because the theory is not available to describe these early life stages, but because researchers have focused on the potential interactions of man with the marine ecosystem. With this limitation, one is restricted to features that are significant and can be measured.

It is evident that in some situations we have sufficient information on which to base management policy at the level of the ecosystem. This does not mean that this will happen immediately, mainly because of the conservative nature of human thinking and reactions. Furthermore, this knowledge must be made available to fisheries managers in a digestible and convincing form. As we consider the effects of our management policies on the ecosystem, we also have to accept responsibility for our actions. At the moment, using the single-species models in fisheries management, we are usually able to restrict our attention to the value of the catch in monetary units with which we are comfortable. The advent of ecosystem management will provide estimates of the direct impact of each proposed intervention on all components of the ecosystem management. This will force us to consider the many and diverse effects of our interventions, and to evaluate effects for which we have, as yet, no common currency and which are sometimes termed 'ethical considerations'.

The marine ecosystem management question has often been raised in connection with marine mammals, where public attitudes have been driven by environmentalists ('greens'). This question has been best reviewed by Harwood (1987), who found that many people now consider a healthy seal population in its own right, rather than its harvesting, as a valuable resource. As a result commercial seal harvesting has ceased and some herds have increased substantially. In turn increasing seal populations now threatens the livelihood of fishermen. The two interested groups, 'greens' and fishermen, exercise different influences on governments.

If a marine mammal population is to be reduced, straightforward calculations should be made of the possible biological gains. These biological gains (benefits) are not necessarily immediate, but might take a few years to materialize. Unfortunately, marine mammal management and protection measures are often decided on political and emotional grounds. Harwood (1987) pointed out that the most constructive contribution that marine biologists can make in this decision-making process is to clarify the interactions of marine mammals and the fish ecosystem and calculate the results of prospective measures.

Flaaten (1988) analysed the ecosystem in the Barents Sea with emphasis on the management of commercial fishery resources and obtaining optimal yields. He found that one of the best means to this end is to overexploit the predators. The greatest predators in the Barents Sea are marine mammals. Consequently, Flaaten arrived at the following conclusion:

'Sea mammals should be heavily depleted to increase the surplus production of fish resources for man. Controversial it might be, our findings are nevertheless rational from an economic point of view.'

Five basic kinds of information are required for marine ecosystem management:

(1) The determination of the present state of the ecosystem. This is an expensive and nearly continuous process, involving trawl and acoustic surveys, collection of various data from fisheries and population evaluations, the use of VPA, ecosystem simulation and other models, various samplings (e.g. benthos, plankton) and collecting of appropriate environmental data for assessment of recent and present environmental anomalies.

(2) The need to know quantitatively the processes affecting the natural fluctuations of the components of the ecosystem and the possible magnitude and direction of these fluctuations (i.e. possible limits of natural changes without the effect of man) (Table 8.2). The processes affecting the fluctuations of the fish biomasses are interconnected (Fig. 8.7) and can be evaluated with numerical ecosystem simulations (e.g. Laevastu & Larkins 1981, Fig. 8.8). These numerical ecosystem simulations also allow the evaluation of the effects of different fishing intensities on the targeted species as well as on the fish ecosystems at large.

(3) The evaluation of a variety of economical aspects of fishing concurrently with the examination of the effects of different fishing intensities assigned in the numerical simulations. This evaluation also pertains to the future aspect of fisheries, i.e. what remains in the sea and how much could be removed in the

Table 8.2   External and internal factors causing fluctuation in the marine ecosystem.

Factors	Effects
External	
Temperature anomalies	Affecting metabolic rate, growth rate and food uptake, and thus predation rate of prey.
Fishing	Affecting abundance of older biomass, thus affecting predation, cannibalism, and recruitment to exploitable stock.
	Affecting total biomass, including spawning biomass.
Internal	
Predation (including cannibalism)	Affecting recruitment to exploitable biomass; main mechanism in interspecies interaction in predator/prey system.
Competition	Interspecies interaction in predator/prey system; can also affect starvation, which in turn affects growth.
Migration	Affecting predator/prey system by changing predator/prey overlap (local density).
	Affecting prey availability and thus predation in space and time.

future, under various fishing strategies, dictated mainly by socio-economic considerations.

(4)   The evaluation from the biological point of view of the state of the ecosystem resulting from the selected plausible exploitation strategies, especially in respect of its future potential evolution under different potential levels of recruitment.

(5)   The determination of management criteria, such as TACs, their allocation, management measures (e.g. gear restriction, closed seasons) and enforcement. This usually involves several regulatory and administrative bodies.

8.2.2   *Unmanaged ecosystems*

Some marine ecosystems show little, if any, effect of man's harvesting of the resources of the sea. We may not need to consider the management of these ecosystems. However, we must be aware of these ecosystems and especially of the more profound natural changes in them, because they are connected in various ways, e.g. as a food source, to the fish ecosystems of interest to man. These ecosystems were described in Chapter 2, and below we present only a few summary notes pertaining to their relationship to fish ecosystems and to man.

Plankton

Phytoplankton communities and production cannot be changed or influenced by man in any noticeable degree, except for local eutrophication. There can, however, be

$$\text{BIOMASS}_{i,t} = \text{Biomass}_{i,t-1} + \text{Growth} - \text{Mortalities} - \text{Emigration (immigration)}$$

i species
t time (month)
+ increase
− decrease

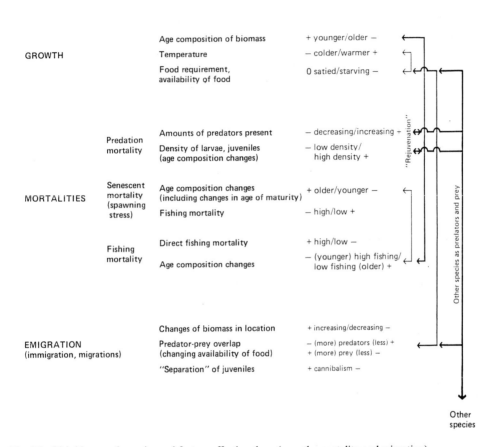

Fig. 8.7   Fish biomass dynamics and factors affecting them (growth, mortality and migration).

considerable natural seasonal and annual fluctuations of phytoplankton production, especially in upwelling regions and occasionally at high latitudes near the seasonal ice cover regions.

Eutrophication has occurred in some semi-closed bays and brackish water seas, which were oligotrophic in their earlier natural state. Eutrophication has had some positive effects from man's point of view, by increasing the basic organic production in these areas and ultimately increasing fish production. The negative side-effects of eutrophication have been minor, compared with some harmful local industrial pollution. Red tides, sometimes connected with eutrophication, are natural phenomena and there is no proof that man has influenced their frequency of occurrence.

Man cannot generally influence zooplankton communities or zooplankton pro-

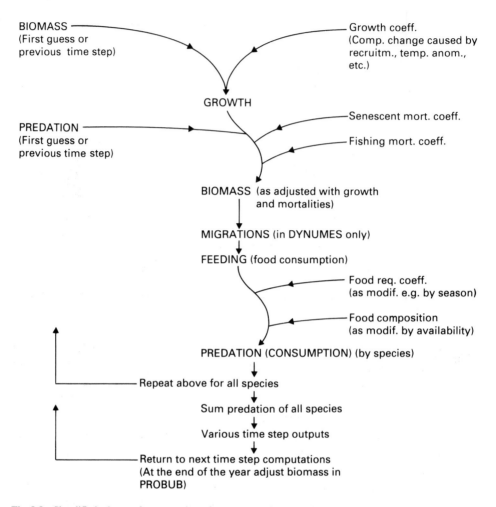

Fig. 8.8 Simplified scheme of computation of processes and state variables in ecosystem simulation. In the model used in the present study, only growth and food consumption were affected by temperature.

duction. The only commercial utilization of zooplankton by man is the past limited harvest of Antarctic krill. To date, this harvest appears to be minor considering the large production of krill and its consumption by other elements of the Antarctic ecosystem. The natural fluctuations of the zooplankton production on a large scale may limit fish production, and although badly known, the possibility of these fluctuations must be included in holistic ecosystem simulation.

The effects of the possible changes in the standing stocks of zooplankton and benthos, epifauna and infauna, on the carrying capacity of the fish ecosystem components can be studied with holistic numerical ecosystem simulations. The results of one such study on demersal and semi-demersal species in the Bering Sea are shown in Fig. 8.9. In this simulation the availability of zooplankton and benthos components was changed to 0.75, 1.0 and 1.35 times the input average value. The effect of changes

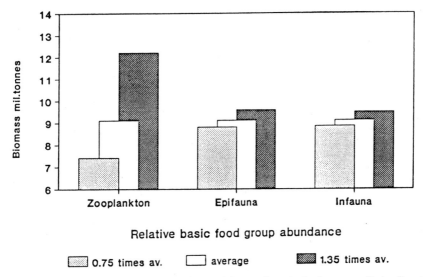

Fig. 8.9 Annual mean biomasses of demersal and semi-demersal species in the eastern Bering Sea at different availability of basic food groups.

in zooplankton standing stock on the demersal and semi-demersal species biomass is approximately proportional to the change of zooplankton abundance, but the effect of change in the benthos components is small. The main reasons for this disparity are twofold. Firstly, the zooplankton constitutes about 60% of the food of these fish components, whereas the infauna contributes only 6.5% and epifauna 8.2%. Secondly, if benthos feeders need to switch food, they switch predominantly to zooplankton, whereas plankton feeders, e.g. younger fish with pelagic feeding habits, do not easily switch to benthos.

The possible annual anomalies in basic organic production and in standing stocks of zooplankton can affect the carrying capacity of regional ecosystems. The organic production and standing stocks of zooplankton can be affected by surface weather anomalies, especially wind, and by local pollution. Thus we need to consider recent and present anomalies in plankton abundance when evaluating the state of the ecosystem and its fluctuations.

Benthos

The amount of suitable fish food in the benthos can affect the production of demersal fish in some regions, and so the estimated benthos biomass and its production should be included in holistic ecosystem simulations. Part of the benthic ecosystem is harvested by man, e.g. crabs, shrimps and clams. Juvenile stages of these species are also important prey to the demersal fish ecosystem. These harvestable species could be included in holistic ecosystem simulation as entities separate from other benthic species. Benthic organisms can be a considerable part of the by-catch in otter trawling, which might need to be considered quantitatively. By-catch problems and management of less migratory benthic animals might necessitate some restrictions on area, season and gear.

Although bottom trawling (beam and otter trawl) will affect benthos (see [6.1]), over the past decades no alarming broad-scale changes in benthos caused by trawling have been reported. It is assumed that in most intensively-fished areas, e.g. the North Sea, the benthos has adjusted to the effect of trawling. Some past industrial waste disposal might have affected benthos in very limited areas.

The standing stocks of benthos and its species composition in any given area can be only approximate, whereas little is known regarding seasonal and long-term fluctuations and annual production of benthos. In general, benthos has been much less studied quantitatively than zooplankton.

In consideration of benthos and fish ecosystem relationships we have to bear in mind that not all benthos is suitable as fish food or available to fish. Furthermore, benthos also contains predatory animals, e.g. starfish, which prey on other benthic organisms as well as on demersal fish eggs and young.

Nekton

Nekton consists mainly of fish and squids and is the most exploited and managed marine ecosystem. Fisheries resources are motile in space and time and their abundance fluctuates considerably even without the effects of fishing. Moderate fishing can remove variable amounts of the species biomass each year without materially affecting the ecosystems. If the fishery does not harvest part of the resource, a portion of that potential harvest will die anyway or will be eaten. The moral requirement of full utilization arises from the latter condition and could be considered in each ecosystem management, especially in the international context of providing food for starving people of the world.

Most of the nektonic ecosystem, i.e. the fish ecosystems, are exploited by man, and their exploitation is the subject of management. However, short-lived smaller species, such as sardines and squids, whose abundance is largely controlled by large variations in recruitment, in turn caused by local environmental fluctuations, might require different management strategies. For example, the Japanese sardine, although yielding high total catches, is not managed by catch limitations because its abundance is controlled by recruitment variations. Assessment of short-lived pelagic species is done mostly with acoustic surveys. Many short-lived species are important forage fish in the local ecosystems. Thus in estimating their TACs the need of the ecosystem for forage should also be taken into account.

Although some specific large-scale environmentally-defined ecosystems cannot be considered as management units, some of their special characteristics should be considered in the management process and are therefore briefly discussed below.

COASTAL AND CONTINENTAL SHELF ECOSYSTEMS

The most intensively-exploited ecosystems are located on the continental shelves. These areas might be affected locally near coasts by pollution and eutrophication.

Aggregation of fish on the shelf occurs for spawning as well as for feeding. Total biomass removal by predation, fishing and senescent mortality is usually larger on the continental shelf than biomass production because of the immigration of the juvenile age groups of many species, e.g. hake, which have been feeding and growing offshore.

Coastal and shelf ecosystems, defined by their location, and neritic, pelagic and demersal fish ecosystems are the main subjects of marine ecosystem management. However, the degree of influence of management on the different ecosystems of the shelf regime is variable, depending partly on the interactions between the ecosystems directly affected by man's actions, mainly fishing, and those which are only indirectly affected by the interactions. Thus ecosystem management requires that we can quantify the large-scale interactions.

Among the large-scale anomalies in continental shelf ecosystems which might be of concern in ecosystem management are environmental anomalies, especially in currents, and the distribution of pollutants. There can also be large-scale changes in migrations and distributions of the fishery resources. The knowledge of these anomalies and changes is applicable to smaller area units of ecosystems subject to management.

OFFSHORE ECOSYSTEMS

Although the plankton ecosystems in offshore waters are in many ways similar to neritic plankton ecosystems except for the absence of some neritic forms, the production is usually lower than on the shelf. Oceanic nekton ecosystems are dispersed and their main characteristic is extensive migration, usually in constant search for food. Some of the coastal-spawning fish spend part of their pelagic life feeding offshore, where the predation mortality is lower than on continental shelf. The main subject of management in offshore ecosystems are the large migratory species such as salmon, tuna and squids. These species pass through several ecosystems. Straddling parts of the shelf fish ecosystems, they also require some exploitation control.

SPECIAL ECOSYSTEMS

Brackish-water and estuarine ecosystems contain fewer species than marine ecosystems. These ecosystems are also the most affected by man's actions, especially by pollution (see [6.2]). The nature of the brackish-water ecosystems is varied and largely determined by local fresh water inflow and its variations.

Coral reef ecosystems have special characteristics making them more sensitive to exploitation than other ecosystems (see [2.3.3]). Coral reefs are also subjected to natural hazards such as hurricanes, periodic coral diseases and predator (e.g. starfish) invasions. Man-made hazards are heavy fishing, the use of poisons and dynamite, and pollution, which promote algal overgrowth. Artificial reef ecosystems are created by man, but their species composition and dynamics are not controllable to any high degree.

8.2.3   *Fisheries ecosystem management principles and procedures*

Two basic prerequisites for ecosystem management are paramount. The first pre-requisite is to assemble a team of people who are collectively knowledgeable of the quantitative composition of the ecosystem under consideration and its functioning; of the nature of the present fishery and its economic and social setting, its capacity and requirements; of the desires of society at large pertaining to the state of the particular ecosystems; and of the methods of fisheries management and their effects. The task of this team is to delineate the ecosystems of concern and to assemble all essential information. There might be a need to present some of this information in a semi-popular but factual form, accessible to the concerned and affected society at large, who have the say about the future state of the ecosystem.

The second prerequisites is that the team must consider the ecosystem as a whole rather than on a species-by-species basis. Obviously, some emphasis should be given to the present and potential future commercial species and the effects of their exploitation on the future state of the ecosystem and its productivity. The team should also be cognizant of the time- dependent numerical ecosystem simulations in order to forecast the future consequences of any effects of fishing and the possible effects of proposed management measures.

The fisheries ecosystem management procedures can be divided into ten tasks (Table 8.3). Most of these tasks and their results form an 'environmental impact statement' and management procedure for the particular ecosystem. The ecosystem management process, outlined below and in Table 8.3, conforms to the traditional pattern of management activity (Fig. 8.10), with the exception that the ecosystem as a whole and interactions within it are considered and the future of the ecosystem is predicted in a numerical and reproducible fashion. Also in ecosystem management more attention can be given to the socio-economic considerations than in some of the existing complex management processes (Fig. 8.11), and reduction of various special interest committees and sometimes overwhelming capital-intensive fishing interests is possible.

The first task in management is to establish and review the general objectives pertaining to the specific regional ecosystem under consideration [7.1] and to assess priorities and modify the objectives as local conditions require. These objectives might include some limits of acceptable changes in the affected ecosystem that might be caused by exploitation of given resources.

The objectives of management for local ecosystems and especially the requirement of the state of the exploited ecosystem are normally established with inputs of requirements from the society concerned and possibly affected, especially coastal and fishing communities, fishing industry, conservation and environmental groups, and national economic interests at large.

The second task is to give priority to specific regional economic, management objectives, considering the socio-economic constraints, including the nature of the existing fishery and fishing fleets. These socio-economic considerations should include the sustainability of a resource, income, and the potential utilization of

Table 8.3 Example of ecosystem management and environmental impact considerations.

Task No.	Action and/or consideration	Remarks
(1) Principles of management and impacts	General management objectives and priorities Environmental effects, including common usage (fishing, oil development, shipping) Species to target and desirable sustained levels; optimal biology and economic utilization Resource allocation and prior rights Other environmental usage Establishment of acceptable limits of changes in ecosystem caused by man	Management principles, priorities and their 'weights' Definitions, limits & boundaries
(2) Local environment impacts	Prioritization of regional objectives General state of the ecosystem (including mammals and birds) Regional allocation demands (onshore/offshore, sports fisheries) Other local requirements (including waste disposal, sanctuaries, environment laws) Sustainability of resources and income and possible alternative resources	Local environmental impacts statement (incl. sustainability and alternative resources)

(3) State of ecosystem	Resource and ecosystem assessment
	Results and evaluation
	  Trawl and acoustic surveys
	  VPA (commercial species)
	  Holistic ecosystem simulation
	  Other resource and ecosystem indices
	Auxiliary information
	  (Recruitment, growth rates, migration, spawning peculiarities, etc.)
		Initial status of environment and stocks; alternatives
		Annual updating

(4) Ecosystem processes (present & near past)	Fluctuation of ecosystem and stocks Past catches Past recruitment fluctuations Observed ecosystem composition changes (effects of fishery and other factors) Past effects of environmental anomalies Predation conditions & changes	Assessment of fluctuations, limits and causes
(5) Assessment of environmental anomalies	Recent and near past environmental anomalies Recent wind anomalies Temperature, ice etc. anomalies Expected effects on stocks and ecosystems Prognoses of recent environment-caused changes	Prognosis of natural anomaly effects
(6) Economics of fishing	Fishing fleet requirements and capabilities Fleet composition & gear mixture Minimum annual catch required (economics, fleets) Catches req. of targeted species Seasonality, alternative resources, their markets Possibilities of gear mixture etc. change	Economic assessment and regimes

Table 8.3 *(Continued)*   Example of ecosystem management and environmental impact considerations.

Task No.	Action and/or consideration	Remarks
(7) Determination of ecosystem response and sustainability of catches	Ecosystem simulation and determination of effect of catch on stocks Use of ecosystem simulation, MSVPA, effect of fishing numerical models) First guess inputs from subjective evaluation from Task 3 Inputs of selected fishing regimes from Tasks 3 and 6 Assessment of all biomasses and their future (5-year) dynamics (runs also with plausible extremes of uncertain inputs) Assessment of by-catch quantities Evaluation of present and future exploitation economics Sustainability of reasonable state of ecosystem and resources	Effects of fishing on ecosystem (minimization of uncertainties) Sustainability of catches and of ecosystem
(8) Selection of biological TACs	Ecosystem evaluation with selected scenarios Use time-dependent ecosystem simulations, compute minimum 5 years What would be the quantitative ecosystem composition in next 3–5 years with selected TACs? Is any biological or economic collapse of stock imminent? Is selective harvesting and/or change of gear mixture recommended? What will other regulation effects be? Determination of most limiting constraints	Testing of about three selected regimes Biological limitations Effects of selected exploitation levels
(9) Economic realities and possible future	Economics of selected scenarios Profitability for different fleets Flexibility to change, target species, gear type Taxation, market demands Fuel savings, other economic effects Future economic outlook of fisheries	Consequences of realities and future outlook
(10) Restrictions and control acceptance by populus	Fisheries management proper Final TACs, their legalization Allocations to fleets Regulation methods determination and legalization Enforcement and control measures Reactions of fisheries and of concerned public	Execution and its applicability

alternative resources. The economic objectives are normally established by responsible government offices and pertain to long-range economic planning.

The third task is to assess the present state of the ecosystems. In this task, the present stock sizes and age compositions of the important commercial species must be determined, taking into account the possible biases and errors in these assessments.

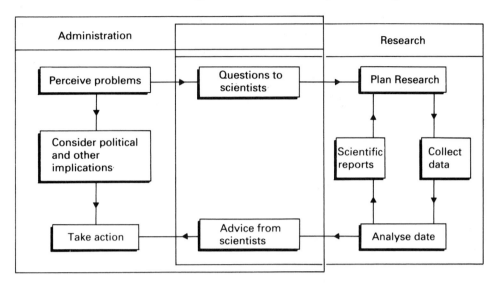

Fig. 8.10 Pattern of fisheries management activity with close links between administration and research (Gulland 1983, in Parsons 1993).

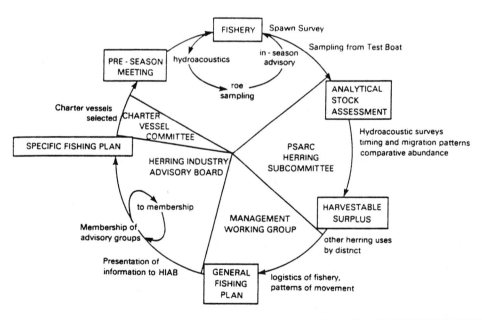

Fig. 8.11 Processes for management of the British Columbia herring fishery (Stocker 1993, in Parsons 1993).

Many essential parameters may be uncertain and biased (e.g. mortalities, age-composition, etc.) caused by recent changes in fishing patterns and by existing regulations. One must also be aware of the inherent natural variability in the fish ecosystem. Best estimates should also be at hand for the abundance of non-commercial species with reference to predator/prey systems. Other species-specific information could be established in a species thesaurus of summaries of information, e.g. migrations, life history, growth rates, estimations of past recruitment and past history of catches. The composition and dynamics of the given ecosystem are established with diverse and long-range research. Annual updating of selected and essential components of this system is necessary.

In the past this task has been the main activity for resource assessment and management advice. A multitude of different approaches are used, compared and evaluated in the assessment of commercially important stocks, e.g. fisheries trawl and acoustic surveys, computations with VPA and MSVPA models, and various indices and observations of catch such as CPUE. For more information see resource assessment in Gunderson (1993) and Laevastu & Favorite (1988).

The fourth task is to evaluate the possible natural fluctuations of the stocks of commercially important species as well as the concomitant changes in various components of the ecosystem. The assessment of the effects of past fishing on the stocks and on the ecosystem is included in this evaluation. In most cases only semi-quantitative pictures within this evaluation task can be obtained even for well-investigated areas. Use can be made of various indices of abundance, including CPUE. Condition factor and growth rate changes are also useful indices for the condition of some stocks. Effects of past fishing can be ascertained with numerical models.

In most areas some information is available on the possible effects of large-scale environmental anomalies on some stocks and recruitment. Therefore the fifth task is to evaluate the recent large-scale environmental (e.g. wind and temperature) anomalies in the region under consideration and their effects on the ecosystem. Some of the possible ecosystem changes caused by environmental factors might be hypothetical only, but could be taken into consideration in task seven, numerical ecosystem simulations. Some of the environmental anomalies from the past might have influenced the present ecosystem, e.g. as regards recruitment, and might affect the future availability of exploitable resources. The state of the pollution and its possible effects must be considered in some coastal areas.

The sixth task in the management decision preparation is to evaluate the existing size and properties of the fishing fleet and the socio-economics of coastal fisheries and fishing communities. Attention should also be given to such conditions as traditional fishing areas, and gear, mixtures of catches, possible species, changes in fishing area and season, and exploitation of alternative resources. These socio-economic considerations pertain to the present conditions, whereas task two pertains to long-term considerations. Additional socio-economic considerations are listed in task nine. The socio-economic consideration in international fisheries is usually based on past fair and equitable distribution problems.

The seventh task is to evaluate the total ecosystem quantitatively with numerical simulations (see examples in Appendix) under different exploitation regimes that are considered reasonable and possible based on the previous six tasks, e.g. different catch levels (including bycatches), possible effects of environmental anomalies, recruitment variations, etc. Figure 8.8 shows a schematic flow diagram of the stepwise computations. Computation in each regime must be carried in seasonal time steps up to 5 years into the future in order to establish trends and possible stabilization with a given regime, as well as possible future sustainability of resources on some estimated lower level. This task is in essence the digestion of data from the previous six tasks. It might be necessary to return to this task several times before deciding final management measures.

The various selected regimes should also be computed with plausible maximum and minimum values for various uncertain, but important, parameters in the simulations, such as recruitment affecting predation, growth rates and immigration/emigration where applicable. Such computations would establish quantitative considerations for such terms as 'uncertainty' and 'risk'.

Several available simulations and numerical models can and should be utilized in this task that pertain to commercial species, such as MSVPA with available latest and best data, gear selectivity effect programmes, and any other species-specific programme. The results of the computations with different regimes can be presented for the decision-making process by authorities and by the society concerned.

The eighth task is to evaluate the results from task seven (i.e., the evaluation of the resulting catches and ecosystem composition now and into the near future under different exploitation regimes). In many cases, the evaluation of the effects of *status quo* catches of the present and near past is the first step in this task. Specific questions to ask are:

(1) At which level will the biomasses stabilize in a few years with different exploitation rates and what would be the corresponding mixture of catches of all exploited species (considering also the species interactions)?
(2) Will some of the biomasses fall to a very low level so that the spawning biomass would become and remain low?
(3) What type of selective harvesting (and gear mixture) would be necessary and possible with realistic and effective regulations?
(4) What are the economics of the possible selected (see task nine)?
(5) What is the most limiting constraint of catch levels?

Tasks seven and eight will diminish and/or eliminate the use of any artificial biological reference point at present in use.

The ninth task is to provide economic projections for the selected possible regimes remaining as plausible TACs for the exploited species considering the ecosystem constraints in task eight.

Many diverse questions pertaining the economics of fishing can be asked, some of which are:

(1) What is the profitability of a given catch level for a species with given densities (biomasses) for the present fleet (vessel types) with prevailing prices and operating costs? The same question can be asked relating to a projected TAC and its allocation.
(2) What is the flexibility of the fleet to change its target species and species mixture and gear type?

(3)   What will be the fishing economics for the next few years at the prognosticated resource and exploitation levels?

In this task the desires (tasks one and two) are weighed against the realities.

The tenth and final task would be to set the TAC limits, allocate the resources to different fleets, and select methods of regulation and enforcement. This task is usually executed by administrators and bureaucrats with technical and scientific advice. The main question in this task is: how to achieve the goals established in task nine and which methods to use, e.g. quotas, subsidies, area and season closures or gear specifications.

The fisheries regulations to be decided in task ten attempt to control:

(1)   The size of fish caught, size/age at first capture and the species of fish caught. These controls use minimum size of fish landed, mesh size and other gear specifications.
(2)   The amount of fish caught or the fishing effort, effected by quotas, vessel licenses and effort limits by fishing time, including closed seasons.
(3)   The special protection measures, e.g. spawning stock protection, minimization of by-catch of given species, effected by gear specifications, area closures and catch quotas on by-catch.

Fisheries regulation effects have recently been reviewed by Parsons (1993). The application of regulations varies considerably from one ecosystem to another and depends on management objectives and their assigned priorities. A few generalities of the properties of the regulations and their side-effects follow.

Gear specifications include mesh size and shape in trawls and gill-nets; hook size in longlines; and the addition of devices such as tickler chains, sorting devices and turtle excluders in shrimp trawls. Certain gears might be forbidden or their use limited to given areas to avoid gear conflicts. The gear specification depends on the objective and what is feasible from a local socio-economic point of view. Usually the use of more efficient gear is limited and it is often uncertain who benefits from gear regulations. Vessel size limitation and the amount of gear allowed per vessel usually limits the fishing effort.

Minimum size limits and by-catch restriction are attempted through gear specifications. Once popular, the attempt to regulate the size at first capture has had questionable effects and is not economically beneficial. The capture of males only, e.g. in a crab fishery with pots, can also be questioned as to whether it is biologically justified.

Closed area regulation can be beneficial in some local fisheries, especially for less migratory species or for the protection of spawning runs of salmon. This type of regulation has little or no effect on controlling catches and landings in most other fisheries, because fishing outside closed area usually intensifies. Using closed areas and seasons to protect spawning stocks has little or no effect on recruitment.

Closed seasons are easier to enforce than closed areas, and can be used to control fishing effort and, in special cases, also to protect a spawning stock in the rare event of this protection being advisable and possible. The amount of fishing is most often controlled by a TAC, which is divided into quotas between user groups. Quotas allow equitable distribution, but must be adjusted from year to year, depending on the status of the stocks. Quotas on major species are usually unable to regulate by-catch, which must be regulated by various other means, such as gear specification, closed areas and quotas on the by-catch. Quotas can also promote under-reporting of catches and discards.

Effort control is necessary in some fisheries and is especially useful as a means of regulating catches of short-lived species. Effort control is often defined by time limits, e.g. hours' fishing or hours on the grounds. In some cases fishing effort is difficult to enforce and its efficiency varies with different vessels and gear.

The regulation of a mixed fishery is difficult, especially if the TAC of one species can become limiting. Effort limitation is most useful in a mixed fishery, although closed areas and specification of gear properties can also be used.

All fisheries management measures, especially those pertaining to effort limitations, have side-effects and other shortcomings and difficulties to enforce.

The quota system of management, especially the individual vessel quota system (ITQ, see Chapter 7) has recently become popular with economists because it has some national and industrial economic benefits. The introduction of an ITQ can reduce the existing over-capacity of fishing fleets, the investment in fishing capital, the number of operating vessels in some fisheries, and fishing effort (e.g. in Iceland, Arnason 1993). Economic rents generated by this system result in some economic benefits. However, the ITQ system cannot be implemented in all fisheries, especially those of an international nature, such as in the European Union (EU).

Some simple management rules can be established for shortlived pelagic species if the fishery is managed on a short-term basis without the need to consider future consequences. This subjective management has three prerequisites:

(1) The managers must be aware that fishing one species, which, if short-lived, is an important prey species, can also affect the biomasses of predator species, thereby requiring a holistic ecosystem approach for the evaluation of the consequences of management options.
(2) Heavy fishing on dominant species in the ecosystem often suppresses the fluctuations of this species, but not necessarily the fluctuations in other species that are affected by fishery through inter-species interactions, i.e. predation.
(3) Fishing can seldom be fully prevented on any species. There will always be by-catches and it is difficult to effectively eliminate or minimize by-catch.

Problems arise in the determination of an annual TAC in species with a widely fluctuating abundance and in species with a short life span where the exploitable stock is not buffered by the presence of several age groups. Furthermore, a rapid, preliminary estimate of TAC is often required, which is based on some biologically and economically reasonable assumptions.

The following estimation procedure could be used to obtain a preliminary estimate of the annual allowable catch in fluctuating stocks and short-lived species:

(1)   Determine the sign of change (+ or –) of the biomass in the previous year, or in the last few years, and extrapolate the near-future trend of the biomass changes, also considering any recruitment indices and the effects of possible environmental anomalies on the recruitment.

(2)   Determine the state of the stock in the previous or present year in relation to its long-term mean biomass, using ecosystem simulation, and determine the plausible allowable catch and the catch of the species as by-catch using the same model.

(3)   Adjust the estimated allowable catch under the following simple considerations:
   (a)   if the biomass is declining, but is near the apex of its past abundance, the *status quo* could be continued;
   (b)   if the decline of biomass is in its second or third year, the allowable catch with reference to long-term mean biomass should be lowered to correspond to the biomass at the lower level of fluctuation;
   (c)   if the biomass is at a low level of fluctuation, a low allowable catch might be required for 1 or 2 years;
   (d)   if the biomass has been increasing in the previous 2 years, heavy fishing (above the allowable catch for long term mean biomass) for next 1 or 2 years might be possible.

(4)   If the target species is a pelagic forage species, its biomass will respond not only to fishing, but also to the biomasses of its principal predator. Thus, the state of the predator's biomasses must be considered in determining allowable catches. If a predator biomass is high, the allowable catch should be lowered. If the predator biomasses are low, increased fishing effort could be allowed. Some pelagic and semi-pelagic species respond also to environmental anomalies such as large-scale changes of zooplankton abundance or changes in surface currents, which must be accounted for when setting the allowable catch, provided that the environmental anomalies and their effects are known.

Total ecosystem management and especially ecosystem steering (see [9.2.1]) require the potential use of a variety of regulatory measures, some of which have rarely been used in marine fisheries. These additional measures might include subsidized fishing for selected species with the purpose of removing predators such as rays, dogfish and other 'scrape fish' and starfish, and also larger, older, highly piscivorous cod and haddock. These measures might also include subsidized fishing on cheaper pelagic fish, e.g. mackerel, which are considered predators on juveniles of more valuable commercial species.

Total ecosystem management and steering might also involve detailed specifications for the use of area- and season-specific gear, all being adjusted to the objectives of the management and tailored to the specific ecosystems and existing local fisheries conditions, including economic considerations. There are no management measures which would not be controversial to some special interest groups and segments of the population. Therefore it is imperative that all the management objectives and measures be investigated and adjusted using accumulated and available knowledge of the particular marine ecosystem and its functioning. The numerical holistic ecosystem simulations would become important tools in this task.

An International Code of Conduct for Responsible Fishing (ICCRF) is under development in FAO. This code is intended as a follow-up of the UN Conference on Straddling Fish Stocks and Highly Migratory Fish Stocks (August 1993, New York). A draft of this code emphasizes the ecosystem approach in the utilization and management of fisheries resources. It emphasizes that management should be based on the best scientific knowledge and evidence available. The code calls for the development of more selective gear and practices and the minimization of waste (discards) and incidental catch of non-target species. The code also asks for the protection of the interests of small-scale and indigenous fisheries.

8.3   The role of scientific and technical knowledge in management

The application of scientific and technological knowledge to fisheries and fisheries management is necessary and economically justified. To conduct pure scientific research has become a moral requirement, mainly with the hope that its results will ultimately benefit mankind as it slowly unlocks the secrets of nature. Ultimately the acquired knowledge will find application in the various aspects of man's activities.

There is an urgent need to apply all available pertinent knowledge to the management of marine natural resources, because a fear has been created in the populace that irreparable harm will be done to them and to marine ecosystems by the present forms of resource exploitation.

Rational fisheries management requires detailed knowledge of:

(1)   The magnitude of the resource at any given time.
(2)   Natural fluctuations of the resource and their causes.
(3)   The amount of the species which can be taken in any given year without unduly diminishing its productivity, i.e. the response of the resource to exploitation.
(4)   Other peculiarities of the resource, e.g. its reproductive and growth potential, its migrations, and other factors affecting availability, which are pertinent for obtaining the highest possible yield and future sustained production.
(5)   Economic and social forces that influence the operation of a fishery and how they are affected by stock fluctuations.

The knowledge listed above must be provided through the collaborative effort of scientists with different specialities. The main tasks of fisheries services and research are listed in Table 8.4. The general tasks of science in fisheries management are depicted in Figure 8.12.

It should be emphasized that fisheries management is done by humans who have inherent limitations of decision making (Downs 1967). Three of the six limitations listed by Downs apply especially to complex fish ecosystem management:

Table 8.4 Main tasks of fisheries services and research.

Determination of and research on	Means and methods
Abundance of resources (stock)	Population dynamics Ecosystem models Surveys (acoustic and trawl) Catch per unit effort
Fluctuations of stocks	Effects of environment Ecosystem internal effects Recruitment and predation
Behaviour of stocks	Migrations Availability Response to environment

Effects of fishing	Effects on:
	    Age composition changes and abundance
	    Recruitment and predation
	    Non-target species

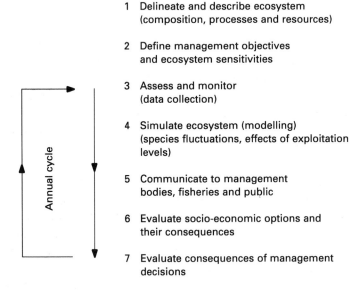

1 Delineate and describe ecosystem
   (composition, processes and resources)

2 Define management objectives
   and ecosystem sensitivities

3 Assess and monitor
   (data collection)

4 Simulate ecosystem (modelling)
   (species fluctuations, effects of exploitation
   levels)

5 Communicate to management
   bodies, fisheries and public

6 Evaluate socio-economic options and
   their consequences

7 Evaluate consequences of management
   decisions

Annual cycle

Fig. 8.12 The tasks of science in fisheries management.

(1) Only a limited amount of information can at one time be mentally considered by each manager. Ecosystem management involves, however, a relatively great amount of diverse information.

(2) The amount of information available to managers is only a small fraction of information needed and potentially available.

(3) Information on future aspects of events has not been available to managers, who have had to make management decisions without them.

These three limiting aspects, especially the third, can be at least partly remedied through the use of ecosystem simulation as reviewed in this and other chapters. These time-dependent ecosystem simulations have a predictive phase and can project events and their consequences several years into the future, but the reliability of results decreases with time.

Because man cannot see through water as easily as he can see through air, two basic problems confronting fisheries biologists are the temporal and spatial distribution and abundance of stocks. It is only through accurate assessments of the resources in the ocean that one can proceed to the more difficult questions of why fish are when and where they are, and how they interact, the basic knowledge required for implementation of ecosystem management and resource predictions based on environmental factors and the use of holistic ecosystem simulations. Total stock assessments are a vast problem and can be made through the extensive data obtained from commercial fishing and from research surveys, including acoustic surveys. These techniques have been used for decades with varying degrees of success, and one must constantly ask whether or not man will ever be able to assess and comprehend nature with sufficient accuracy to permit forecasting of conditions and events. Although forecasting may be a desirable target or the ultimate goal for resource/environment studies, this does not imply that available knowledge, which is not complete, does not have prognostic value.

Effective management requires timely and accurate estimates of fish abundance and the prediction of the effects of alternative management options on the fish ecosystem. Thus scientists must map the fluctuations of the fish ecosystem under different levels of mortalities with reasonable accuracy. This involves time-dependent predictions. The better these predictions, the less the uncertainty of objective management decisions. The past uncertainty can be diminished, and its causes localized and ascertained if and when the understanding of the dynamic processes in the fish ecosystem is improved.

Elegant mathematical models would not contribute much to our knowledge on the dynamics of ecosystems, but numerical ecosystem simulations using empirical knowledge would. Fisheries ecosystems are complex and cannot be explained nor simulated with simple mathematical models. However, using our accumulated empirical knowledge in a step-wise fashion in composing ecosystem simulations, we can achieve a complex, interacting picture of the ecosystem.

There is some question as to whether the expanding marine sciences have adequate continuity to provide the extensive background information required to plan integrated marine studies without 'reinventing the wheel'. Much of the knowledge is gained from the retrospective study of the causes of events in the ecosystem. Knowledge and science also involve the interpretation of observations and data, which is always complex and includes alternative solutions. For example, the age composition not only indicates the effect of fishing, but also includes recruitment by immigration and 'mortality' by emigration, e.g. movement of larger/older specimens to deeper water or to higher latitudes. Because predation is one of the main mechanisms of interaction within the ecosystem, one has to be aware of its complexities. Any species can be predator as well as prey, and this can change with time and space and age of the species. For example, cod is a predator of juvenile and adult herring, but herring is a predator of cod eggs and fry.

For the evaluation of the effects of natural processes in the marine fisheries ecosystem, quantitative knowledge of the effects of various processes is necessary (Fig. 8.13). Obviously, few specific data from a given ecosystem are available and considerable generalization is necessary whenever possible.

Because the components of fisheries ecology are highly variable and fisheries behavioural problems are difficult to define scientifically, we cannot make final categorical statements in fisheries ecology, as is the custom in some other branches of science. We can, however, condense our knowledge of ecological phenomena in simulation models, which are derived from experience as well as experiments, and we must use a mixture of deterministic, causal and probabilistic approaches.

A frequent complaint about ecosystem simulation models has been that the models require lots of data which are not available. However, applicable ecosystem simulations can be programmed with routinely-available data. Below we review the basic and essential data requirements and computations in the Andersen–Ursin and the DYNUMES and PROBUB models.

MAIN PROCESS	MAJOR AFFECTING FACTORS
(a) Biomass abundance affecting processes	
SPAWNING	Spawning biomass size (fecundity)
(reproduction, larval recruitment)	Predation on eggs and larvae
	Availability of food (starvation)
	Environmental factors (advection)
GROWTH	Age
	Temperature (anomalies)
	Food availability
PREDATION	Vulnerability
(mortality)	Predator abundance
MORTALITIES	Senescent mortality
	Spawning stress mortality
	Disease mortality
(b) Biomass distribution affecting processes	
SOURCE-SINK AREAS	Growth
(spatio-temporal differences in	Predation
abundance affecting factors)	Other mortalities
MIGRATIONS	Feeding migrations
Seasonal	Search for optimum environment
Life-cycle dependent	Spawning migrations
	Predation avoidance migrations
	Feeding migrations
Environment-dependent	Search for optimum environment
	Advection by currents
APEX PREDATORS AND FISHERY	

Fig. 8.13 Major dynamic processes in the marine fish ecosystem.

Growth rates are computed from empirical weight/age data. However, these data on juveniles, especially in the first and second years of life, are scarce and deficient for nearly all species and some extrapolation is necessary.

In biomass-based models the growth coefficient of the biomass of any species is dependent on the growth rate at any given time and the relative abundance of biomass at this time. The growth coefficient should vary with trophic and environmental effects and also with the variation of recruitment and fishing, which affect the age distribution of the biomass. Thus it is desirable to account for the age distribution change either by dividing the species into several age groups or by computing the age distribution with another model or sub-routine. In some cases, the growth rate of a given biomass should be considered as a function of fishing intensity, as heavy fishing can cause rejuvenation of the population and consequent increase of growth rate.

The empirical knowledge of distribution of numbers or biomass of fish in pre-fishery juveniles and their mortalities is lacking in all species. These parameters are derived with auxiliary models with some plausible assumptions.

The effects of water temperature on growth are treated in the models according to the best available knowledge. However, few empirical data are available on the acclimatization of different species to different temperature ranges. The effects of the availability of food and related partial starvation on growth are treated in a manner consistent with available knowledge.

In trophodynamics both models (Anderson–Ursin and DYNUMES) can partition the food requirement and utilization between growth and maintenance. Although the temporal dependence of the feeding rate on food density is simulated in both models, the simulation of the spatial food density dependence is possible only in gridded models such as DYNUMES. The fundamental differences between single-species models and the ecosystem model is that predation and trophodynamics in general are included in the latter, thus quantitatively connecting the dynamics of all species in the ecosystem.

Changes in relative age and size composition occur in all species, from a variety of causes, e.g. variation in recruitment, fishery, etc., which in turn induce other changes in the ecosystem and its dynamics (e.g. with reference to size-dependent feeding). The effects of different spatial distributions of juveniles and adults on predation can be simulated only in gridded models such as DYNUMES.

The turnover rates of most fish biomasses can be computed with bulk biomass or 'box' models such as PROBUB. However, more empirical data on annual turnover rates of zooplankton and benthos from ecologically different locations would be highly desirable. These data could be used for determining more accurate carrying capabilities for different regions.

The changes in age composition of the target species caused by fishing are computed directly in the Andersen–Ursin model where all species are divided into a number of age groups. In the DYNUMES model only one or two species at a time are divided into different age groups. In non-divided species, the changes in age composition are depicted in a number of parameters, notably in biomass growth rate.

These age (or size) composition-dependent changes are indirectly depicted in trophodynamics (size-dependent feeding, composition of food, etc.).

In species in which schooling is pronounced, the fishery is computed as constant catch per time step, whereas in more dispersed species subject to a bottom trawl fishery the fishing mortality coefficient can be used.

Spawning stress mortality is used in both models. There is a non-linear interaction between fishing mortality, constant over fully-exploited year classes, and spawning stress mortality, which increase about 10% per year. The predation mortality must be quantitatively limited (density dependent) when the density of prey becomes low. In the Andersen–Ursin model it is limited indirectly via a vulnerability coefficient. In the DYNUMES model it is limited by two factors; a prescribed monthly maximum percentage of a given biomass allowed to be consumed, and a predation level factor. The effects of spatially and seasonally changing predator–prey distributions on the predation rates can be included and studied only in models with spatial resolution (DYNUMES).

The gridded models (DYNUMES) have several advantages, allowing the presentation of space resolution, migrations, differences in space and time of predator–prey relations, etc. The box models can, however, be satisfactorily used for many fisheries research and management problems.

Whereas the Andersen–Ursin model takes its initial input from various available assessments, the DYNUMES model needs an initial analysis of input biomasses. This initial analysis consists of computing a unique solution to the biomass equations with predetermined food composition and food requirements. With this approach the mean carrying capacity, or equilibrium biomass, can be computed.

The ecosystem models require that all the components of the biota be presented quantitatively in the simulation. This presentation is not always possible for every species, but can be done for ecologically similar groups of species. Feeding habits and food composition are the main criteria for grouping species into ecological groups.

The migrations in box models (PRORUB) can be described as 'boundary values', i.e. the fluxes through the boundaries. Growths and mortalities outside the boxes cannot be computed but must be estimated. The migrations are treated in detail in gridded models, provided that some prior information on migrations, e.g. seasonal and spawning migrations, is available. Furthermore, the dispersal, aggregation and passive transport by currents can be computed in gridded models.

The accuracy of the prognoses of the fisheries resources is largely dependent on recruitment. In the Andersen–Ursin model the spawning products are released in a given month of spawning. The number of larvae surviving is controlled by an empirical formulation that allows lower survival from a high number of spawners and high survival from a low number of spawners.

In the present DYNUMES and PRORUB models recruitment is a function of the biomass size and the growth coefficient of the biomass. In these models, recruitment to the exploitable part of the population becomes a function of predation mortality on the juveniles.

Several ecosystem models deal exclusively with plankton production. However, previous attempts to compute fishery production from plankton production have not been successful, mainly because the pathways of plankton utilization are diverse and highly variable in space and time. One of the main tasks of the fisheries-orientated ecosystem models is to determine quantitatively the abundance of the species and ecological groups in the ecosystem and the resulting utilization of available food resources. Consequently the fisheries-orientated ecosystem models also need the average standing crops or production of plankton and benthos as input.

The standing crop data of zooplankton are simulated in DYNUMES using the available empirical data, and the consumption of zooplankton is computed in detail in the models. However, the data on turnover rates of zooplankton, needed for a more accurate determination of carrying capacity, are generally poorly known.

The sensitivity analysis, verification and validation of large ecosystem models pose many problems which have not been addressed properly by conventional means. In some cases the error limits of the outputs can be estimated by considering the possible error limits in the input data and the formulae used in the model where these inputs are used as parameters. This will also lead to verification of the results of the individual processes.

Consideration of risk in management should not be based on one-sided mathematical and statistical definition, but could be evaluated and demonstrated with the calculation of alternative measures, which include different assumptions on uncertain basic inputs, such as recruitment failures in several consecutive years. Uncertainty can be evaluated with different assumptions and alternative measures and would include the preparation of management contingency plans.

In this book we have presented a summary of fisheries ecology and fisheries management problems, mostly in a deterministic and causal manner. In Chapter 4 we summarized the causality of fish behaviour and physiology in respect of selected environmental parameters. However, we have also described the probabilistic nature of fisheries ecology throughout this book by reporting the variability of a fish's response to changes in the ecosystem, and have emphasized that many factors might be operative simultaneously in environment/species interactions. Thus we cannot explain all the interactions on the basis of cause and effect, i.e. causality, but have to admit chance, i.e. probability. This is partly because not all phenomena are yet adequately described and explained, and partly because fish have some freedom of choice in their behaviour, i.e. have their own 'mind and will', although rudimentary from a human point of view.

8.4   What successes or failures to expect when we use different management measures

The possible benefits of fisheries management are in the future and it is not always clear what the benefits will be and who the beneficiaries will be. The maximization of economic return, by reducing the TAC and hoping that the stock will respond to it is often an illusion. Equally illusory can be the hope of economic maximization by fleet reduction; the remaining fleet will usually increase its efficiency. Fishing effort reduction can be achieved by specification of the gear used, and economic benefits can be improved by specifying fishing methods and gear that require less fuel. Both alternatives are drastic measures. The introduction of the ITQ as a management measure is possible in some national fisheries and is advocated by many economists. It reduces the capital investment in fisheries and can reduce the existing fleet size. The use of an ITQ in an international fishery is, however, highly questionable.

Fisheries managers attempt to regulate harvest levels, but the outcomes are usually uncertain. If the harvest is constrained at a lower level than necessary, some of the economic opportunity is forgone. The basic question is always how much harvesting to allow to prevent depletion of a stock below some level at which harvesting becomes uneconomic. This harvest level depends on the nature of the fishery and on the types of fleets harvesting the resource. However, it is to some extent possible to predict the effects of different fishing intensities on the targeted stocks as well as other affected species using numerical ecosystem simulations (see [5.4] and Appendix).

It is often desirable to modify the proportion of a species in a given ecosystem, where the beneficiary would be the ecosystem itself. This could be effected through ecosystem steering (see [8.2.1]), using various management means such as gear specification and subsidized fishing on non-food species for industrial use. As the traditional fishery resources are nearly fully exploited and the need for food keeps rising, other changes can be expected in fish resource utilization, such as the shift from non-food to food use of abundant species, the development of substitutes (e.g. surimi, a fish paste), and the utilization of discards.

It would normally take a few years before a given management measure would have effects on a stock or an ecosystem at large. Meanwhile, several other measures have usually been taken and unanticipated changes have occurred. World-wide attempts to identify possible environmental causes of recruitment fluctuations have had little success. Even if such causes could be identified, they would remain largely unpredictable and uncontrollable. Variations in marine ecosystems remain endemic.

Rebuilding of stocks has been attempted in the past by the drastic reduction and termination of fisheries on the depleted species. Often the stocks have 'bounced back', but not necessarily to their previous high levels. In other cases the rebuilding attempts have been fruitless. Where there has been an apparent successful recovery of the stock, it is not always possible to determine whether this happened because of the rebuilding measures (cessation of fishing) or because of natural fluctuations in recruitment.

From some points of view, fisheries management is as risky as the fisheries itself. Potential returns can be expected, but they do not necessarily materialize. Thus, both management and fishery must build in alternatives, anticipate changes and adjust strategies in a relatively short time interval. The task of fisheries scientists is to see that the gambling task in this process is carried out intelligently.

Some offshore fisheries, e.g. oceanic gill-net fisheries for tuna and squids, have been terminated because of by-catch problems. Other fisheries have been constrained by by-catch levels or limits on the time and area of operation, mainly because of the by-catch. It seems that many of these decisions and actions have been based on emotions and political considerations rather than on logic and scientific facts.

Good information is lacking on the seasonal distribution of most targeted species and of those species which it is desirable to avoid as by-catch. Consequently, there is a need for good fisheries charts and information systems.

It is not easy to make an accurate diagnosis or assessment of marine ecosystems and prescribe treatments for their changes. Risks and uncertainties and their implications should be evaluated and explained by cause. Ecosystem simulations offer one of the means of evaluating some of the existing uncertainties. Ecosystem and multi-species management is not easily welcomed because it is considered as complicating management. Thus education on this subject is needed.

Squire (1993) characterized the pelagic resources as being 'in a state of chaotic fluctuations relative to apparent abundance'. Even short-term predictions seem not to be feasible in short-lived pelagic species. Here management must rely on near real-time estimates, based on the measurement of a small fraction of the stock. More than one method for estimating abundance must be used.

It has become apparent that advice on theoretical marine fisheries ecosystem management is plentiful. Much of it is, however, coloured by personal desires. Thus much more education on the marine ecosystem and its functioning is needed, together with co-operation between different specialists and different advisory and management bodies in sharing their experiences.

Fishermen at large are quite a heterogeneous group, and so are the fishing fleets, but in some economic models they are assumed to be homogeneous units. Economic rationales of profit and loss do not apply to all groups and categories of fishermen. Preservation of the life styles from earlier generations is often a driving motive of fishing. The existing fisheries economics models are not adapted to an ecosystem consideration. However, various economic considerations must be applied in marine ecosystem management. We cannot use rigid mathematical 'canned' models, because the conditions of ecosystems and their exploitation by man vary in space and time, and subjective adjustments are needed. Ecosystem management also needs the consideration of social effects and thus requires social impact analysis. Various indirect socio-economic management measures are becoming more popular. Among these measures are distribution of benefits and subsidies instead of allocation of stock quota, and bilateral agreements on resource allocation between coastal communities and large-scale offshore fishing.

Most fisheries and ecosystem management objectives pertain to the future, such as maintenance of an adequate resource level and the possible effects of management actions on the ecosystem. Consequently, it is necessary to predict functioning of the ecosystem with the present and prospective future exploitation, as well as the effects of natural forces acting now and possibly in the future on the ecosystem. These predictions are now possible using numerical ecosystem simulations tailored to particular local and regional ecosystems, utilizing available data and knowledge.

8.5   National management and international co-operation

The first fisheries management systems, started at the beginning of the twentieth century, were national. However, international systems evolved shortly before and after World War II in ICES and later in ICNAF. Now most management systems have again become largely national, often tailored to local social and economic objectives. Sometimes these systems are influenced by special interests and also by local fishing rights.

ICES has been an effective international forum for deliberation of fisheries biological and related marine science problems and giving biological advice to the European management bodies. It has not embarked on economic issues because of complex and varying economic considerations in its member countries. ICES management advice has also been short of alternative measures. Biological advice is provided by national fisheries research institutions with little economic expertise. Furthermore, the enforcement methods differ from one nation to another.

Despite the EEZ regime, international fisheries conflicts still exist, especially with regard to widely migrating species. In addition, despite the existence of inter-governmental bodies, such as NAFO, fisheries disagreements, e.g. between Canada and the European Union, have existed and calls have been made for international discussion and settlement of fisheries disputes through the United Nations.

The international fisheries management and politics prior to the EEZ regime in the North Atlantic have been summarized by Underdal (1980), who also foreshadowed the development of international fishery regulations (Table 8.5).

Hinds (1991) pointed out that international fisheries commissions have been good for scientific and technical discussions but not effective in fisheries management, mainly because they have not been able to acquire the authority for enforcement. Thus these bodies are only advisory groups with limited effects on national policy-making bodies. Distant-water fleets, though decreasing because of EEZ regimes and economics, are still in existence. These fleets are difficult to manage, nationally as well as internationally.

Fisheries management may be considered a co-operative undertaking by its nature, involving scientists, industry and government; the government is supposed to represent the populace.

Recently the role of fisheries science in fisheries management and policy has become better defined. It is the function of scientists to determine the size of the stocks, their distributions, fluctuations caused by a multitude of factors, and their response to present and future exploitation.

It is up to industry representatives and the populace, together with appointed or elected managers, with the advice of economists, to determine which fish to harvest and how. For example, should man heavily exploit the gadoids, at the same time relieving predation pressure on herring, and later harvest herring in higher quantities as a consequence; or should he harvest only herring roe to the extent that recruitment would not suffer greatly, or harvest only a little or nothing, leaving the marine mammals to control the food resources of the sea. In the past

Table 8.5   Hypothetical development of international fishery regulations (Underdal 1980).

Hypothetical order of introduction	Substantive need	Kind of externality problem	Political function of regulations	Regulatory techniques	Most likely level of conflict
1	Defining fishing rights, jurisdiction	Incompatible demands	Distributive	Agreement on fishing limits, jurisdiction	Moderate to high
2	Rules for conduct at sea, policing	Deprivation or collective inefficiency, depending on symmetry of the situation	Distributive or determinate combination, depending on situation	Rules of navigation, marking of gear, prohibiting mobile gears in certain areas, rules for liability	Low (if symmetrical), high (if asymmetrical)
3	Protecting young fish	Collective inefficiency	Sum of regulations, possible combinations	Mesh-size, minimal landing size, closed areas/ seasons, taxes	Low to moderate
4	Reducing overall fishing mortality	Collective inefficiency	Sum of regulations or combinations	TACs, effort limits, quotas, closed areas/ seasons, taxes	Low to moderate

the fisheries management practices have paid little attention to scientific findings (Underdal 1980).

A variety of methods are in use in fisheries management, most of which serve to check and limit the catches and landings by fishermen. These include quotas by time, area and fleets; vessel sizes; limited entry, mesh size of gear; closed seasons and areas. The effectiveness of these methods has been variable. Unfortunately, few reliable comparative results from the studies of the suitability and effectiveness of the management methods are available, and the results vary from one region to another. Thus one has to come to the conclusion that the fisheries management practice is neither a science nor an art, although attempts to apply both have been made; but the management is an undefinable political pot-pourri (see Underdal 1980).

There is no universal approach to fisheries management, although most of the principles and methods might be used universally. There will always be political considerations and 'give and take' negotiations which are not based on scientific or technical knowledge and facts (Underdal 1980). It has also become painstakingly clear that the people involved in management decisions, as well as the fishing industry, which is affected, must become more familiar with the processes involved in the marine ecosystem and the possible effects of various management methods.

References and Further Reading

Adams, J.A. (1987) The primary ecological sub-divisions of the North Sea: some aspects of their plankton communities. In: *Developments in Fisheries Research in Scotland* (eds R.S. Bailey and B.B. Parrish), pp. 165–81. Fishing News Books, Oxford.

Adams, N.J., Seddor, P.J. and Van Hezile, Y.M. (1992) Monitoring of seabirds in the Benguela upwelling systems: Can seabirds be used as indicators and predictors of change in the marine environment? *S. Afr. J. Mar. Sci.*, **12**, 959–74.

Alexander, V. (1981) Ice-biota interactions: an overview. In: *The Eastern Bering Sea Shelf: Oceanography and Resources*, Volume 2 (eds D.W. Hood and J.A. Calder), pp. 757–61. University of Washington Press, Seattle.

Alton, M.S. (1973) Bering Sea benthos as a food resource for demersal fish populations. In: *Oceanography of the Bering Sea* (eds D.W. Hood and E.J. Kelley), *Univ. Alaska Occ. Publ.*, **2**, 257–78.

Alverson, D.L. (1967) Distribution and behaviour of Pacific hake as related to design of fishing strategy and harvest rationale. *Conference on Fish Behaviour in Relation to Fishing Techniques and Tactics*. Bergen. Paper E/11.

Alverson, D.L. (1992) A review of commercial fisheries and the Steller sea lion (*Eumetopia jubatus*): The conflict arena. *Rev. Aquat. Sci.*, **6** (3,4), 203–56.

Alverson, D.L., and Larkin, P. (1994) Fisheries: Fisheries science and management, Century 21. In: *The State of the World's Fisheries Resources* (ed. C.W. Voigtlander), pp. 150–67. Oxford & J.B.H. Publishing Co. Pvt. Ltd, New Delhi.

Alverson, D.L., Freeberg, M.H., Pope, J.G. and Murawski, S.A. (1994) *Bycatch discards in world fisheries: Quantities, impacts and the philosophic bases for their management.* Report from Natural Resources Consultants Inc., Seattle, 91 pp.

Andersen, K.P. and Ursin, E. (1977) A multispecies extension to the Beverton and Holt theory of fishing with accounts of phosphorus circulation and primary production. *Medd. Danm. Fisk.-og Havunders*, N.S., **7**, 319–435.

Andersen, V., Sardou, J. and Nival, P. (1992) The dual migrations and vertical distributions of zooplankton and micronekton in the Northwestern Mediterranean Sea. *J. Plankton Res.*, **14**(8), 1129–69.

Anderson, G.C. (1969) Subsurface chlorophyll maximum in the north-east Pacific Ocean. *Liminology and Oceanography*, **14**(3), 386–91.

Anderson, I. (1993) Aliens slip through international safety net'. *New Scientist*, 3 July 1993, p. 5.

Anderson, J.T. (1988) A review of size dependent survival during pre-recruit stages of fishes in relation to recruitment. *J. Northw. Atl. Fish. Sci.*, **8**, 55–66.

Anderson, L.G. (1986) *The Economics of Fisheries Management.* The John Hopkins University Press, Baltimore.

Annala, J.H., Sullivan, K.J. and Hore, A. (1989) Management of multispecies fisheries in New Zealand by individual transferrable quotas. *ICES MSM Symposium*, Paper 41.

Anon. (1991) A review of world salmon culture. *US Mar. Fish. Review*, **53**(1), 27–40.

Anon. (1993) Status of fishery resources off the Northeastern United States for 1993. *NOAA Techn. Memo.* NMFS – F/NEC-101.

Appollov. A.B. (1992) Biocoenoses of bottom fauna in the eastern part of the Baltic Sea. *Biologiya Morya*, **5–6**, 62–69.

Armstrong, M.J., Chapman, P., Dudley, S.F.J., Hampton, I. and Malan, P.E. (1991) Occurrence and population structure of pilchard *Sardinops ocellatus*, round herring, *Etrumeus whiteheadi* and anchovy *Engraulis capensis* off the east coast of South Africa. *S. Afr. J. Mar. Sci.*, **11**, 227–49.

Arnason, R. (1993) Ocean fisheries management: recent international developments. *Marine Policy*, **17** (5), 334–9.

Arntz, W.E. (1979) Predation by demersal fish and its impact on the dynamics of macro-benthos. In: *Marine Benthic Dynamics* (eds K.R. Tenore and B.C. Coull), pp. 121–49. University of South Caroline Press, Columbia.

Arntz, W.E. and Fahrbach, E. (1991) *El Niño. Klimaexperiment der Natur physikalische Ursachen und biologische Folgen.* Birkhauser Verlag, Basel.

Arntz, W.E., Brey, T., Tarazona, J. and Robles, A. (1987) Changes in the structure of a shallow sandy-beach community in Peru during an El Niño event. *S. Afr. J. Mar. Sci.*, **5**, 645–58.

Asada, Y., Hirasawa, Y., & Nagasaki, F. (1983) Fishery Management in Japan. *FAO Fisheries Tech. Paper* **238**, 26 pp.

Atema, J. (1980) Chemical senses, chemical signals and feeding behaviour in fishes. In: *Fish Behaviour and its Use in the Capture and Culture of Fishes. ICLARM Conf. Proc.*, **5**, 57–101.

Bader, R.G. (1954) The role of organic matter in determining the distribution of polychaetes in marine sediments. *J. Mar. Res.*, **13**(1), 32–47.

Bainbridge, V. and Corlett, J. (1968). The zooplankton of the NORWESTLANT surveys. *ICNAF Spec. Publ.*, **7**, 101–122.

Bainbridge, V. and McKay, B.J. (1968) The feeding of cod and redfish larvae. *ICNAF Spec. Publ.*, **7**, 187–205.

Baker, J.T. and Murphy, P.T. (1991) Tropical marine biodiversity. *International Marine Bio-technology Conference, Baltimore*, 13–16.

Bakken, E. (1983) Recent history of Atlanto–Scandian herring stock. *FAO Fish. Rep.*, **291**(2), 521–36.

Baldwin, J.R. and Louvorn, J.R. (1994) Expansion of seagrass habitat by the exotic *Zostera japonica*, and its use by dabbling ducks and brant in Boundary Bay, British Columbia. *Mar. Ecol. Prog. Ser.*, **103**, 119–27.

Barber, R.T and Smith, R.L. (1981) Coastal upwelling ecosystem. In: *Analysis of Marine Ecosystems* (ed A.R. Longhurst), pp. 31–68. Academic Press, London.

Barnes, R.S.K. and Mann, K.H. (1991) *Fundamentals of Aquatic Ecology.* Blackwell Science, Oxford.

Bartell, S.M., Gardner, R.H. and O'Neil, R.V. (1992) *Ecological Risk Estimation.* Lewis Publications, Boca Raton.

Bax, N. (1991) A comparison of the fish biomass flow to fish fisheries and mammals in six marine ecosystems. *ICES Mar. Sci. Symp*, **193**, 217–24.

Bax, N.J. and Laevastu, T. (1989) Biomass potential of large marine ecosystems, a system approach. In: *Large Marine Ecosystems* (eds K. Sherman, L.M. Alexander and B.D. Gold), pp. 188–205. American Association for the Advancement of Science, Washington, DC.

Bax, N.J., Mehl, S., Godø, O.R. and Sunnanå, K. (1989) Transparent multispecies analysis. An exploration of fisheries and survey data off the Norwegian coast and Barents Sea. *ICES MSM Symposium*, Paper 18.

Bercy, C. and Bordeau, B. (1985) Physiological and ethological reactions of fish to low frequency noise radiated by sounders and sonars. *International Symposium on Fishery Acoustics*, Seattle.

Bergstad, O.A., Jogrenson, T. and Dragesund, O. (1985) Life history features and ecology of the gadoid resources of the Barents Sea. Paper presented at the Gadoid Workshop, Seattle, Washington, June 1985.

Beukema, J.J. (1982) Annual variation in reproductive success and biomass of the major macrozoobenthic species living in a tidal flat area of the Wadden Sea. *Netherlands J. Sea Res.*, **16**, 37–45.

Bevan, D.E., Brocher, C. and Alverson, D.L. (1992) Regional fisheries management councils in need of reform. *Fisheries*, **17**(5), 35–7.

Beverton, R.J.H. and Holt, S.J. (1957) On the dynamics of exploited fish populations. *Fishery Invest.*, London, Ser. **2**(19), 533 pp.

Bigford, T.E. (1991) Sea-level rise, nearshore fisheries and the fishing industry. *Coastal Management*, **19**, 417–37.

Bjøordal, A. (1989) Recent developments in longline fishing – catching performance and conservation aspects. *Proceedings of the World Symposium on Fishing Gear and Fishing Vessel Design*, St. John's, November 1988, pp. 19–24.

Bjørdal, A. and Laevastu, T. (1990) Effect of trawling and long lining on the yield and biomass of cod-stock – numerically simulated. ICES, C.M. 1990/G:32, 31 pp.

Bjørge, A., Christensen, I. and Oritsland, T. (1981) Current problems and research related to interactions between marine mammals and fisheries in Norwegian coastal and adjacent waters. ICES, C.M. 1981/N:18.

Blackburn, M. (1981) Low latitude gyral regions. In: *Analysis of Marine Ecosystems* (ed A.R. Longhurst), pp. 3–29. Academic Press, London.

Blacker, R.W. (1965) Recent changes in the benthos of the west Spitsbergen fishing ground. *ICNAF Spec. Publ.*, **6**, 791–4.

Blanchet, J., Rodier, M. and LeBouteiller, A. (1992) Effect of El Niño Southern oscillation events on the distribution and abundance of phytoplankton in the western Tropical Pacific Ocean along 165 degrees E. *J. Plankton Res.*, **14**(1), 137–56.

Blanton, J.O. (1986) Coastal frontal zones as barriers to offshore fluxes of contaminants. *Rapp. P.-v. Réun. Cons. Int. Explor. Mer*, **186**, 18–30.

Blaxter, J.H.S. (1965) Effect of change of light intensity on fish. *ICNAF Spec. Publ.*, **6**, 647–61.

Blaxter, J.H.S. (1969) Swimming speeds of fish. *FAO Fisheries Rpt*, **62**(2), 69–100.

Blaxter, J.H.S. (1980) Vision and feeding of fishes. In: *Fish Behaviour and its Use in the Capture and Culture of Fishes. ICLARM Conf. Proc.*, **5**, 32–56.

Blaxter, J.H.S. (1988) Sensory performance, behaviour and ecology of fish. In: *Sensory Biology of Aquatic Animals* (ed J. Atema), pp. 203–232. Springer-Verlag, New York and Berlin.

Blaxter, J.H.S. and Hunter, J.R. (1982) The biology of clupeoid fishes. *Adv. Mar. Biol.*, **20**, 1–199.

Blix, A.S. (1991) On marine mammals and fish of the Northwest Atlantic Ocean. *Rep. Benguela Ecol. Prog. S. Afr.*, **22**, 39–40.

Bonner, W.N. and Laws, R.M. (1964) Seals and sealing. In: *Antarctic Research* (eds R. Prestly, R.J. Adie and G. de O. Robin), pp. 63–90. Butterworth, London.

Bragg, J. (1991) El Niño. Uncertainty in the ocean ecosystem. *Pacific Fishing*, **12**(7), 47–53.

Branch, G.M., Barkai, A. and Hockey, P.A.R. (1987) Biological interactions: Causes or effects of variability in the Benguela ecosystem? *S. Afr. Mar. Sci.*, **5**, 425–45.

Brander, K.M. and Dickson, R.R. (1984) An investigation of the low level of fish production in the Irish Sea. *Rapp. P.-v. Réun. Cons. Int. Explor. Mer.* **183**, 234–42.

von Brandt, A. (1959) Classification of fishing gear. In: *Modern Fishing Gear of the World* (ed H. Kristjonsson), pp. 274–96. Fishing News Books, Oxford.

von Brandt, A. (1984) *Fish catching methods of the world.* Fishing News Books, Oxford.

Brown, P.C., Painting, S.J. and Cochrane, K.L. (1991) Estimates of phytoplankton and bacterial biomass and production in the northern and southern Benguela ecosystems. *S. Afr. J. Mar. Sci.*, **11**, 537–64.

Butterworth, D.S. (1991) A brief review of population trends for the South African (Cape) fur seal, and the potential impacts of a continued increase in population. *Rep. Benguela Ecol. Prog. S. Afr.*, **22**, 24–5.

Butterworth, D.S., Duffy, D.C., Best, P.B. and Bergh, M.O. (1988) On the scientific basis for reducing the South African seal population. *S. Afr. J. Sci.*, **84**, 179–88.

Butterworth, D.S., Punt, A.E., Bergh, M.O. and Borchers, D.L. (1992) Assessment and management of South African marine resources during the period of the Benguela ecology programme: key lessons and future directions. *S. Afr. J. Mar. Sci.*, **12**, 989–1004.

Caddy, J.F. (1991) Towards a comparative evaluation of human impacts on fishery ecosystems of enclosed and semienclosed seas. *Reviews in Fisheries Science*, **1**(1), 57–95.

Carr, A.H. and Robinson, W. (1992) Survival of juvenile cod and American plaice in the Northwestern Atlantic trawl fishery. *ICES Symposium on Fish Behaviour in Relation to Fishing Operations*, Bergen, Paper No. 7.

Christensen, V. (1983) Predation by sand eel on herring larvae. ICES, C.M. 1983/L:27.

Christy, F.T. and Scott, A. (1965) *The Common Wealth in Ocean Fisheries*. Johns Hopkins University Press, Baltimore.

Colebrook, J.M. (1985) Continuous plankton records: overwintering and annual fluctuations in the abundance of zooplankton. *Mar. Biol.*, **84**, 261–5.

Colijn, F. (1991) Changes in plankton communities: when, where, and why? ICES Variability Symposium. Paper 13.

Conan, G. (1982) The long-term effects of the *Amoco Cadiz* oil spill. *Phil. Trans. R. Soc. London*, **B297**, 323–33.

Connell, D.W. and Muller, G.J. (1980a) Petroleum hydrocarbons in aquatic ecosystem – Behaviour and effects of sublethal concentrations: Part 2. *CRC Critical Reviews in Environmental Control*, **11**(2), 105–161.

Connell, D.W. and Miller, G.J. (1980b) Petroleum hydrocarbons in aquatic ecosystems – Behaviour and effects of sublethal concentrations: Part 1. *CRC Critical Reviews in Environmental Control*, **11**(1), 37–104.

Constanza, R. (ed) (1991) *Ecological Economics: The Science and Management of Sustainability*. Columbia Univ. Press, New York.

Cooney, T.R. (1981) Bering Sea zooplankton and micronekton communities with emphasis on annual production. In: *The Eastern Bering Sea Shelf. Oceanography and Resources*, Volume 2 (eds D.W. Hood and J.A. Calder), pp. 947–74. University of Washington Press, Seattle.

Corlett, J. (1965) Wind, currents, plankton and the year-class strength of cod in the western Barents Sea. *ICNAF Spec. Publ.*, **6**, 373–8.

Corten, A. (1983) Predation on herring larvae by the copepod *Candacia armata*. ICES C.M. 1983/H:20.

Corten, A. and van der Kamp, G. (1991) Natural changes in pelagic fish stocks of the North Sea. *ICES Variability Symposium*. Paper 27.

Cramer, S. and Daan, N. (1986) Consumption of benthos by North Sea cod and haddock in 1981. ICES C.M. 1986/G:56.

Crawford, R.J.M., Shannon, L.V. and Pollock, D.E. (1987) The Benguela ecosystem. Part IV. The major fish and invertebrate resources. *Oceanogr. Mar. Biol. Ann. Rev.*, **25**, 353–505.

Crawford, R.J.M., Ryan, P.G. and Williams, A.J. (1991) Seabird consumption and production in the Benguela and western Agulhas ecosystems. *S. Afr. J. Mar. Sci.*, **11**, 357–75.

Crowley, R.W. and Palsson, H. (1991) Rights based fisheries management in Canada. *Mar. Res. Econ.*, **7**, 1–21.

Cushing, D.H. (1968) *Fisheries Biology, A Study of Population Dynamics*. University of Wisconsin Press, Madison.

Daan, N. (1976) Some preliminary investigations into predation on fish eggs and larvae in the southern North Sea. ICES, C.M. 1976/L:15.

Daan, N. (1980) Review of replacement of depleted stocks by other species and the mechanisms underlying such replacement. *Rapp. P.-V. Cons. Int. Explor. Mer.* **177**, 405–21.

Daan, N. (1987) Multispecies versus single-species assessment of North Sea fish stocks. *Can. J. Fish. Aquat. Sci.*, **44**, 360–70.

Daan, N., Rijnsdorp, A.D. and van Overbeeke, G.R. (1985) Predation by North Sea herring *Clupea harengus* on eggs of plaice *Pleuronectes platessa* and cod *Gadus morhua. Trans. Am. Fish. Soc.*, **114**, 499–506.

Daniulyte, G. and Malukina, G. (1969) The reaction of some fishes in an electric field. *FAO Fisheries Rpt.*, **62**(2), 775–80.

Davenport, J. (1982) Oil and planktonic ecosystems. *Trans. R. Soc. London*, **B297**, 369–84.

David, P.M. (1967) Illustrations of oceanic neuston. In: *Aspects of Marine Zoology* ed N.B. Marshall), *Symp. Zool. Soc. London*, **19**, 211–13.

DeBlois, E.M. and Leggett, W.C. (1993) Impact of amphipod predation on the benthic eggs of marine fish: an analysis of *Calliopius laeviusculus* bioenergetic demands and predation on the eggs of a beach spawning osmeriid (*Mallotus villosus*). *Mar. Ecol. Prog. Ser.*, **93**, 205–16.

Demel, K. and Kulikowsky, J. (1965) *Fishing Oceanography.* Polish Scientific Publishers, Warszawa.

Dementeva, T.F. and Mankovich, E.M. (1966) Fluctuations in the growth of cod in the Barents Sea according to external conditions. In: *Biologicheskie i okeanograficheskie usloviya obrazovaniya promyslovykh skoplenii ryb.*, pp. 383–95. Pishevaya Promyshlennost, Moskva.

Dethlefsen, V. (1984) Diseases in North Sea fishes. *Helgolander Meeresunters.*, **37**, 353–74.

Dethlefsen, V. and von Westernhagen, H. (1983) Oxygen deficiency and effects on bottom fauna in the eastern German Bight 1982. *Meeresforsch.*, **30**, 42–53.

Dickie, L.M. (1979a) Perspectives on fisheries biology and implications for management. *J. Fish. Res. Board Can.*, **36**, 838–44.

Dickie, L.M. (1979b) Predatory–prey models for fisheries management. In: *Predator–Prey Systems in Fisheries Management* (ed H. Clepper), pp. 281–92. Sport Fishing Institute, Washington DC.

Dickson, J.O. (1993) Deep sea fish aggregating devices for commercial fisheries in the Philippines. *INFOFISH Intern.* **4**, 51–6.

Doubleday, W.G. (1980) Coping with variability in fisheries. *FAO Fish. Rep.* **236**, 131–9.

Downs, A. (2967) *Inside Bureaucracy.* Little, Brown & Co, Boston.

Duarte, C.M., Cebrian, J. and Marba, N. (1992) Uncertainty of detecting sea changes. *Nature*, **356**, 190.

Duval, W.S. and Fink, R.P. (1981) The sublethal effects of water-soluble hydrocarbons on the physiology and behaviour of selected marine fauna. MS report EE-16 to Environment Canada, 86 pp.

Dye, A.H. (1992) Experimental studies of succession and stability in rocky intertidal communities subject to artisanal shellfish gathering. *Netherlands J. Sea Res.*, **30**, 209–17.

Einarson, H. (1955) Nokkur ord um samanorraenu sildarrannsoknirnar a 'Dana', 'G.O. Sars', og 'Aegir'. *Aegir*, **48**(12), 179–81.

Eleftheriou, A. and Robertson, M.R. (1992) The effects of experimental scallop dredging on the fauna and physical environment of a shallow sandy community. *Netherlands J. Sea Res.*, **30**, 289–99.

Elmgren, R. (1984) Trophic dynamics in the enclosed, brackish Baltic Sea. *Rapp. P.-v. Réun. Cons. Int. Explor. Mer.***188**, 152–69.

Elmgren, R. (1989) Man's impact on the ecosystem of the Baltic Sea: Energy flows today and at the turn of the century. *Ambio*, **18**(6), 326–32.

Endal, A. (1980) Fuel saving potential in Norwegian fisheries. ICES C.M. 1980/B:14.

Engås, A. and Ona, E. (1990) Day and night fish distribution pattern in the net mouth area of the Norwegian bottom-sampling trawl. *Rapp. P.-v. Réun. Cons. Int. Explor. Mer.* **189**, 123–7.

Engås, A., Soldal, A.V. and Ovredal, J.T. (1991) Avoidance reactions of ultrasonic tagged cod during bottom trawling in shallow water. ICES, C.M. 1991/B:41.

Enger, P.S., Karlser, H.E. and Knudsen, F.R. (1992) Perception and reaction of fish to infrasound. *ICES Symposium on Fish Behaviour in Relation to Fishing Operations*, Bergen, Paper No. 8

Estep, K.W. (1989) Creating and using taxonomic keys with Hyper-Card. *Fisken Hav.*, **1989**(1), 1–13 & App.

Fager, E.W. (1963) Communities of organisms. In: *The Sea*, Volume 2 (ed M.N. Hill), pp. 415–437. Interscience Publishers, New York.

FAO (1986) Technical consultation on open sea shellfish culture in association with artificial reefs. *FAO Fish Rep.*, **357**, 1–175.

FAO (1992) Marine fisheries and the Law of the Sea: a decade of change. *FAO Fisheries Circular*, **853**.

Favorite, F. and Laevastu, T. (1979) A study of the ocean migrations of sockeye salmon and estimation of the carrying-capacity of the North Pacific Ocean using a dynamical numerical salmon ecosystem model (NOPASA). *NWAFC Processed Rpt.*, 79–16.

Favorite, F. and Laevastu, T. (1981) Finfish and the environment. In: *The Eastern Bering Sea Shelf: Oceanography and Resources*, Volume 1 (eds D.W. Hood and J.A. Calder), pp. 597–610. University of Washington Press, Seattle.

Feder, H.M. and Jewett, S.C. (1981) Feeding interactions in the Eastern Bering Sea with emphasis on the benthos. In: *The Eastern Bering Sea Shelf: Oceanography and Resources*, Volume 2 (eds D.W. Hood and J.A. Calder), pp. 1229–61. University Washington Press, Seattle.

Fernø, A. (1993) Advances in understanding of basic behaviour; consequences for fish capture studies. *ICES Mar. Sci. Symp.*, **196**, 5–11.

Fiscus, C.H. (1978) Marine mammal–salmanoid interactions: A review Paper presented at a Symposium on Salmanoid Ecosystems of the North Pacific Ocean, Newport, Oregon, May 1978.

Fiscus, C.H. and Mercer, R.W. (1982) Squids taken in surface gill-nets in the North Pacific Ocean by the Pacific Salmon Investigation Program 1955–1972. *NOAA Techn. Memo NMFS F/NWC-28.*

Flaaten, O. (1988) *The Economics of Multispecies Harvesting.* Springer-Verlag, Berlin.

Fleming, R.H. and Laevastu, T. (1956) The influence of hydrographical conditions on the behaviour of fish. *FAO Fish. Bull.*, **9**(4), 181–96.

Fogarty, M.J., Rosenberg, A.A. and Sissenwine, M.P. (1992) Fisheries risk assessment, sources of uncertainty. *Environ. Sci. Technol.*, **26**(3), 441–7.

Folsom, W., Altman, D., Manuar, A., Nielsen, F., Revord, T., Sandborn, E. and Wildman, M. (1992) World Salmon Culture. *NOAA Techn. Memo. NMFS- f/SPO-3.*

Fransz, H.G. and Gieskes, W.W.C. (1984) The unbalance of phytoplankton and copepods in the North Sea. *Rapp. P.-v. Réun. Cons. Int. Explor. Mer.* **183**, 218–25.

Franz, D.R. (1976) Distribution and abundance of inshore populations of the surf clam *Spisula solidisima*. *Am. Soc. Limnol. Oceanogr. Spec. Symp.*, **2**, 404–13.

Fraser, J.H. (1952) Hydrobiological correlations at the entrance to the North Sea in 1947. *Rapp. P.-v. Réun. Cons. Int. Explor. Mer.* **131**, 38–43.

Fraser, J.H. (1961) The oceanic and bathypelagic plankton of the north-east Atlantic. *Res. Rep. Scot.*, **4**.

Furness, R.W. (1982) Competition between fisheries and seabird communities. *Adv. Mar. Biol.*, **20**, 225–307.

Furness, R.W. (1989) Estimation of fish consumption by seabirds and the influence of changes in fish stock size on seabird predation. *ICES, MSM Symposium*. Paper No. 36.

Garrod, D.J. (1987) The scientific essentials of fisheries management and regulations. *Min. Agric. Food Fish., Lowestoft Lab. Leaf.*, **60**.

Gauldie, R.W. (1991) Taking stock of genetic concepts in fisheries management. *Can. J. Fish. Aquat. Sci.*, **48**, 722–31.

Gearing, J.N. and Gearing, P.J. (1983) Suspended load and solubility affect sedimentation of petroleum hydrocarbons in controlled estuarine ecosystems. *Can. J. Fish. Aquat. Sci.*, **40** (Suppl. 2), 54–62.

Gillbricht, M. (1988) Phytoplankton and nutrients in the Helgoland region. *Helgolander Meeresunters*, **42**, 435–67.

Gislason, H. (1983) A preliminary estimate of the yearly intake of fish by saithe in the North Sea. ICES CM 1983/G:52, 10 pp.

Glemarec, M. and Hussenot, E. (1982) A three-year ecological survey in Benoit and Wrac'h abers following the *Amoco Cadiz* oil spill. *Netherlands J. Sea Res.*, **16**, 483–90.

Godø, R. (1985) Dispersion and mingling of cod from various nursery and feeding areas along the Norwegian coast and in the Barents Sea. Paper presented at the Gadoid Workshop, Seattle, Washington, June 1985.

Godø, R. and Moksness, E. (1985) Growth and maturation of Norwegian coastal cod and Arcto–Norwegian cod under different conditions. Paper presented at the Gadoid workshop, Seattle, Washington, June 1985.

Goering, J.J. and Iverson, R.L. (1981) Phytoplankton distribution on the southern Bering Sea shelf. In: *The Eastern Bering Sea Shelf: Oceanography and Resources*, Volume 2 (eds D.W. Hood and J.A. Calder), pp. 933–946. University of Washington Press, Seattle.

Gonzalez, J.M. and Suttle, C.A. (1993) Grazing by marine nanoflagellates on viruses and virus-sized particles: ingestion and digestion. *Mar. Ecol. Prog. Ser.*, **94**, 1–10.

Gordon, H.S. (1954) The economic theory of a common property resource: the fishery. *J. Political Economy*, **62**, 124–42.

Gore, A. (1992) *Earth in the Balance: Ecology and the Human Spirit*. Houghton Miffin, Co., New York.

Gorman, T. (1994) Longlining, jigging and trolling. *Fish. Boat World*, **5**(11), 9–13.

Gorschkova, T. (1938) Organischer Stoff in der Sedimenten des Motovskij Busens. *Trans. Inst. Mar. Fish.*, **5**, 71–84.

Gotceitas, V. and Brown, J.A. (1993) Risk of predation to fish larvae in the presence of alternate prey: effect of prey size and number. *Mar. Ecol. Prog. Ser.*, **98**, 215–22.

Graham, M. (1943) *The Fish Gate*. Faber and Faber, London.

Grahl-Nielsen, O., Neppelberg, T., Palmork, K.H., Westrheim, K. and Wilhelmsen, S. (1976) The *Drupa* oil spill, investigation concerning oil, water and fish. ICES, C.M. 1976/E:34.

Grebmeier, J.M. (1992) Benthic processes on the shallow continental shelf. In: *Results of the Third Joint US–USSR Bering and Chukshe Sea Expedition* (ed J.F. Turner), pp. 243–251. US Fish and Wildlife Service, Washington DC.

Greenwood, P.H. (1992) Are the major fish faunas well-known? *Netherlands J. Zool.*, **42**(2–3), 131–8.

Greer Walker, M., Jones, Harden, F.R. and Arnold, G.P. (1978) The movement of plaice (*Pleuronectes platessa* L.) tracked in the open sea. *J. Cons Int. Explor. Mer.* **38**(1), 58–86.

Grey, J.S. (1982) Effects of pollutants on marine ecosystems. *Netherlands J. Sea Res.*, **16**, 424–43.

Grigg, R.W. (1994) Effects of sewage discharge, fishing pressure and habitat complexity on coral ecosystems and reef fishes in Hawaii. *Mar. Col. Prog. Ser.*, **103**, 25–43.

de Groot, S.J. (1974) The impact of bottom trawling on benthic fauna of the North Sea. *Ocean Man.*, **9** 177– 90.

Grosslein, M.D., Langton, R.W. and Sissenwine, M.P. (1980) Recent fluctuations in pelagic fish stocks of the Northwest Atlantic, Georges Bank region, in relationship to species interaction. *Rapp. P.-v. Réun. Cons. Int. Explor. Mer*, **177**, 374–404.

Gulland, J.A. (1987a) The impact of seals on fisheries. *Marine Policy*, July 1987, 196–204.

Gulland, J.A. (1987b) The effects of fishing on community structure. *S. Afr. J. Mar. Sci.*, **5**, 839–49.

Gulland, J.A. (1987c) The management of North Sea fisheries. Looking towards the 21st century. *Marine Policy*, October 1987, 259–72.

Gunderson, D.R. (1993) *Surveys of Fisheries Resources*. John Wiley & Sons, New York.

Hagmeier, A. (1951) Die Nahrung der Meerestiere. Die Bodenfauna. *Handb. Seefisch. Nordeurop.*, **1**(5b).

Hamre, J. (1988) Some aspects of the interrelation between the herring in the Norwegian Sea and the stocks of capelin and cod in the Barents Sea. ICES C.M. 1988/H:42.

Hannesson, R. (1993) *Bioeconomic Analysis of Fisheries*. Fishing News Books, Oxford.

Hara, J. (1984) *Fish Chemoreception*. Chapman and Hall, London and New York.

Harden Jones, F.R. (1968) *Fish Migration*. Edward Arnold Ltd, London.

Harden Jones, F.R. (1974) Objectives and problems related to research into fish behaviour. In: *Fisheries Research* (ed. F.R. Harden Jones), pp. 261–75. John Wiley & sons, New York.

Hardgrave, B.T. (1991) Impacts of man's activities on aquatic systems. In: *Fundamentals of Aquatic Ecology* (eds R.S.K. Barnes and K.H. Mann), pp. 245–64. Blackwell Science, Oxford.

Harwood, J. (1984) Seals and fisheries. *Mar. Poll. Bull.*, **15**(12), 426–9.

Harwood, J. (1987) Competition between seals and fisheries. *Sci. Prog., Oxf.*, **71**, 429–37.

Harwood, J.W. and Cushing, D.H. (1972) Spatial distribution and ecology of pelagic fish. In: *Spatial Pattern in Plankton Communities* (ed J.H. Steele), pp. 355–83. Plenum Press, New York.

Hashimoto, T. and Maniwa, Y. (1971) Research on the luring of fish schools by underwater sounds. In: *Modern Fishing Gear of the Word 3* (ed H. Kristjonsson), pp. 501–3. Fishing News Books, Oxford.

Hawkins, A.D. (1993) Underwater sound and fish behaviour. In: *Behaviour of Teleost Fishes* (ed T.J. Pitcher), pp. 129–69. Chapman & Hall.

He, P. (1993) Swimming speeds of marine fish in relation to fishing gear. *ICES Mar. Sci. Symp*, **196**, 183–9.

He, P. and Xu, G. (1992) Underwater observations of the behaviour of cod towards different fishing bait. *ICES Symposium on Fish Behaviour in Relation to Fishing Operations*, Bergen. Paper No. 18.

Heinrich, A.K. (1962a) The life histories of plankton animals and seasonal cycles of plankton communities in the oceans. *J. Cons. Int. Explor. Mer.* **27**(1), 15–24.

Heinrich, A.K. (1962b) On the production of copepods in the Bering Sea. *Int. Revue. Ges. Hydro-biol.*, **47**(3), 465–9.

Heip, C. (1992) Benthic studies: summary and conclusions. *Mar. Ecol. Prog. Ser.*, **91**, 265– 8.

Hela, I. and Laevastu, T. (1962) *Fisheries Hydrography*. Fishing News Books, Oxford.

Helbing, E.W., Villafane, V., Ferrario, M. and Holm-Hansen, O. (1992) Impact of natural ultraviolet radiation on rates of photosynthesis and on specific marine phytoplankton species. *Mar. Ecol. Prog. Ser.*, **80**, 89–100.

Hey, E. (1992) A healthy North Sea ecosystem and healthy North Sea fishery: two sides of the same regulation: *Ocean Dev. Int. Law*, **23**, 217–38.

Hinds, L. (1992) Management and development problems. *Marine Policy*, September 1992, 394–403.

Hughes, R.N. (1991) Reefs. In: *Fundamentals of Aquatic Ecology* (eds R.S.K. Barnes and K.H. Mann), pp. 213–29. Blackwell Science, Oxford.

Hunt, G.L. Jr, Burgeson, B. and Sanger, G.A. (1981) Feeding ecology of seabirds of the Eastern Bering Sea. In: *The Eastern Bering Sea Shelf: Oceanography and Resources*, Volume 2. (Eds D.W. Hood and J.A. Calder), pp. 629–48. University of Washington Press, Seattle.

ICES (1984) Reports of the ICES Advisory Committee on Fishery Management, 1983. *ICES Coop. Res. Rep.*, 128.

ICES (1989) Report of the Multispecies Assessment Working Group. ICES, C.M. 1989/Assess:20.

ICES (1991) Report of the study groups on ecosystem effects of fishing activities. ICES, C.M. 1991/G:7.

ICES (1992) Report of the ICES Working Group on the Effects of Extraction of Marine Sediments on Fisheries. *ICES Coop. Res. Rep.*, 182.

ICES (1993) Report of the Working Group on Methods of Fish Stock Assessment. *ICES Coop. Res. Rep.*, 191.

Idelson, M.S. (1934) Materials for the quantitative evaluation of the bottom fauna of the Barents, White and Kara Seas. No. 7: Distribution of the benthos biomass in the southern Barents Sea. The influence of different factors on the density of the sea bottom population. *Trans. Oceanogr. Inst. Moscow*, **3**(4).

Inoue, Y., Matsushita, Y. and Arimoto, T. (1993) The reaction behaviour of walleye pollock (*Theraga chalcogramma*) in a deep/low-temperature trawl fishing ground. *ICES Mar. Sci. Symp*, **196**, 77–9.

Jakobsen, T. and Nedreaas, K. (1985) A model for simulating the effects on management advice of different strategies for sampling of demersal fish in Norway. Paper presented at Gadoid Workshops, Seattle, June 1985.

Jakobsen, T. and Olsen S. (1985) Variation in rates of migration of saithe from Norwegian waters to Icelandic and other areas. Paper presented at the Gadoid Workshop, Seattle, June 1985.

Jakobsson, J. (1978) The North Icelandic herring fishery and environmental conditions 1960–1968. Paper 30, *ICES Symposium on the Biological Basis of Pelagic Fish Stock Management*. Mimeo.

Jakobsson, J. (1991) Recent variability in fisheries of the North Atlantic. *ICES Variability Symposium*. 21.

Jensen, P.B. (1919) Valuation of the Limfjord. I. Studies on the fish food in the Limfjord, 1909–1917: its quantity, variations and annual production. *Rep. Danish Biol. Sta.*, **26**, 1–44.

Jewett, S.C. and Feder, H.M. (1976) Distribution and abundance of some epibenthic invertebrates of the northeast Gulf of Alaska with notes on feeding biology. *Science in Alaska, Proc. 27th Alaska Sci. Conf. 1976*, **2**, 377–431.

Johnston, R. (1977) What North Sea oil might cost fisheries. *Rapp. P.-v. Réun Cons. Int. Explor. Mer*. **171**, 212–23.

Jones, R. (1984) Some observations on energy transfer through the North Sea and Georges Bank food webs. *Rapp P.-v. Reun. Cons. Int. Rxplor. Mer*, **183**, 204–17.

Josefson, A.B. (1990) Increase of benthic biomass in the Skagerrak–Kattegat during the 1970s and 1980s – effects of organic enrichment? *Mar. Col. Prog. Ser.*, **66**, 117–30.

Kalejs, M. and Ojaveer, E. (1989) Long-term fluctuations in environmental conditions and fish stocks in the Baltic. *Rapp. P.-v. Réun. Cons. Int. Explor. Mer*, **190**, 153–8.

Kallio-Nyberg, I. and Ikonen, E. (1992) Migration patterns of two salmon stocks in the Baltic Sea. *ICES J. Mar Sci.*, **49**, 191–8.

Karohji, K. (1972) Regional distribution of phytoplankton in the Bering Sea and western and northern subarctic regions of the North Pacific Ocean in summer. In: *Biological Oceanography of the Northern North Pacific Ocean*, pp. 99–115. Idemista Shoten, Tokyo.

Kato, M. (1969) Longlining for swordfish in the eastern Pacific. *Comm. Fish. Rev.*, **31**(4), 30–2.

Kepler, A.K., Kepler, C.B. and Ellis, D.H. (1992) Ecological studies of Caroline Atoll, Republic of Kiribati, south-central Pacific Ocean. In: *Results of the First Joint US–USSR Central Pacific Expedition (BERPAC)*, Autumn 1988, 5–137. US Fish and Wildlife Service, Washington DC.

Ketchum, B.H. (1957) The effect of the biological system on the transport of elements in the sea. *US Nat. Acad. Sciences, Nat. Res. Couns.*, Publ. 551, 53–9.

Khaldinova, N.A. (1966) Fluctuations in the growth of haddock in the Barents Sea with the food supply. In: *Biologicheskie i Okeanograficheskie Usloviya Obrazovaniya Promyslovykh Skoplenii Ryb*, pp. 401–15. Pishehvaya Promyshlennost, Moskva.

Kikuchi, T. and Tsajita, T. (1977) Some relations between spawning of pelagic fishes in economic importance and the oceanographic structure in the northwestern North Pacific Ocean. In: *Fisheries Biological Production in the Subarctic Pacific Region*, pp. 397–438. Hokkaido University, Hakodate.

Konstantinov, K.G. (1966) Duration of existence of schools of cod, their size, composition and distribution on the fishing grounds of the Barents Sea. In *Biologicheskie i Okeanograficheskie Usloviya Obrazovaniya Promyslovykh Skoplenii Ryb*, pp. 223–35. Pishehvaya Promyshlennost, Moskva.

Korringa, P. (1967) Estuarine fisheries in Europe as affected by man's multiple activities. In: *Estuaries.* (ed G.H. Lauff), pp. 658–63. American Association for the Advancement of Science, Washington, DC.

Korringa, P. (1968) Biological consequences of marine pollution with special reference to the North Sea fisheries. *Helgolander wiss. Meeresunters.*, **17**, 126–40.

Krumholz, L.A. and Goldberg, E.D. (1957) Ecological factors involved in the uptake, accumulation, and loss of radionuclides by aquatic organisms. In: *The Effects of Atomic Radiation on Oceanography and Fisheries.* US Nat. Acad. Sci., Nat. Res. Council, Publication 551, 69–79.

Laevastu, T. (1966) Effects of ocean thermal structure on fish finding with sonar. *FiskDir. Skr. Ser. Havunders*, **15**, 202–209.

Laevastu, T. (1992) Interactions of size-selective fishing with variations in growth rates and effects on fish stocks. ICES, G.M. 1992/G:5.

Laevastu, T. (1993) *Marine Climate, Weather and Fisheries.* Fishing News Books, Oxford.

Laevastu, T. and Bax, N. (1991 Predation-controlled recruitment in the Bering Sea fish ecosystem. *ICES Mar. Sci. Symp.*, **193**, 147–52.

Laevastu, T. and Favorite, F. (1978a) Numerical evaluation of marine ecosystems. Part I. Deterministic Bulk Biomass Model (BBM). *Northwest & Alaska Fish. Cen., Seattle, Wash., Proc. Rep.*

Laevastu, T. and Favorite, F. (1978b) Numerical evaluation of marine ecosystem. Part II. Dynamical Numerical Marine Ecosystem Model (DYNUMES III) for evaluation of fishery resources. *Northwest and Alaska Fish. Cen., Seattle, Wash. Proc. Rep.*

Laevastu, T. and Favorite, F. (1981) Holistic simulation of marine ecosystem. In: *Analysis of Marine Ecosystems* (ed A.R. Longhurst), pp. 702–27. Academic Press, London.

Laevastu, T. and Favorite, F. (1988) *Fishing and Stock Fluctuations.* Fishing News Books, Oxford.

Laevastu, T. and Hayes, M.L. (1981) *Fisheries Oceanography and Ecology*. Fishing News Books, Oxford.

Laevastu, T. and Larkins, H.A. (1981) *Marine Fish Ecosystem*. Fishing News Books, Oxford.

Laevastu, T. and Marasco, R. (1982) Fluctuations of fish stocks and the consequences of the fluctuations of fishery and its management. *NOAA Techn. Memo NMFS F/NWC-27*.

Laevastu, T. and Marasco, R. (1985) Evaluation of the effects of oil development on commercial fisheries in the Eastern Bering Sea. *NWAFC Processed Rep. (Seattle)*.

Laevastu, T. and Rosa, H. (1962) Distribution and relative abundance of tunas in relation to their environment. *FAO World Sci. Meeting on the Biol. of Tunas and Related Species*, La Jolla, Paper No. 2.

Larkin, P.A. (1977) An epitaph for the concept of maximum sustained yield. *Trans. Am. Fish. Soc.*, **106**(1), 1–11.

Larkin, P.A. (1980) Objectives of management. In: *Fisheries Management*. (ed R.T. Lackey and L.A. Nielson), pp. 245–61. John Wiley & Sons, New York.

Larsen, R.B. and Isaksen, B. (1993) Size selectivity of rigid sorting grids in bottom trawls for Atlantic cod (*Gadus morhua*) and haddock (*Melanogrammus aeglefinus*). *ICES Mar. Sci. Symp.*, **196**, 12–16.

Larson, U.R., Elmgren, R. and Wulff, F. (1985) Eutrophication and the Baltic Sea: causes and consequences. *Ambio*, **14**(1), 9–14.

LeBrasseur, R.J. (1972) Utilization of herbivore zooplankton by maturing salmon. In: *Biological Oceanography of the Northern North Pacific Ocean* (ed A.Y. Takonouti), pp. 581–8. Idemitsu Shoten, Tokyo.

Lee, A. (1978) Effects of man on the fish resources of the North Sea. *Rapp. P.-v. Réun. Cons. Int. Explor. Mer*, **173**, 231–40.

Lewis, J.B. (1981) Coral reef ecosystems. In: *Analysis of Marine Ecosystems* (ed A.R. Longhurst), pp. 127–58. Academic Press, London.

Lie, U. (1965) Quantities of zooplankton and propogation of *Calanus finmarchicus* at permanent stations on the Norwegian coast and at Spitsbergen, 1959–1962. *FiskDir. Sk., Ser. Havunders.*, **13**(8), 5–19.

Lie, U. (1969) Standing crop of benthos infauna in Puget Sound and off the coast of Washington. *J. Fish. Res. Bd Canada*, **26**, 55–62.

Livingston, P.A. (1989) Key fish species, northern fur seals *Callorhinus ursinus*, and fisheries interactions involving walleye pollack *Theraga chalcogramma* in the Eastern Bering Sea. *J. Fish. Biol.*, **35**(Suppl. A), 179–86.

Livingston, P.A. (ed) (1991) Groundfish food habitats and predation on important prey species in the Eastern Bering Sea from 1984 to 1986. *US Dept. Commer. NOAA Tech. Memo NMFS F/NWC-206*.

Lockyer, E. (1993) Dead orca tells grim tale for sea lion. *Alaska Fishermans Journal*, **16**(2), 16.

Løkkeborg, S., Bjordal, A. and Fernø, A. (1989) Responses of cod (*Gadus morhua*) and haddock (*Melanogrammus aeglefinus*) to baited hooks in the natural environment. *Can J. Fish. Aquat. Sci.*, **46**, 1478–83.

Løkkeborg, S. and Soldal, A.V. (1993) The influence of seismic exploration with airguns on cod (*Gadus morhua*) behaviour and catch rates. *ICES Mar. Sci. Symp*, **169**, 62–7.

Longhurst, A.R. (1957) Density of marine benthic communities off West Africa. *Natur*, **179**(4558), 542–3.

Longhurst, A.R. (ed) (1981) *Analysis of Marine Ecosystems*. Academic Press, London.

Loshbaugh, S. (1992) Will sea lions deliver the next blow to the North Pacific groundfisheries? *National Fisherman*, **73**(7), 17.

Lough, R.G. and Laurence, G.C. (1982) Larval haddock and cod survival studies on George Bank. In: *Fisheries Ecosystem Study News. Narrangsett Laboratory*. Ref. 8206, 176–92.

Lucas, C.E. (1947) The ecological effects of external metabolites. *Biol. Rev.*, **22**(3), 220–95.

Lyles, C.H. (1968) Historical statistics. The squid fishery. US Dept Interior, Bureau of Commercial Fisheries CFS No. 4833.

McAllister, C.D. (1961) Zooplankton studies at ocean weather station 'p' in the northeast Pacific Ocean. *J. Fish. Res. Bd Canada*, **18**(1), 1–29.

McConnaughey, T. and McRoy, P. (1976) Food-web structure and the fraction of carbon isotopes in the Bering Sea. *Science in Alaska 1976*, Alaska Div. of AAAS, 296–316.

McIntyre, A.D. (1982) Oil pollution and fisheries. *Phil Trans. R. Soc. Lond.*, **B297**, 401–11.

McIntyre, A.D. (1992) The current state of the oceans. *Mar. Pollution Bull.*, **25**(3/4), 28–31.

McRoy, P.C., Goering, J.J. and Shields, W.E. (1972). Studies of primary production in the eastern Bering Sea. In: *Biological Oceanography of the Northern North Pacific Ocean.* (ed A.Y. Takonouti), pp. 199–216. Idemitsu Shoten, Tokyo.

Magnusson, K.G. and Palsson, O.K. (1989) Trophic ecological relationships of Icelandic cod. *Rapp. P.-v. Réun. Cons. Int. Explor. Mer*, **188**, 206–24.

Malins, D.C. and 16 other co-authors (1982) Sublethal effects of petroleum hydrocarbons and trace metals, including biotransformations, as reflected by morphological, chemical, physiological, pathological, and behavioral indices. MS Report, NWAFC, Env. Cons div.

Mamaeva, N.V. (1992) Ciliate protozoa in plankton. In: *Results of the Third Joint US–USSR Bering and Chukchi Sea Expedition.* US Fish & Wildlife Service, Washington DC, pp. 155–61.

Mann, K.H. and Lazier, J.R.N. (1991) *Dynamics of Marine Ecosystems.* Blackwell Science, Oxford.

Manzer, J.I., Ishida, T., Peterson, A.E. and Hanavan, M.G. (1965) Salmon of the North Pacific Ocean – Part V, offshore distribution of salmon. *Bull. Int. North Pacific Fish Comm.*, **15**, 1–247.

Marchand, M. and Caprais, M.P. (1982) Suivi de la pollution de l'*Amoco Cadiz* dans l'eau de mer et les sédiments marins. *Amoco Cadiz*, consequences d'une pollution accidentelle par les hydrocarburas. *Cont. Natl. Explor. Oceans*, Paris, 23–54.

Marques, J.C. and Bellan-Santini, D. (1993) Biodiversity in the ecosystem of the Portuguese continental shelf: distributional ecology and the role of benthic amphipods. *Mar. Biol.*, **115**, 555–64.

Mesnil, B. (1986) Towards more consistent advice on North Sea demersal stock. ICES, C.M. 1986/G:40.

Minoda, T. (1972) Characteristics of the vertical distribution of copepods in the Bering Sea and south of the Aleutian Chain, May–June 1962. In: *Biological Oceanography of the Northern North Pacific Ocean* (ed A.Y. Takonouti), pp. 323–31. Idemitsu Shoten, Tokyo.

Misund, O.A. (1991) Swimming behaviour of school related to fish capture and acoustic abundance estimation. *University of Bergen, Department of Fisheries and Marine Biology.* Thesis.

Mohr, H. (1971) Behaviour patterns of different herring stocks in relation to ships and mid-water trawls. In: *Modern Fishing Gear of the World 3* (ed H. Kristjonsson), pp. 368–71, Fishing News Books, Oxford.

Morioka, Y. (1972) The vertical distribution of calanoid copepods off the southern coast of Hokkaido. In: *Biological Oceanography of the Northern North Pacific Ocean* (ed A.Y. Takonouki), pp. 309–21. Idemitsu, Shoten, Tokyo.

Morita, R.Y. (1981) Microbiology of the Eastern Bering Sea. In: *The Eastern Bering Sea Shelf: Oceanography and Resources*, Volume 2 (ed D.W. Hood and J.A Calder), pp. 903–18. University of Washington Press, Seattle.

Motoda, S. and Minoda, T. (1973) Plankton of the Bering Sea. In: *Oceanography of the Bering Sea* (eds D.W. Hood and E.J. Kelley), *Univ. Alaska Occ. Publ*, **2**, 207–30.

Murphy, C.I. (1980) Schooling and the ecology and management of marine fish. In: *Fish Behaviour and its Use in the Capture and Culture of Fishes*, pp. 400–12. ICLARM, Manila.

Murphy, R.C. (1972) The oceanic life of the Antarctic. In *Oceanography*, pp. 287–200. Scientific American, W.H. Freeman & Co., San Francisco.

Nakajima, K. and Nishizawa, S. (1972) Exponential decrease in particulate carbon concentration in a limited depth interval in the surface layer of the Bering Sea. In: *Biological Oceanography of the Northern Pacific Ocean* (ed A.Y. Takonouti, pp. 496–505. Idemitsu Shoten, Tokyo.

Nakken, O. and Raknes, A. (1985) The horizontal distribution of Arctic cod in relation to the distribution of bottom temperature in the Barents Sea, 1978–1984. Paper presented at the Gadoid Workshop, Seattle, June 1985.

Nedreaas, K. (1987) Food and feeding habits of young saiths *Pollachius virens* (L.) on the coast of western Norway. *FiskDir. Skr. Ser. Havunders.*, **18**, 263–301.

Nellen, W. and Schadt, J. (1991) Year to year variability in plankton community on the spawning ground of the Herbides–Buchan herring. *ICES Variability Symposium*. Paper 25.

Nemoto, T. and Harrison, G. (1981) High latitude ecosystems. In: *Analysis of Marine Ecosystems* (ed A.R. Longhurst), pp. 95–126. Academic Press, London.

Nesis, K.N. (1977) General ecological concepts as applied to marine communities. The community as continuum. In: *Biological Productivity of the Oceans*. Translation in NOAA Technical Memo NMFS-F/NEC-34, 1–11.

Nilsson, E.M. and Hopkins, C.C.E. (1991) Regional variability in fish–prawn communities and catches in the Barents Sea, and their relationship to the environment. *ICES Variability Symp*, Paper 29.

Nilssen, E.M., Pedersen, T., Hopkins, C.C.E., Thyholdt, K. and Pope, J.G. (1993) Recruitment variability and growth of northeast Arctic cod: Influence of physical environment, demography, and predator–prey energetics. *ICES Cod Symposium Reijkjavik*, Paper 30.

Nomura, M. (1980) Influence of fish behavior on use and design of setnets. In: *Fish Behavior and its Use in the Capture and Culture of Fishes*, pp. 446–71. ICLARM, Manila.

Norges Offentlige Utredninger NOCI 1980:25 (1980) *Muligheter og Konsekvenser ved Petroleums funn Nord for 62 deg*. Norwegian Government Publications, Oslo.

O'Boyle, R., Sinclair, A. and Hurley, P. (1989) A bioeconomic model of an age-structured groundfish resources, exploited by a multi-gear fishing fleet. *ICES MSM Symposium*. Paper 23.

Ojaveer, E. (1981) Influence of temperature, salinity and reproductive mixing of Baltic herring groups on its embryonal development. *Rapp. P.-v. Réun. Cons. Int. Explor. Mer.***178**, 409–15.

Ojaveer, E. (1988) *Baltic Herring* Agropromizdat, Moscow (in Russian).

Ojaveer, E. (1989) Population structure of pelagic fishes in the Baltic. *Rapp. P.-v. Réun. Cons. Int. Explor. Mer*, **190**, 17–21.

Olsen, K. (1969) A comparison of acoustic threshold in cod with recordings of ship-noise. *FAO Fish. Rep.*, **62**(2), 431–5.

Olsen, K. (1971) Influence of vessel noise on behaviour of herring. In: *Modern Fishing Gear of the World 3*. (ed H. Kristjonsson), pp. 291–4. Fishing News Books, Oxford.

Olsen, K. (1981) The significance of fish behaviour in the evaluation of hydroacoustic survey data. ICES C.M. 1981/B:22.

Olsen, S. and Laevastu, T. (1983) Fish attraction to baits and effects of currents on the distribution of smell from baits. *NW & Alaska Fisheries Center Report*, 83–05.

Olsen, K., Angell, J., Pettersen, F. and Lovik, A. (1982) Observed fish reactions to a surveying vessel with special reference to herring, cod, capelin and polar cod. *ICES Symposium on Fish Acoustics*, Bergen, 1982. Paper 48.

Ona, E. and Godø, O.R. (1990) Fish reaction to trawling noise: the significance for trawl sampling. *Rapp. P.-v. Réun. Cons. Int. Explor. Mer*, **189**, 159–66.

Østvedt, O.J. (1965) The migration of Norwegian herring to Icelandic waters and the environmental conditions in May–June 1961–1964. *FiskDir. Skr. Ser. Havunders.*, **13**(8), 27–47.

Overholtz, W.J. and Tyler, A.V. (1985) Long-term responses of the demersal fish assemblages of Georges Bank. *Fish. Bull.*, **83**(4), 507–20.

Palacin, C., Gili, J.M. and Martin, D. (1992) Evidence of meiofauna spatial heterogenity with eutrophication processes in a shallow-water Mediterranean bay. *Estuarine, Coastal and Shelf Science*, **35**, 1–16.

Palsson, O.K. (1983) The feeding habits of demersal fish species in Icelandic waters. *J. Mar. Res. Inst. Reijkjavik*, **7**(1), 1–60.

Parrish, B.B. (1969) A review of some experimental studies of fish reactions to stationary and moving objects of relevance to fish capture processes. *FAO Fish. Rep.*, **62**(2), 233–45.

Parrish, B.B. and Blaxter, J.H.S. (1964) Notes on the importance of biological factors in fishing operations. In: *Modern Fishing Gear of the World 2.* (ed H. Kristjonsson), pp. 557–60. Fishing News Books, Oxford.

Parsons, L.S. (1993) Management of marine fisheries in Canada. *Can. Bull. Fish. Aquat. Sci.* **225**, 1–763.

Petersen, C.G.J. (1918) The sea bottom and its production of fish food. *Rep. Danish Biol. Sta.*, **25**, 1–62.

Petersen, C.G.J. and Jensen, P.B. (1911) Valuation of the sea. I. Animal life of the sea bottom, its food and quantity. *Rep. Danish Biol. St.*, **20**, 3–79.

Pihu, E. (1987) *Matk kalariiki.* Valgus, Tallinn.

Pitcher, T.J. (1986) Fluctuations of shoaling behaviour in teleosts. In: *The Behavior of Teleost Fishes* (ed T.J. Pitcher), pp. 294–337. Johns Hopkins University Press, Baltimore.

Pope, J.G. (1972) An investigation of the accuracy of Virtual Population analysis using Cohort Analysis. *Res. Bull. Int. Comm. NW Atlant. Fish.*, **9**, 65–74.

Pope, J.G. and Knights, B.J. (1982) Comparison of length distributions of combined catches of all demersal fishes in surveys in the North Sea and at Faroe Bank. In: *Multispecies Approaches to Fisheries Management Advice* (ed M.C. Mercer), pp. 116–118.

Pope, J.G. and Shepherd, J.G. (1982) A simple method for the consistent interpretation of catch-at-age data. *J. Cons. Int. Explor. Mer*, **40**(2), 176–84.

Prenski, L.B. and Angelescu, V. (1993) Biologia trofica de la merluza (*Merluccius hubbsi*) del mar Argentino. Parte 3: Consumo anual de alimento a nivel poblacional y su relacion con la explotacion des las pesquerias multispecificas. *INIDEP Docum. Cient*, 7, Mar de La Plata, Argentina.

Preston, A., Jefferies, D.F. and Mitchell, N.T. (1971). Experience gained from the controlled introduction of liquid radioactive waste to coastal waters. *Nuclear Techniques in Environmental Pollution.* International Atomic Energy Agency, IAEA-SM- 142a/40, 629–644.

Pyanov, A.I. (1993) Fish learning in response to trawl fishing. *ICES Mar. Sci. Symp.*, **196**, 12–16.

Quast, J.C. and Hall, E.L. (1972) List of all fishes of Alaska and adjacent waters with a guide to some of their literature. *NOAA Techn. Rep.*, NMFS SSRF-658, 47 pp.

Rass, T.S. (1979) The biographical basis of regionalization of the fish producing areas of the world ocean. In: *Biological Resources of the World Ocean*, pp. 48–83. Nauka Publishing House, Moscow.

Redant, F. (1991) An updated bibliography of the effects of bottom fishing gear and harvesting techniques on sea bed and benthic biota. *Working Document to the Study Group on Ecosystem Effects of Fishing Activities.* ICES.

Redford, K.H. (1994) Science and the nature conservancy. *Nature Conservancy*, **44**(1), 15.

Ricker, W.E. (1975) Computation and interpretation of biological statistics of fish populations. *Bull. Fish. Res. Bd Canada*, **191**, 1–382.

Riemann, B. and Hoffman, E. (1991) Ecological consequences of dredging and bottom trawling in the Limfjord, Denmark. *Mar. Ecol. Prog. Ser.*, **69**, 171–8.

Rosenfeld, A. (1976) Infectious diseases in commercial shellfish on the middle Atlantic coast. *Am. Soc. Limnol. Oceanogr. Spec. Symp.*, **2**, 414–23.

Rowe, G.T. (1981) The deep-sea ecosystem. In: *Analysis of Marine Ecosystems.* (ed A.R. Longhurst), pp. 235–67. Academic Press, London.

Rowe, G.T., Smith, K.L. and Clifford, C.H. (1976) Benthic-pelagic coupling in the New York Bight. *Am. Soc. Limnol. Oceanogr. Spec. Symp.*, **2**, 370–76.

Royce, W.F., Smith, L.S. and Hartt, A.C. (1968) Models of oceanic migrations of Pacific salmon and comments on guidance mechanisms. *Fish. Bull.*, **66**(3), 441–62.

Rumohr, H. and Krost, P. (1991) Experimental evidence of damage to benthos by bottom trawling with special reference to *Arctica islandica. Meeresforsch.*, **33**, 340–5.

Saetersdal, G. (1969) Review of information on the behaviour of gadoid fish. *FAO Fisheries Rep.*, **62**(2), 201–15.

Saetersdal, G. and Loeng, H. (1983) Ecological adaption of reproduction in Arctic cod. Contribution to *PINRO/IMR Symposium on Arctic Cod*, St Petersburg 1983. Mimeo.

Sahrhage, D. (1967) Ueber die Verbreitung der Fischarten in der Nordsee. Teil I, Januar 1962 und 1963. *Ber. Dt Wiss. Komm. Meeresforsh.*, **19**(3), 16–179.

Sampson, D.B. (1992) Fishing technology and fleet dynamics: Prediction from bioeconomic model. *Mar. Resource Econ.*, **7**, 37–58.

Sanger, G.A. (1972) Preliminary standing stock and biomass estimates of seabirds in the subarctic Pacific region. In: *Biological Oceanography of the Northern North Pacific Ocean.* (ed A.Y. Takenouti), pp. 589–611. Idemitsu Shoten, Tokyo.

Sato, M. (1977) Prospects of fisheries in the 200 mile fishery zone of Japan. In: *Fisheries Biological Production in the Subarctic Pacific Region*, pp. 7–68. Hokkaido University, Hakodate.

Saville, A. (1965) Factors controlling dispersal of the pelagic stages of fish and their influence on survival. *ICNAF Spec. Publication*, **6**, 335–48.

SCAR (Scientific Committee for Antarctic Research) (1976) Report of Meeting, Woods Hole, Massachusetts, 23–24 August 1976.

Schott, G. (1935) *Geographie des Indischen und Stillen Ozeans.* Boysen, Hamburg.

Schott, G. (1944) *Geographie des Atlantischen Ozeans.* Boysen, Hamburg.

Scott, G.P., Kenney, R.D., Thompson, T.J. and Winn, H.E. (1983) Functional roles and ecological impacts of the cetacean community in the waters of the northeastern US continental shelf. ICES, C.M. 1983/N:12.

Seckel, G.R. (1972) Hawaii-caught skipjack tuna and their physical environment. *Fish Bull.*, **70**(3), 763–87.

Seki, M. (1970) Microbial biomass in the euphotic zone of the North Pacific subarctic waters. *Pac Sci.*, **24**(2), 269–74.

Sharp, G,D. and McLain, D.R. (1993) Fisheries, El Niño-southern oscillation and upper-ocean temperature records: an Eastern Pacific example. *Oceanography*, **6**(1), 13–22.

Shaw, E. (1962) The schooling of fishes. *Oceanography*, pp. 235–43. W.H. Freeman & Co., San Francisco.

Shepherd, J.G. (1993) Key issues in the conservation of fisheries. *Laboratory Leaflet*, **72** Lowestoft.

Sherman, K. and Jones, C. (1980) The zooplankton component of a Northwest Atlantic eco-system. ICES, C.M. 1980/L:67.

Sherman, K., Green, J.R., Goulet, J.R. and Fofonoff, P.W. (1982a) Coherence in the zoo-plankton component of the marine ecosystem off the northwest coast of the United States. ICES, C.M. 1982/L:51.

Sherman, K., Jones, C., Sullivan, L., Smith, W., Berrien, P. and Ejsymonth, L. (1982b) Congruent shifts in sand eel abundance in western and eastern North Atlantic ecosystems. In: *Fisheries Ecosystem Study News. Narragansett Lab. Ref. 82–6*, 32–35.

Sherman, K., Grosslein, M., Mountain, D., Busch, D., O'Reilly, J. and Theroux, R. (1988) The continental shelf ecosystem off the northeast coast of the United States. In: *Ecosystems of the World 7*. (ed H. Postma and J.J. Zijlstra), pp. 279– 337. Elsevier, Amsterdam and Oxford.

Sherman, K., Alexander, L.M. and Gold, B.D. (1990) *Large Marine Ecosystems: Patterns, Processes and Yields*. American Association for the Advancement of Science, Washington, DC.

Sinclair, A. (1992) Fish distribution and partial recruitment: The case of eastern Scotian Shelf cod. *J. Northw. Atl. Fish. Sci.*, 13, 15–24.

Sinclair, A.D., Gascom, D., O'Boyle, R., Rivard, D. and Gavaris, S. (1991) Consistency of some Northwest Atlantic groundfish stock assessments. *NAFO Sci. Coun. Studies*, **16**, 59–77.

Sissenwine, M.P. (1990) Perturbation of a predator-controlled continental shelf system. In: *Variability and Management of a large Marine Ecosystem*. (ed K. Sherman and L.M. Alexander), pp. 55–85. Westview Press, Boulder.

Sissenwine, M.P., Cohen, E.B. and Grosslein, M.D. (1982) Structure of the Georges Bank ecosystem. In: *Fisheries Ecosystem Study News, Narragansett Lab. Ref. 82–6*, 260–84.

Skud, B.E. (1977) Drift, migration and intermingling of Pacific halibut stocks. *Int. Pacific Halibut Comm., Rep.*, 63.

Skud, B.E. (1982) Dominance in fishes: The relation between environmental factors and abundance. *Science*, **216**, 144–9.

Soldal, A.V. Engås, A. and Isaksen, B. (1993) Survival of gadoids that escape from demersal trawl. *ICES Mar. Sci. Symp.*, **196**, 183–9.

Soria, M. Gerlotto, F. and Freon, P. (1993) Study of learning capabilities of tropical clupeoids using an artificial stimulus. *ICES Mar. Sci. Symp*, **196**, 17–20.

Sorokin, Y.I. (1977a) Production of microflora. In *Oceanography: Biology of the Ocean* Vol. 2. *Biological Productivity of the Ocean*. Nauka Press, Moscow. (Translation *NOAA Techn. Memo NMFS-F/NEC- 34*, 256–85).

Sorokin, Yu. I. (1977b) The communities of coral reefs. In *Oceanography: Biology of the Ocean*, Vol. 2, Biological Productivity of the Ocean. Nauka Press, Moscow. (translation *NOAA Techn. Memo NMFS-F/NEC-34*, 161–89).

Soutar, A. and Isaacs, J.D. (1974) Abundance of pelagic fish during the 19th and 20th centuries as recorded in anaerobic sediment off the Californias. *Fishery Bulletin*, **72**(2), 257–73.

Southward, A.J. and Boalch, G.T. (1986) Aspects of long term changes in the ecosystem of the western English Channel in relation to fish populations. *International Symposium on Long Term Changes in Marine Fish Populations* (eds T. Wyatt and M.G. Larrañeta), pp. 415–447. Vigo.

Spärck, R. (1935) On the importance of quantitative investigation of the bottom fauna in marine biology. *J. Cons. Int. Explor. Mer.*, **10**(1), 3–19.

Sparholt, H. (1990) An estimate of the total biomass of fish in the North Sea. *J. Cons. Int. Explor. Mer*, **46**, 200–10.

Sparre, P. & Willmann, R. (1992) BEAM 4 a bio-economic multispecies, multi-fleet, multi-plant, multi-area extension of the traditional forecast model. ICES, C.M. 1992/D:2.

Squire, J.L., Jr (1993) Relative abundance of pelagic resources utilized by the Californian purse-seine fishery: Results of an airborne monitoring program, 1962–90. *Fish. Bull. US*, **13**, 348–61.

Steele, J.H. (1974) *The Structure of Marine Ecosystems*. Harvard University Press, Cambridge.

Stocker, S. (1981) Bentic invertebrate macrofauna of the eastern Bering/Chukchi continental shelf. In: *The Eastern Bering Sea Shelf: Oceanography and Resources*, Vol. 2 (eds D.W. Hood and J.A. Calder), pp. 1069–90. University of Washington Press, Seattle.

Sunnanå, K., Bax, N.J. and Godø, O.R. (1986) Further analysis of the ecosystem model-NORFISK-for studies of fish stocks off western Norway. Gadoid Workshop, Seattle, June 1985.

Thomas, J.P., Phoel, W.C., Steimle, F.W., O'Reilly, J.E. and Evans, C.A. (1976) Seabed oxygen consumption – New York Bight apex. *Am. Soc. Limnol. Oceanogr. Spec. Symp.*, **2**, 354–69.

Thomsen, B. (1993) Selective flatfish trawling. *ICES mar. Sci. Symp*, 196, 161–4.

Thorsen, G. (1955) Modern aspects of marine level-bottom animal communities. *J. Mar. Res.*, **14**(4), 387–97.

Thorsen, G. (1957) Bottom communities (sublittoral or shallow shelf). In: *Treatise of Marine Ecology and Paleoecology*, Vol. 1. *Ecology*. (ed J.W. Hedgpeth), pp. 461–534. *Geol. Soc. Am. Mem.*, **67**, 461–534.

Tiews, K. (1983) Über die Veranderungen in Auftreten von Fischen und Krebsen im Beifang der deutschen Garnelenfischerei während der Jahre 1954–1981 – Ein Beitrag zur Oekologie des deutschen Wattenmeers und zum biologischen Monitoring von Oekosystemen in Meer. *Ark. Fisch. Wiss.*, **34**(Beih.1), 1–156.

Tsyban, A.V., Kudryatsev, V.M., Mamaev, N.V. and Sukhanova, N.V. (1992) Total number of biomass and activity of bacteria. In: *Results of the Third Joint US–USSR Bering and Chukchi Seas Expedition* (ed J.F. Turner), pp. 55–60. US Fish & Wildlife Service,. Washington, DC.

Ulltang, Ø (1977) Sources of errors in and limitations of Virtual Population Analysis (Cohort Analysis). *J. Cons. Int. Explor. Mer.*, **37**(3), 249–60.

Ulltang, Ø. (1980) Factors affecting the reaction of pelagic fish stocks to exploitation and requiring a new approach to assessment and management. *Rapp. P.-v. Réun. Cons. Int. Explor. Mer*, **177**, 489–504.

Ulltang, Ø (1989) Current ICES approaches to fish stock assessment and management advice. (MS)

Underdal, A. (1980) *The Politics of International Fisheries Management*. Universitetsforlaget, Oslo.

Ursin, E. (1982) Stability and variability in the marine ecosystem. *Dana*, **2**, 51–67.

Ursin, E. (1984) The tropical, the temperate and the arctic seas as media for fish production. *Dana*, **3**, 43–60.

US National Academy of Sciences (1985). *Petroleum in the Marine Environment*. US National Research Council, Washington, DC.

van der Valk, L. (1992) Estimated amount of physical disturbance of the seabed in the shallow southern North Sea due to natural causes. ICES, CM 1992/E:39.

Vethaak, A.D. (1992) Gross pathology and histopathology in fish: summary. *Mar. Ecol. Prog. Ser.*, **91**, 171–2.

Vethaak, A.D., Bucke, D., Lang, T., Wester, P.W., Jol, J. and Carr, M. (1992) Fish disease monitoring along a pollution transect: a case study using dab *Limanda liminda* in the German Bight. *Mar. Ecol. Prog. Ser.*, **91**, 173–92.

Vinogradov, M.E. (1981) Ecosystems of equatorial upwellings. In: *Analysis of Marine Eco-systems.* (ed A.R. Longhurst), pp. 69–93. Academic Press, London.

Walford, L. and Wickland, R. (1973) Contribution to a world wide inventory of exotic marine and anadromous organisms. *FAO Fish. Techn. Paper,* 121.

Walker, D.R. and Pitcher, G.C. (1991) The dynamics of phytoplankton populations, including a red-tide bloom, during a quiescent period in St. Helena Bay, South Africa. *S. Afr. J. mar. Sci.,* **10**, 61–70.

Walsh, J.J. (1981) Self-Sea ecosystems. In: *Analysis of Marine Ecosystems.* (ed A.R,. Long-hurst), pp. 159–96. Academic Press, London.

Wardle, C.S. (1993) Fish behaviour and fishing gear. In: *Behaviour of Teleost Fishes.* (ed. T.J. Pitcher), pp. 609–43.

Waterman, B. and Kranz, H. (1992) Pollution and fish diseases in the North Sea. *Mar. Pollution Bull.,* **24**(3), 131–8.

Weber, W. (1989) A review of stock assessment of cod in the Baltic. *Rapp. P.-v. Réun. Cons. Int. Explor. Mer,* **190**, 224–34.

Wespestad, V.G. (1993) Status of Bering Sea pollock and the effect of the 'Donut Hole' fishery. *Fisheries,* **18**(3), 18–24.

Wiens, J.A. and Scott, J.M. (1975) Model estimation of energy flow in Oregon coastal seabird populations. *The Condor,* **77**, 439–52.

Wilder, R.J. (1993) Is this holistic ecology or just muddling through? The theory and practice of marine policy. *Coastal Management,* **21**, 209–24.

Williams, H. and Pullen, G. (1993) Schooling behaviour of jack mackerel. *Trachurus declivis* (Jenyns), observed in the Tasmanian purse seine fishery. *Aust. J. Mar. Freshw. Res.,* **44**, 577–87.

Winters, G.H., Wheeler, J.P. and Dalley, E.L. (1986) Survival of a herring stock subjected to a catastrophic event and fluctuating environmental conditions. *J. Cons. Int. Explor. Mer.,* **43**, 26–42.

Wolotira, R.J., Sample, T.M., Noel, S.F. and Iten, C.R. (1993) Geographic and batymetric distributions for many commercially important fishes and shellfishes off the West Coast of North America based on research survey and commercial catch data, 1912–1984. *NOAA Tech. Memo NMFS- AFSC-6.*

Woodhead, P.M.J. (1965) Effects of light upon behaviour and distribution of demersal fishes in the North Atlantic. *ICNAF Spec. Publ.,* 267–87.

Woodley, S., Kay, J. and Francis, G. (1993) *Ecological Integrity and the Management of Ecosystems.* St Lucie Press, Delrey Beach, Florida.

Wyman, E. (1994) Deep water fishery resource assessment and management strategies. *Fish. Boat World,* **5**(11), 35.

Young, de B. and Rose, G.A. (1993) On recruitment and distribution of Atlantic cod (*Gadus morhua*) off Newfoundland. *Can. J. Fish. Aquat. Sci.,* **50**, 2729–41.

Yusa, T., Kyushin, K., and Forrester, C.R. (1977) Outline of life history information on some marine fish. In: *Fisheries Biological Production in the Subarctic Pacific Region,* pp. 123–74. Hokkaido University, Hakodate.

Zusser, S.G. (1966) Reaction of horse mackerel to light according to the amount of food consumed. In: *Biologicheskie i Okeanograficheskie Usloviya Obrazovaniya Promyslovykh Skoplenii Ryb,* pp. 325–45. Pishehvaya Promyshlennost, Moskva.

Glossary of Acronyms and Terms

Special terms used in this book are usually defined and/or described in the text, usually where they first occur. The locations of these definitions and/or descriptions can be found with the help of the index.

Acronyms

ACFM	Advisory Committee on Fisheries Management (International Council for the Exploration of the Sea).
BEAM	A bio-economic multi- species, multi-fleet fisheries forecasting model (Sparre & Willman 1992).
CPUE	Catch per unit effort (an index of relative stock abundance).
DDT	A pesticide (dichlorodiphenyltrichloroethane).
DYNUMES	Dynamical numerical ecosystem simulation (a holistic ecosystem simulation with spatial resolution).
EEZ	Exclusive economic zone.
EU	European Union.
F	Fishing mortality (coefficient).
FAD	Fish attraction device.
FAO	Food and Agriculture Organization.
ICES	International Council for the Exploration of the Sea.
ICNAF	International Council for Northwest Atlantic Fisheries (now NAFO).
ITQ	Individual transferable quota.
IUCN	International Union for Conservation of Nature.
LTPY	Long-term potential yield.
M	Natural mortality (coefficient).
MEY	Maximum economic yield.
MSVPA	Multi-species virtual population analysis.
MSY	Maximum sustainable yield.
NAFO	Northwest Atlantic Fisheries Organization.
NEAFC	Northeast Atlantic Fisheries Council.
NORFISK	A fisheries ecosystem simulation model for the Norwegian coastal areas.
OSY	Optimum sustainable yield.
OY	Optimum yield.
PCB	A pesticide.
PROBUB	Prognostic bulk biomass simulation model.
SKEBUB	Skeleton bulk biomass simulation model.
SR	Spawner–recruit relation.
TAC	Total allowable catch.
TARS	Weathered hydrocarbons.
UN	United Nations.
UNEP	United Nations Environmental Programme.
UUSDYNE	An ecosystem simulation programme.
VPA	Virtual population analysis.
WSF	Water soluble fraction of hydrocarbons.

Explanation of terms

Acclimatization temperature   Annual mean temperature (also range of temperature) of the normal habitat of a species or a fish stock.

Basic organic production   Also primary production. Production of organic matter by phytoplankton and by sessile plants (seaweeds).

Biocoenosis   Biological community; animals and plants living in a biotape and dependent on each other in various degree.

Boreal   Northern (latitudes).

Brackish water   Salt water with salinity less than 20%.

Carrying capacity   The amount of biota (e.g. finfish) that a defined ocean can sustain.

Community   Animals which live in the same locality under the influence of similar environmental factors and affect the existence of each other through their activities. The communities are usually named by the dominant species in them.

Ecosystem   Plants and animals in a specified environment, depending on and interacting with this environment and with each other.

El Niño (EN)   A condition off the coasts of Peru and Ecuador when there is a cessation of upwelling. EN is usually accompanied by a northward shift of surface mean low and high pressure centres. This term has lately been misused.

Equilibrium biomass   Long-term mean biomass of a given species which can be sustained in a given region under defined and balanced conditions.

Eutrophication   Increasing the productivity of a body of water by adding limiting nutrient salts (nitrates and phosphates) usually with domestic and agricultural wastes.

Exploitable biomass   The biomass of fish in a given stock that is available to the fisheries.

Food chain   The sequence of using plants and animals as food. The food chain starts with plants (primary producers), followed successively by plant-eating animals (herbivores) and animals that feed on herbivores (first stage carnivores), etc. However, this sequence does not occur in nature in this pure form, as the food composition of most animals varies in space and time and also changes with age (size) of the organisms.

Heterotrophs   Bacteria, saprophytic plants and ciliates that obtain nourishment from mostly dead organic matter.

Holistic ecosystem (simulation)   Numerical simulation of (marine) ecosystem which contains all biota, environment and all essential processes within this ecosystem.

Hypoxia   Low oxygen content or lack of oxygen.

Mare liberum   Freedom of the seas. Freedom to fish and navigate. A concept initiated in the sixteenth and seventeenth centuries.

Neritic   Pertaining to waters and ecosystems above the continental shelf.

Phototaxis   A movement, stimulated by light, either toward the light, photopositively, or away from the light, photonegatively.

Population dynamics   Quantitative change occurring in a population with time. The term is often applied to signify mathematical approaches to simulate the dynamics of single-species populations.

Recruitment   Process of movement of age (size) classes towards older age classes measured quantitatively by the amounts of the species biomass moving up to older age classes per unit time (usually a year). Usually applied to the amount of younger fish moving annually into the exploitable population.

Rejuvenation of population   Decrease of average age of a given population, either by the decrease of exploitable biomass or increase of juvenile biomass due to good recruitment.

Red tide   A term applied to the mass development of plankton in coastal waters.

Semi-pelagic (semidemeral)   Fish who spend part of their life on the bottom and part in the water mass above and obtain their food from both regimes.

Seston   Suspended matter in the water mass. Bioseston – living suspended matter; abioseston – dead suspended matter.

Spawning stress mortality   Mortality during and shortly after spawning.

Synecology   Ecology pertaining to relationships between communities and their environment.

Target species   Species which is the main subject for fishing by a given vessel with a given type of gear.

Thermocline   Also 'temperature discontinuity layer' is a horizontal layer below the surface with limited thickness at which the temperature changes abruptly vertically.

Trophodynamics   Dynamics of feeding and quantitative considerations of conversion of food to tissue and for activity (maintenance).

Appendix: Numerical Dynamical Ecosystem Simulations SKEBUB and UUSDYNE

A.1 Description of the SKEBUB fish ecosystem simulation programme

The SKEBUB fish ecosystem simulation is a 'box' model and is the simplest multi-species ecosystem model based on biomass, in contrast to conventional number-based models. One of the purposes of the description of this model here is to explain the essentials of fisheries ecosystem simulation.

The equations presented here can be applied, with proper data, to any fish species or ecological group of species. The biomass and trophodynamic equations can also be applied, with some modifications, to a single cohort of any species. The numerical behaviour of the individual formulae is well known and so is not described here. Plankton and benthos biomasses and their annual changes are prescribed (input) with harmonic formulae. The consumption by marine mammals is computed in another model, e.g. with a spreadsheet, and input in each seasonal time step.

The biomass growth and mortality is computed in discrete seasonal time steps. The biomass growth rate is computed from the empirical data of annual growth rates and the distribution of biomass with age. The latter is computed with an auxiliary model.

The biomass (B) of a cohort, species, or group of species (i) at the end of a given time step (t) (seasonal time step is normally used) is computed with a well-known equation (1), using biomass from the previous time step ($t - 1$) and growth rate (coefficient) (g) minus total mortality rate (Z) for this time step.

$$B_{i,t} = B_{i,t-1} * e^{gi(t)-Zi(t)} \tag{1}$$

The yield (Y) is computed with a prescribed fishing mortality coefficient ϕ_i. It should be noted that all the instantaneous coefficients (growth, mortality, fishery) are different from the corresponding conventional coefficient for number-based models that use an annual time step. Thus all these coefficients have to be computed on biomass base and for the time step used in the model.

$$Y_{i,t} = B_{i,t} - B_{i,t} * e^{-\phi i(t)} \tag{2}$$

As a relatively short time step is used in the computation, the second-order terms, such as non-linearities in growth and mortality during the time step, can be ignored.

The growth coefficient is computed in each time step, accounting for the effects of starvation in the previous time step:

$$g_{i(t)} = g_i^0 * ((R_{i,t-1} - S_{i,t-1})/R_{i,t-1}) \tag{3}$$

If there was no starvation in the previous time step ($S_{i,t-1} = 0$), the rate of growth ($g_{i(t)}$) will take the prescribed value g_i^0, but if the species was not able to get all the food required ($R_{i(t)}$) for maximum growth rate, the prescribed growth rate will be reduced by the ratio of the amount of food which the species was not able to get during the previous time step ($R_{i,t-1} - S_{i,t-1}$) over the total amount of food required by the biomass to grow under unlimited conditions ($R_{i,t-1}$). Both values are available from the previous time step and the possible error caused by this necessary back-stepping choice is again minimized by the use of a short time step in the computations.

The initial (prescribed) growth rate can be presented as a harmonic function over time to take account of seasonal differences in growth $(g_i^0 = \gamma_i + \sigma_i * \text{COS}\,(\alpha_i t - \kappa_i))$ where γ_i is the annual mean growth coefficient, σ_i is half of the magnitude of its annual change, α_i is the phase speed, and κ_i is the time lag to reach the maximum. Furthermore, the growth rate can be made a function of either surface or bottom temperature. Growth rate is also a recruitment parameter in biomass-based models (see below).

The mortality rate $(Z_{i(t)})$ is the addition of all negative rates of changes, thus representing the total mortality rate:

$$Z_{i(t)} = \phi_{i(t)} + \mu_i + \beta_{i,t-1} \tag{4}$$

All rates of change are presented as instantaneous coefficients and are therefore additive. Fishing mortality $(\phi_{i(t)})$ and natural mortality from old age and diseases, including also spawning stress mortality (μ_i), are prescribed, but the predation mortality coefficient $(\beta_{i,t-1})$ is computed trophodynamically in the previous time step from the ratio of consumption of the species over its biomass $(\beta_{i,t-1} = \ln(1 - (C_{i,t-1}/B_{i,t-1})))$.

The amount of food eaten by a species $(Ri(t))$ with unlimited food availability is:

$$R_{i(t)} = B_{i,t} * r_i * \tau \tag{5}$$

where r_i is the prescribed daily ration (in fraction of body weight daily) and τ is the length of the time step in days. If the growth rate (g_i^0) is made a harmonic function over the year, r_i must also be made a harmonic function $(r_i = \rho_i + \partial_i * \text{COS}\,(\alpha_i t - \kappa_i))$.

If the food supply of all food items for a given species would be unlimited, we could compute the consumption of each food item (e.g. the consumption of species j by species i $(C_{j,i})$) from the food requirement (R_i) and the fraction of species j (prey) in the food of species i (predator $(\pi_{i,j})$):

$$C_{j,i} = R_{i,t} * \pi_{i,j} \tag{6}$$

In this case the total consumption of species i would be:

$$C_i = \Sigma_j\, C_{i,j} \tag{7}$$

and the starvation would be 0. However, some food might be in limited supply and only part of the biomass of a prey is usually accessible as suitable food because feeding is size-dependent. The vulnerability of one species (prey) to another species (predator) is prescribed by average composition of the food of the predator. Therefore the fraction of each species that is allowed to be consumed in each time step is prescribed in the model (ρ_j), considering mainly the size composition of the biomasses of individual species. Furthermore, substitution of low-availability food items with high-availability items must be used. However, conditions can arise where full substitution is unrealistic and partial starvation will occur. There are various ways of computing the actual consumption with the above limitations.

The recruitment is usually depicted in number-based models as a discontinuity relating it to a discrete spawning period. In our biomass-based model we have treated it as a continuous process. This treatment is acceptable if we think in terms of size groups rather than age groups and consider variations in growth of individuals belonging otherwise to the same age groups, and assume a longer spawning period, as is the rule with most species.

Considering a continuous recruitment to all size groups and assuming that there are no exceptionally strong or weak year classes of post-larval juveniles, the recruitment would be proportional to the biomass present. The variations in post-larval recruitment would be depicted in the biomass-based model by the variations of the growth coefficient of the species biomass.

However, large spawning biomasses are known to produce proportionally small year classes and small spawning biomasses are known to produce proportionally large recruitment year classes. Therefore the recruitment could be controlled in biomass-based models, making the growth coefficient inversely proportional to biomass present:

$$g_c^0 = g_i^0 \, B_i^E / B_{i,t-1} \tag{8}$$

Where B_t^E is the equilibrium or mean biomass of species i. This computation can be done in the models in prognostic mode after the determination of the equilibrium biomasses.

If the biomasses of all species in the ecosystem do not change over a year (i.e. the previous January biomass is equal to the present January biomass), then we can say that the biomasses are in equilibrium. This implies that the growth of the biomass equals its removal by mortalities, specially by predation. If we want to achieve this equilibrium, we can change the growth rate, the mortality rate, or the biomass level itself. The growth rate is determined by empirical data and the other factors, such as temperature, are assumed, in equilibrium, to be the same from one year to another, although seasonal changes can occur. Fishing and other mortality rates are also assumed to remain the same from one year to another. The predation mortality (consumption), together with other mortalities that remain unchanged, must then balance the growth rate.

This balance can be achieved if the biomass levels of the predators are adjusted so that the biomasses remain constant from one January to the next. This adjustment can be done by finding a unique solution to the biomass equations of all species or groups of species in the ecosystem. This unique solution exists when one of the biomasses and consumption by it are predetermined, assumed to be known and fixed. In this case an iterative solution can be applied to adjust the biomasses of other species once after each year's computation:

$$B_{i,t12,0} = B_{i,t12,a} + \frac{(B_{i,b} - B_{i,a})}{k} \tag{9}$$

where $B_{i,t12,0}$ is the new (adjusted biomass for December, $B_{i,t12,a}$ is the biomass of the previous December, $B_{i,b}$ is the biomass of the previous January (computed as the next step from $B_{i,t12,a}$), $B_{i,a}$ is the computed biomass in January 1 year later, and k is an iteration constant (2.5–10, depending on the state of convergence). Twenty years or more of computation is needed before the solution converges to a unique equilibrium solution.

The model requires as input a number of species-specific constants. Besides these, the biomass of at least one species must be prescribed as known, i.e. not altered in iterative adjustment. The biomasses of other species must be initially prescribed as the best first guesses. The first guess values of the consumption (C) can be computed by assuming C_l to be 7% of B_i per month.

To determine the carrying capacities of given ocean regions with the model and to obtain realistic equilibrium biomasses, the model must include all species. Computer capacity as well as basic information available does not usually allow the specification of all species separately, but many species must be grouped ecologically, using food composition and feeding habits as the main criteria.

A.2   Symbols for constants, calculated variables, and state vectors

Constants

α_i	Phase speed, time-step dependent (e.g. 30° per month, radiants)
β_i	Predation mortality
γ_i	Annual average instantaneous growth rate
ϕ_i	Instantaneous fishing mortality coefficient
∂_i	Half amplitude of annual change of food requirement (fraction of body weight daily)
σ_i	Half amplitude of annual change of growth rate
ρ_i	Annual average food requirement (fraction of body weight daily)

κ_i Phase lag (in radians)

μ_i Instantaneous rate of mortality (other than predation mortality, including spawning stress mortality)

$\pi_{i,j}$ Fraction composition of prey j in the food of predator i

p_j Fraction of biomass j allowed to be taken in one time step (month)

r_i Prescribed ration (food requirement, fraction of body weight daily)

g_i^0 Prescribed instantaneous growth rate

Note: The last two parameters (r_i and g_i^0) can also be computed if γ_i and ρ_i and related constants are prescribed.

Dynamically calculated parameters

$g_{i(t)}$ Calculated instantaneous growth rate

$Z_{i(t)}$ Calculated total instantaneous total mortality rate

B_i^E Equilibrium biomass

State vectors

$B_{i,t}$ Biomass of species i at time t

$C_{i(t)}$ Consumption of species i (predation) during time step t

$C_{j,i}$ Consumption of species j by predator i

$R_{i(t)}$ Food requirements

$S_{i(t)}$ Starvation (the amount of food missing from the full food requirement ($R_{i(t)}$))

$Y_{i(t)}$ Yield

A.3   Purpose of the UUSDYNE programme and its capabilities

UUSDYNE is a relatively basic fish ecosystem simulation programme with spatial resolution and seasonal time step. It is a PC version of DYNUMES, programmed in FORTRAN for use on personal computers with a minimum of two disk drives and with 512K or larger memory.

The UUSDYNE programme described here can serve only as an example for other ecosystem simulations for other areas. Any ecosystem simulation must be adapted to specific regional ecosystems, to data availability and, to some extent, to the desired purposes of the simulations.

Simulation programmes similar to UUSDYNE can be used to study quantitatively and determine the effects of fishing on stocks, effects of environmental anomalies, effects of predation and spatial source and sink areas and seasons of species biomasses.

UUSDYNE has both diagnostic (analytical) and prognostic (predictive) capabilities, although its use for diagnostic purposes (resource assessment) is more time consuming than the use of SKEBUB programme. Thus it is advisable to prescribe the internally consistent biomasses, equilibrium biomasses, from SKEBUB runs.

Simulations such as UUSDYNE are essentially time-dependent bookkeeping programmes, keeping track of the changes in biomass as affected by growth, spawning, and various mortalities, containing a number of empirically-derived, tested, and verified simple formulae. Its logic and data flow can be ascertained from its code.

UUSDYNE is composed of three different programmes: (UUSDYNE – initialization and data flow control: PREDAT — computation of predation mortality species by species, including calls for migration sub-routines; and BIOBAL – computation of growth, fishing and other mortalities); each of which contains several other sub-routines. These three main sub-routines can be compiled separately and then linked to one programme.

Computation of feeding (predation mortality) in UUSDYNE is more realistic and more flexible than in SKEBUB. The apex predation (predation by mammals and birds) is computed in a separate programme and read into UUSDYNE as seasonal values of apex predation. Some biomasses, e.g. squids and salmon, are also read in seasonally and are not subject to growth and mortalities.

A.4   Data requirements and data preparation

The time step used in UUSDYNE presented here, is a season, four time steps in a year. The grid size for the eastern Bering Sea (Fig. A.1) is 63.5 km. The area is divided into nine sub-regions and the sub-region number is digitized for each grid point; land points are 0.

In other arrays, depth (m) and previously analysed long-term mean seasonal surface and bottom temperatures are prescribed with input data.

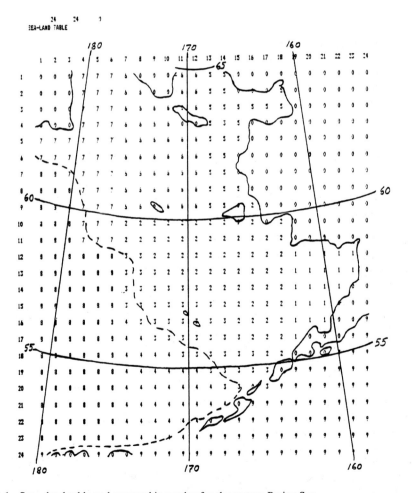

Fig. A.1   Sea – land table and geographic overlay for the eastern Bering Sea.

To input the initial biomasses of species or groups of species at each grid point two steps are necessary. Firstly, the internally consistent biomasses (equilibrium biomasses) must be determined with SKEBUB or PROBUB simulation. Secondly, the relative distribution of the species should be drawn on the map (e.g. abundance indices 0–8 contours), these relative abundances digitized and summed, the factor of this sum and the total biomass determined, and the digitized values at the grid points multiplied by this factor.

The initial food composition of the individual biomasses are adjusted to internally consistent values with spreadsheet programmes and then digitized into an array (FC).

Only one value each of the following parameters is required for each species or group of species: growth rate (GR), food requirement (FR) in body weight daily (BWD), senescent (or spawning stress or residual) mortality rate, and acclimatization temperature. The rate values must be commensurate with the time step used, i.e. seasonal. The growth rate of the biomass is determined with an auxiliary programme, and senescent (residual) mortality, which is a relatively small number, must be estimated. Food requirement, in terms of body weight daily (BWD), is estimated from the available literature, often using data on biomass distribution with age. The acclimatization temperature is estimated as the annual mean temperature of the ecological regime, pelagic or bottom, in the main distribution area of the given stock.

Fishing mortality (catches) is either prescribed with a fishing mortality coefficient, which gives the desired catch, or as a digitized field of actual (or prospective) catches.

Zooplankton and benthos biomasses are simulated with harmonic formulae (see sub-routine PLANKT) only if these biomasses become determinants for starvation.

Seasonal migration is estimated from empirical knowledge and migration speed ('u' and 'v' components), simulated with the sub-routine UNIMIG. Corresponding biomass migrations are computed with the sub-routine RANNAK after separation of migrating fraction using the sub-routine RANPOR.

There are also a few indices and factors input within the programme, such as grid length, migrating fraction, season, and other counters and conversion factors.

A.5   Basic computations of the dynamics of the fish ecosystem

The dynamics of the fish ecosystem and the formulae used in its simulation are described by Laevastu & Larkins (1981). Only a few essential formulae are briefly mentioned here.

The biomasses of the species and various other data, such as growth rate, are read from data file NEWSLDT. Computations are started with the summer season (K = 3) and corresponding seasonal data are read from file SUMDAT (consumption by apex predators in array FO, which continues to serve as the consumption array, surface and bottom temperatures (T1 and T2), salmon (SA) and squid (SQ) biomasses). Thereafter zooplankton generation sub-routine PLANKT could be called.

The sub-routine PREDAT is used for migration and predation (feeding) computation, species by species. First, the species is separated from the 3-D array of the species group. Thereafter the migration speed creation sub-routine UNIMIG is called in seasons 2 and 4, followed by call to RANPOR (migrating portion separation) and the RANNAK (migration) sub-routines. Thereafter the feeding sub-routine SOOK2 is called, wherein the seasonal total consumption of food is made a function of temperature (surface temperature for pelagic and semidemersal species, bottom temperature for demersal species).

The total food consumption is divided into consumption of food item groups according to the prescribed food composition (FC). Seven food item groups are used in UUSDYNE, pelagic, semi-pelagic and demersal fish, squids, salmon, benthos and zooplankton. The separation of the consumption of different species is made in the sub-routine ASVSUR according to the proportion of these species in the corresponding species group.

At the end of sub-routine PREDAT the possible minor consumption of benthos and demersal fish over deep water is replaced with zooplankton consumption. A similar adjustment of squid consumption over the shallower part of shelf is also made.

The main function of the sub-routine BIOBAL is to call growth and mortality computation and fishing sub-routines and to provide various outputs. Before calling the growth and mortality sub-routine ASVSUR, the predation of given species/species group is computed by multiplying the fraction of this species in the ecological group by the consumption of this ecological group. This assumes equal availability of the species as prey in this group.

In the sub-routine ASVSUR the growth of the biomass is made a function of temperature, the same function as for food consumption. The sub-routine VALAND is called for preparation of summary outputs. Outputs can, however, be taken anywhere within the programme.

The species groups as used in the skeletal UUSDYNE programme are given in Table A.1. Species number refers to the corresponding number in the programme (L). The food item number is the food group number in the consumption array (FO(24,24,7)).

A new examples of the potential outputs are given in the following Figs A.2 and A.3 to indicate their possible nature, with reference respectively to source/sink of yellowfin sole biomass in winter (including changes caused by migration), and the distribution of pollack biomass in summer.

Table A.1 Species and groups of species in UUSDYNE.

Species, species group	Species Number	Food Item Number
PB (24, 24, 3) Pelagic species		1
Herring	1	
Atka mackerel	2	
Other pelagic species	3	
SB (24, 24, 4) Semi-pelagic species		2
Pollack	4	
Rockfishes	5	
Cod	6	
Sablefish	7	
DB (24, 24, 5) Demersal species		3
Halibut, turbot	8	
Yellowfin sole	9	
Other flatfish	10	
Other demersal fish	11	
Crabs	12	
Prescribed species (seasonal inputs)		
Salmon	14	5
Squids	13	4
Plankton		7
Benthos		6

304   *Appendix*

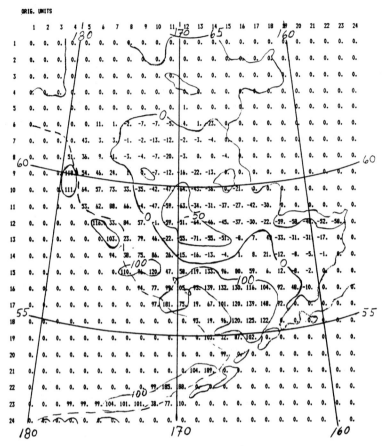

Fig. A.2 Source/decrease of yellowfin sole biomass in winter, including changes caused by migration (kg/km²).

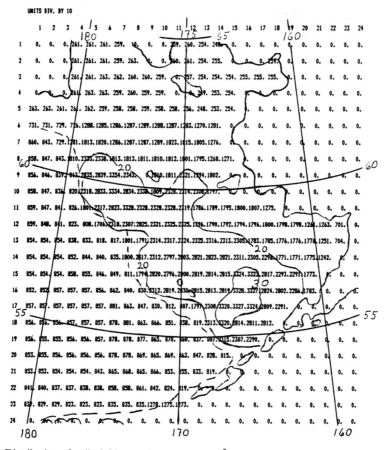

Fig. A.3 Distribution of pollack biomass in summer (t/km²).

A.6 Symbols and abbreviations

(*denotes input; if the symbol is used in sub-routines other than PREDAT and BIOBAL, the routine name is given in parentheses.)

AG	Intermediate (ASVSUR)
ALP	30* RADFA (PLANKT) Phase speed (also smoothing factor in RANNAK)
ALPP	60* RADFA (PLANKT) Phase speed
AS	Area of grid square (63.5 × 63.5) (PUUK)
*AT	Acclimatization temperatures (12)
*AUS	Parameter (0.5) (RANNAK)
AW	Working array (24,24)
*AX	Maximum biomass at grid point (ROMIGR)
B1	DL*DL (Area)
B2	Intermediate (RANNAK)
B5	Intermediate (RANNAK)
BET	Smoothing factor (RANNAK)
BN	Intermediate (PREDAT)
CEU	Speed adjustment factor (UNIMIG)

CEW	Speed adjustment factor (UNIMIG)
CUS	Intermediate (UNIMIG)
CVS	Intermediate (UNIMIG)
CW	Working array (24,24)
*DB	Demersal species (24,24,5)
DCO	Intermediate (UNIMIG)
*DD	Greatest depth of migration (UNIMEG)
DIL	Factor (PRIAFP)
*DK	Speed adjustment factor (UNIMIG)
*DL	Grid size (63.5) (RANNAK)
EP	Species (ROMIGR)
ETN	Intermediate (temperature effect, SOOK2)
F8	Species after migration (RANNAK)
*FC	Food composition (14,7)
FCON	Food consumption in season (SOOK2)
*FIM	Fishing mortality (PUUK)
FLD	Array for output (24,24)
FO	Seasonal food consumption (24,24,7) (initially contains apex predation)
*FR	Food requirement (BWD) (12)
*GR	Growth coefficient (seasonal) (12)
HW	Working array (24,24)
I	Index, counter (also species group number)
IEH	NE-1 (ROMIGR)
IND	Index
*ISL	Sea-land table (24,24)
J	Index, counter (also species number)
JA	Number of species in the group
JC	Number of species in the group
JEH	ME-1 (ROMIGR)
K	Index, counter, (also season)
*KE	Index (1-spring, 2-autumn) (UNIMIG)
KK	Index, counter
KM	Index
KN	Counter
KRC	Number of 2-day periods (RANNAK)
L	Index, counter (also consecutive species number)
M	Index
MA	J-1 (ROMIGR)
ME	Index, size of array
MEH	ME-1
ML	J+1 (ROMIGR)
N	Index
NA	I-1 (ROMIGR)
NE	Index, size of array
NEH	NE-1
NEHH	NE- 2
NL	I+1 (ROMIGR)
*PB	Pelagic species (24,24,3)
PLK	Zooplankton (24,24)
PN	Intermediate (UNIMIG)
PRE	Printing field (PRIAFP)

*RADFA	Conversion factor from degrees to radians (0.0174533) (PLANKT)
*RP	Migrating fraction (RANPOR)
S1	Species (24,24) (RANPOR)
S2	Migrating part of the species (RANPOR)
S3	Non-migrating part of the species (RANPOR)
S8	Species before migration (RANNAK)
*SA	Salmon (24,24)
*SB	Semi-pelagic species (24,24,4)
*SD	Depth, m (24,24)
SH	Intermediate (RANNAK)
*S1	Seasonal catch
SK	Species array (PUUK)
SKH	Intermediate (for SD)
*SM	Spawning stress (or senescent) mortality (12)
SME	Intermediate (ROMIGR)
*SQ	Squids (24,24)
*SS	Shallowest depth of migration (UNIMIG)
STR	Intermediate (UNIMIG)
SU	Sum
SUT	Intermediate (RANNAK)
SV	Intermediate (RANNAK)
SVK	Intermediate (RANNAK)
T	Time (PLANKT)
*T1	Surface temperature
*T2	Bottom temperature
*TD	Time step (2 days) (RANNAK)
U1	Intermediate (RANNAK)
*UK	Basic U component of migration (prescribed) (UNIMIG)
UP	Intermediate (RANNAK)
UR	U component of migration (UNIMIG)
*VAP	Index for printing (RANNAK)
*VK	Basic V component of migration (prescribed) (UNIMIG)
*VM	Minimum biomass of a grid point (ROMIGR)
VP	Intermediate (RANNAK)
VR	V component of migration (UNIMIG)
*ZKAP	Phase lag (see PLANKT)
*ZKAS	Zooplankton parameter (see PLANKT)
ZKP	Intermediate (PLANKT)
ZKPP	Intermediate (PLANKT)
*ZOM	Zooplankton parameter (see PLANKT)
*ZOMS	Zooplankton parameter (see PLANKT)
*ZOO	Zooplankton parameter (see PLANKT)

A.7   Example of UUSDYNE code in FORTRAN®

An example of the UUSDYNE code appears on the following pages. For species and groups of species in UUSDYNE see Table A.1.

```fortran
      PROGRAM UUSDYNE
      COMMON ISL(24,24),SD(24,24),T1(24,24),T2(24,24),SA(24,24),
     1AW(24,24),HW(24,24),CW(24,24),PLK(24,24),UR(24,24),VR(24,24),
     2SQ(24,24),NE,ME,K,RA(24,24)
      COMMON/PLNK/ZOO(10),ZOM(10),ZOMS(10),ZKAP(10),ZKAS(10)
      COMMON/FCO/FO(24,24,7),FCO(24,24,7),FC(14,7),GR(12),FR(12),SM(12),AT(12)
      COMMON/BIM/PB(24,24,3),SB(24,24,4),DB(24,24,5)
C
      OPEN (2,FILE='B:NEWSLDT',FORM='FORMATTED')
      OPEN (3,FILE=' ')
C
C     AT-ACCLIMATIZATION TEMP.
      NE=24
      ME=24
C     READING INPUT DATA
C     SEA-LAND TABLE AND DEPTHS
      READ (2,10)((ISL(N,M),M=1,ME),N=1,NE)
10    FORMAT(24I3)
      READ (2,11)((SD(N,M),M=1,ME),N=1,NE)
11    FORMAT(12F6.0)
C     SPECIES BIOMASSES
21    KN=1
22    READ (2,23)((PB(N,M,KN),M=1,ME),N=1,NE)
23    FORMAT(12F7.0)
      KN=KN+1
      IF(KN-3)22,22,24
24    KN=1
25    READ (2,23)((SB(N,M,KN),M=1,ME),N=1,NE)
      KN=KN+1
      IF(KN-4)25,25,26
26    KN=1
27    READ (2,23)((DB(N,M,KN),M=1,ME),N=1,NE)
      KN=KN+1
      IF(KN-5)27,27,28
28    READ (2,29)((FC(N,M),M=1,7),N=1,14)
29    FORMAT(7F6.2)
72    READ(2,73)(ZOO(N),N=1,10)
73    FORMAT(10F6.0)
C     ZOOPLANKTON PARAMETERS
      READ(2,73)(ZOM(N),N=1,10)
      READ(2,73)(ZOMS(N),N=1,10)
      READ(2,73)(ZKAP(N),N=1,10)
      READ(2,73)(ZKAS(N),N=1,10)
C     SPECIES SPECIFIC PARAMETERS
      READ(2,74)(GR(N),N=1,10)
74    FORMAT(12F6.3)
      READ(2,74)(FR(N),N=1,12)
      READ(2,74)(SM(N),N=1,12)
      READ(2,74)(AT(N),N=1,12)
C     PAUSE 'DYNOUT TO A'
      OPEN (4,FILE='A:DYNOUT',STATUS='NEW',FORM='FORMATTED')
C
C     XXXXXXXXXXXXX
C     SEASONAL COMPUTATION, K=SEASON,START WITH SUMMER
      K=3
30    IF(K-2)33,34,32
31    IF(K-4)35,36,36
32    IF(K-4)35,36,36
C     SEASONAL INPUTS
33    OPEN (5,FILE='C:WINDAT.DAT',FORM='FORMATTED')
39    READ (5,37)((T1(N,M),M=1,24),N=1,24)
      READ (5,37)((T2(N,M),M=1,24),N=1,24)
37    FORMAT(12F6.1)
      READ (5,23)((SA(N,M),M=1,24),N=1,24)
      READ (5,23)((SQ(N,M),M=1,24),N=1,24)
      READ (5,38)((FO(N,M,I),I=1,7),M=1,24),N=1,24)
38    FORMAT(24F5.0)
      CLOSE(5)
      GOTO 40
34    OPEN (5,FILE='C:SPRIDAT.DAT',FORM='FORMATTED')
      GOTO 39
35    OPEN (5,FILE='C:SUMDAT.DAT',FORM='FORMATTED')
      GOTO 39
36    OPEN (5,FILE='C:AUTDAT.DAT',FORM='FORMATTED')
      GOTO 39
40    CALL PLANKT(ISL,SD,PLK,K)
C     WRITE (3,41)K
41    FORMAT(1H1,/,    ZOOPLANKT. KG/KM2, SEASON=     ',I3)
C     CALL PRIAFP(PLK,2)
C     CALL PREDAT
      CALL BIOBAL
C     XXXXXXXXXXXXXXXXXXX SOME OUTPUTS XXXXXXXXXXXXXX
      IF(K-3)800,52,800
52    DO 101 N=1,ME
      DO 101 M=1,ME
      RA(N,M)=FO(N,M,3)
      CW(N,M)=DB(N,M,1)+DB(N,M,2)+DB(N,M,3)+DB(N,M,4)+DB(N,M,5)
      IF(CW(N,M)-5.)101,101,99
99    AW(N,M)=(RA(N,M)/CW(N,M))*100.
101   CONTINUE
103   FORMAT(1H1,/,        PERCENT DEM. FISH CONS.. S=  ',I3/)
      WRITE (3,103)K
      CALL PRIAFP(AW,3)
      DO 111 N=1,ME
      DO 111 M=1,ME
      RA(N,M)=FO(N,M,2)
      CW(N,M)=SB(N,M,1)+SB(N,M,2)+SB(N,M,3)+SB(N,M,4)
      IF(CW(N,M)-5.)111,111,110
110   AW(N,M)=(RA(N,M)/CW(N,M))*100.
111   CONTINUE
112   FORMAT(1H1,/,      PERCENT SEMIPEL. FISH CONS.,  S=  ',I3/)
      WRITE (3,112)K
      CALL PRIAFP(AW,3)
      DO 114 N=1,ME
      DO 114 M=1,ME
      RA(N,M)=FO(N,M,1)
      CW(N,M)=PB(N,M,1)+PB(N,M,2)+PB(N,M,3)
      IF(CW(N,M)-5.)114,114,113
```

```
  113 AW(N,M)=(RA(N,M)/CW(N,M))*100.
  114 CONTINUE
  115 FORMAT(1H1,'    PERCENT PEL. FISH CONS.  S= ',I3/)
      WRITE (3,115)K
      CALL PRIAFP(AW,3)
C     SETTINGS FOR NEXT SEASON
  800 K=K+1
      IF(5-K)810,810,812
  810 K=1
      GO TO 31
  812 IF (K-3)31,900,31
  900 STOP
      END

      SUBROUTINE PLANKT (ISL,SD,PLK,K)
      DIMENSION ISL(24,24),SD(24,24),PLK(24,24)
      COMMON/PLNK/ZOO(10),ZOM(10),ZOMS(10),ZKAP(10),ZKAS(10)
C     UNITS IN (MG/M3 ORIG.) KG/KM2
C     RADFA - CONVERSION FACTOR FROM DEGRESS TO RADIANS
C     ZKAP - PHASE LAGS
C     ZOO ZOM ZOMS - ZOOPLANKTON PARAMETERS
C     ALP  ALPP - PHASE SPEEDS
      T=K
      NE=24
      ME=24
      RADFA=0.0174533
      ALP=30.*RADFA
      ALPP=60.*RADFA
      DO 10 N=1,NE
      DO 10 M=1,ME
      IF(ISL(N,M))12,2,11
   11 I=ISL(N,M)
      ZKP=ZKAP(I)*RADFA
      ZKPP=ZKAS(I)*RADFA
      PLK(N,M)=ZOO(I)*1.8+ZOM(I)*COS(ALP*T-ZKP)+ZOMS(I)*SIN(ALPP*T-ZKPP)
      GO TO 10
    2 PLK(N,M)=0.
   10 CONTINUE
      DO 31 N=1,NE
      DO 31 M=1,ME
      IF (SD(N,M))31,31,32
   32 IF (SD(N,M)-60.) 41,41,42
   42 SKH=60.
      GO TO 49
   41 SKH=SD(N,M)
   49 PLK(N,M)=PLK(N,M)*SKH
   31 CONTINUE
   12 RETURN
      END

      SUBROUTINE SOOK2 (SW,TT,J)
      COMMON/FCO/FO(24,24,7),FC(14,7),GR(12),FR(12),SM(12),AT(12)
C     SW-SPECIES BIOM.
```

```
C     TT-TEMPERATURE (SEASON AND SFC OR BOT)
C     J-NUMBER OF SPECIES (TOT.14)
C     OBS. FR IS IN % BWD, CONV. TO FRACT. (OBS.  .92 DAYS)
C     FC IS IN %, (OBS. * 0.01)
      DIMENSION SW(24,24),TT(24,24)
      NE=24
      ME=24
      DO 250 N=1,NE
      DO 250 M=1,ME
      IF(SW(N,M)-8.)250,250,5
    5 IF(TT(N,M)-0.8)6,6,10
    6 TE=0.8
      GOTO 11
   10 TE=TT(N,M)
   11 ETN=1./AT(J)-(1./TE)
      FCON=SW(N,M)*FR(J)*.92*EXP(ETN)
      DO 230 I=1,7
      FO(N,M,I)=FO(N,M,I)+(FCON*FC(J,I)*0.01)
  230 CONTINUE
  250 CONTINUE
      RETURN
      END

      SUBROUTINE ASVSUR (SP,CP,TT,L)
C     SP-SPECIES BIOM.
C     CP-CONSUMPTION OF THIS SPECIES
C     TT-TEMPERATURE
C     L-NUMBER OF SPECIES (TOT.12)
      COMMON/FCO/FO(24,24,7),FC(14,7),GR(12),FR(12),SM(12),AT(12)
      DIMENSION SP(24,24),CP(24,24),TT(24,24)
      NE=24
      ME=24
      DO 310 N=1,NE
      DO 310 M=1,ME
      IF(SP(N,M)-8.)310,310,8
    8 IF(TT(N,M)-0.8)12,12,14
   12 TE=0.8
      GOTO 15
   14 TE=TT(N,M)
   15 ETN=1./AT(L)-(1./TE)
      AG=GR(L)*EXP(ETN)
      SP(N,M)=SP(N,M)+(AG*SP(N,M))-CP(N,M)-(SM(L)*SP(N,M))
  310 CONTINUE
      RETURN
      END

      SUBROUTINE PUUK (SK,SI,L,FIM,SU)
      DIMENSION SK(24,24),SI(24,24)
C     SI IN MAIN PROGR.=CW. PRINT OUT IN MAIN PROGR.
C     BY CALLING PRIAFP
C     SK-SPECIES
C     SI-SEASONAL CATCH, EITHER INPUT OR OUTPUT
C     IND -INDEX, 1-USE F. 2-USE PRESCRIBED CATCH FIELD
```

```fortran
      NE=24
      ME=24
      SU=0.
      AS=63.5*63.5
1     DO 10 N=1,NE
      DO 10 M=1,ME
6     IF (SK(N,M)-30.) 7,7,8
7     SI(N,M)=0.
      GOTO 10
8     SI(N,M)=SK(N,M)*FIM
12    SU=SU+SI(N,M)*AS/1000.
13    SK(N,M)=SK(N,M)-SI(N,M)
10    CONTINUE
25    SU=SU/1000.
30    RETURN
      END

      SUBROUTINE UNIMIG(ISL,SD,UR,VR,SS,DD,UK,VK,KE)
      DIMENSION ISL(24,24),SD(24,24),UR(24,24),VR(24,24),UR(24,24)
C     SPECIES-ADAPTABLE MIGRATION SPEED SUBROUTINE
C     ISL-SEA-LAND TABLE
C     SD-DEPTH
C     UR-WCOMPONENT
C     VR-V COMPONENT
C     SS-SHALLOWEST DEPTH OF MIGRATION
C     DD-DEEPEST DEPTH OF MIGRATION
C     UK-BASIC U COMPONENT,PRESCRIBED
C     VK-BASIC V COMPONENT,PRESCRIBED
C     (CUS-LATITUDINAL COMPONENT FOR U)
C     (CVS-LAT. COMP. FOR V)
C     DK -SPEED ADJUSTMENT FACTOR
C     KE-INDEX FOR SPRING (1) OR AUTUMN (2)
      NE=24
      ME=24
      DO 9 N=1,NE
      DO 9 M=1,ME
      UR(N,M)=0.0
      VR(N,M)=0.0
9     CONTINUE
      DK=0.31
      CUS=DK*1.1
      CVS=DK*2.60
      CEU=0.4
      CEW=0.12
      NEH=NE-1
      MEH=ME-1
      NEHH=NE-2
      DCO=1.6*DK
      DO 10 N=2,NEHH
      DO 10 M=2,MEH
      IF(ISL(N,M))10,10,105
105   IF (ISL(N,M)-9) 11,10,10
11    IF(SD(N,M)-SS)10,10,13
13    IF(SD(N,M)-DD)14,14,10
14    IF(KE-1)10,15,17
C     SPRING MIGRATION SPEED
15    PN=N
      SM=M
      SV=24-M
      IF (SM-15) 32,31,31
31    SM=M-14
32    UR(N,M)=0.7*UK-CUS*SQRT(PN)+CEU*SQRT(SM)
      VR(N,M)=1.0*VK-CVS*SQRT(PN)+CEW*SQRT(SV)
      IF (UR(N,M)) 33,34,34
33    UR(N,M)=0.
34    IF(VR(N,M)) 35,36,36
35    VR(N,M)=0.
36    IF(SD(N,M)-150.)22,22,21
21    STR=SQRT(0.1*SD(N,M))
      UR(N,M)=UR(N,M)*DCO*SQRT(STR)
      VR(N,M)=VR(N,M)*DCO*SQRT(STR)
22    GO TO 10
C     AUTUMN MIGRATION SPEED
17    PN=N
      SM=M
      SV=24-M
      IF (SM-15) 38,37,37
37    SM=M-14
38    UR(N,M)=-(0.7*UK)+CUS*SQRT(PN)-CEU*SQRT(SM)
      VR(N,M)=-(1.0*VK)+CVS*SQRT(PN)-CEW*SQRT(SV)
      IF (UR(N,M)) 221,221,222
222   UR(N,M)=0.
221   IF (VR(N,M)) 223,10,220
220   VR(N,M)=0.
223   IF(SD(N,M)-150.)10,10,23
23    UR(N,M)=UR(N,M)/(DK*SQRT(0.09*SD(N,M)))
      VR(N,M)=VR(N,M)/(DK*SQRT(0.09*SD(N,M)))
10    CONTINUE
C     ADJUSTMENT OF BOUNDARY VALUES
      DO 101 N=1,NE
      UR(N,2)=UR(N,2)*0.25
      UR(N,3)=UR(N,3)*0.5
      VR(N,2)=VR(N,2)*0.6
      UR(N,23)=UR(N,23)*0.8
101   CONTINUE
      DO 102 M=1,ME
      IF (VR(2,M)) 125,125,111
111   VR(2,M)=VR(2,M)*0.2
      VR(3,M)=VR(3,M)*0.5
      VR(4,M)=VR(4,M)*0.75
      VR(24,M)=0.
      GO TO 121
125   VR(2,M)=VR(2,M)*0.4
      VR(3,M)=VR(3,M)*0.8
121   VR(23,M)=VR(23,M)*0.2
      VR(22,M)=VR(22,M)*0.4
      VR(21,M)=VR(21,M)*0.8
      VR(24,M)=0.
102   CONTINUE
```

```fortran
      RETURN
      END

      SUBROUTINE RANPOR (ISL,SD,S1,S2,S3,RP,SS,DD)
      DIMENSION ISL(24,24),SD(24,24),S1(24,24),S2(24,24),S3(24,24)
C     SEPARATION OF MIGRATING PORTION
C     S1 SPECIES, TOTAL
C     S2 MIGRATING PART OF POPULATION
C     S3 NONMIGRATING PART OF POPULATION
C     RP MIGRATING FRACTION
C     SS SHALLOWEST DEPTH OF NO MIGRATION
C     DD DEEPEST DEPTH OF NO MIGRATION
      NE=24
      ME=24
      DO 10 N=1,NE
      DO 10 M=1,ME
      IF(ISL(N,M))160,160,101
  101 IF (ISL(N,M)-9)11,102,102
   11 IF(SD(N,M)-SS)12,12,13
   12 S2(N,M)=0
      S3(N,M)=S1(N,M)
      GO TO 10
   13 IF(SD(N,M)-DD)14,14,12
   14 S2(N,M)=S1(N,M)*RP
      S3(N,M)=S1(N,M)-S2(N,M)
      GO TO 10
  102 S2(N,M)=0.
      S3(N,M)=S1(N,M)
      GO TO 10
  160 CONTINUE
      S1(N,M)=0.
      S2(N,M)=0.
      S3(N,M)=0.
   10 CONTINUE
      DO 201 M=8,11
      S3(23,M)=S1(23,M)
      S2(23,M)=0.
      S3(24,M)=S1(24,M)
      S2(24,M)=0.
  201 CONTINUE
      RETURN
      END

      SUBROUTINE RANNAK(S8,ISL,UP,VP)
      DIMENSION S8(24,24),ISL(24,24),F8(24,24),UP(24,24),VP(24,24)
C     ISL- SEA-LAND TABLE.
C     S8 - SPECIES (BEFORE MIGRATION, AND UPON EXIT)
C     F8 - SPECIES AFTER MIGRATION
C     UP - (U-COMP.FROM UNIMIG)
C     VP - (V-COMP.FROM UNIMIG)
C     VAP- IND. FOR SPEED PRINTING
C          1-PRINT, 0-NO
C     KRC - NUMBER OF 2-DAY MIGRATIONS

      NE=24
      ME=24
      VAP=1.
      KRC=11
      KK=1
      TD=2.
      NEH=NE-1
      MEH=ME-1
      AUS=0.5
      DL=63.5
      B1=DL*DL
      B2=(2.*TD*AUS)/B1
      B5=(AUS*TD)/B1
      SU=0.
      DO 150  N=2,NEH
      DO 150 M=2,MEH
      IF (ISL(N,M)) 334,334,1335
 1335 IF (ISL(N,M)-9) 335,150,150
  335 SU=SU+S8(N,M)/1000.
      GO TO 150
  334 F8(N,M)=0.
      S8(N,M)=0.
  150 CONTINUE
C     XXXXXXXXXXXXXXXXXXXXXXXX
C     DOUBLE PRESCRIBED SPEED
      DO 321 N=1,NE
      DO 321 M=1,ME
      UP(N,M)=2.*UP(N,M)
      VP(N,M)=2.*VP(N,M)
  321 CONTINUE
  254 US=0.
      DO 133  N=2,NEH
      DO 133 M=2,MEH
      SUT=0.
      SVK=0.
      IF(ISL(N,M))333,333,1221
 1221 IF (ISL(N,M)-9) 221,333,333
  221 IF(UP(N,M))225,1225,231
 1225 SH=0.
      GO TO 232
  225 SH=(S8(N,M)-S8(N,M+1))/DL
      SUT=TD*ABS(UP(N,M))*SH
      GO TO 232
  231 SH=(S8(N,M)-S8(N,M-1))/DL
      SUT=TD*ABS(UP(N,M))*SH
  232 IF(VP(N,M))233,1223,234
 1223 SV=0.
      SVK=0.
      GO TO 235
  233 SV=(S8(N,M)-S8(N-1,M))/DL
      SVK=TD*ABS(VP(N,M))*SV
      GO TO 235
  234 SV=(S8(N,M)-S8(N+1,M))/DL
      SVK=TD*ABS(VP(N,M))*SV
```

```
235  F8(N,M)=S8(N,M)-SUT-SVK
260  US=US*F8(N,M)/1000.
     GO TO 133
333  F8(N,M)=0.
133  CONTINUE
     U1=SU/US
     DO 751 N=2,NEH
     DO 751 M=2,MEH
     IF (ISL(N,M)-9) 1716,751,751
1716 F8(N,M)=F8(N,M)*U1
     IF(F8(N,M)-1.0)716,722,722
716  F8(N,M)=0.
722  IF(ISL(N,M))251,251,751
251  F8(N,M)=0.
751  CONTINUE
     DO 10 N=1,NE
     IF(ISL(N,1))9,9,8
9    F8(N,1)=0.
8    F8(N,1)=S8(N,1)
7    IF(ISL(N,ME))6,6,5
6    F8(N,ME)=0.
     GO TO 10
5    F8(N,ME)=S8(N,ME)
10   CONTINUE
     DO 1810 M=1,24
     F8(24,M)=0.
     S8(24,M)=0.
1810 CONTINUE
     DO 11 M=1,ME
     IF(ISL(1,M))4,4,3
4    F8(1,M)=0.
     GO TO 11
3    F8(1,M)=S8(1,M)
2    F8(NE,M)=S8(NE,M)
11   CONTINUE
     ALP=0.95
     BET=1-ALP
     DO 301 N=2,23
     DO 301 M=2,23
     F8(N,M)=ALP*F8(N,M)+BET*(F8(N-1,M)+F8(N+1,M)
    1+F8(N,M-1)+F8(N,M+1))/4.
301  CONTINUE
     DO 12 N=1,NE
     DO 12 M=1,ME
     IF (ISL(N,M)-9) 1736,151,151
1736 S8(N,M)=F8(N,M)
     IF (ISL(N,M)) 151,151,336
336  IF(F8(N,M))151,12,12
151  F8(N,M)=0.
     S8(N,M)=0.
12   CONTINUE
     DO 1951 M=1,ME
     F8(24,M)=0.
     S8(24,M)=0.

1951 CONTINUE
     KK=KK+1
     IF(KK-KRC)254,254,255
255  KK=0
752  SU=0.
     US=0.
C    XXXXXXXXXXXXXXXX
C    RETURN ORIGINAL SPEED
     DO 322 N=1,NE
     DO 322 M=1,ME
     UP(N,M)=UP(N,M)/2.
     VP(N,M)=VP(N,M)/2.
322  CONTINUE
     DO 270 N=1,NE
     DO 270 M=L,ME
     S8(N,M)=F8(N,M)
270  CONTINUE
     IF(VAP-1)258,259,258
259  WRITE(3,256)
256  FORMAT(1H1,5X,'U COMP. OF MIGR. KM/DAY. DIV.BY 2',//)
     WRITE (3,500)(N,N=1,24)
500  FORMAT(/,5X,24I5)
     WRITE (3,302)(N,(UP(N,M),M=1,24),N=1,24)
302  FORMAT(/,I3,2X,24F5.1)
     WRITE(3,257)
257  FORMAT(1H1,5X,'V COMP. OF MIGR. KM/DAY. DIV.BY 2',//)
     WRITE (3,500)(N,N=1,24)
     WRITE (3,302)(N,(VP(N,M),M=1,24),N=1,24)
258  RETURN
     END

     SUBROUTINE ROMIGR(EP,ISL,AX,VH,NE,ME)
     DIMENSION EP(24,24),ISL(24,24)
C    EP - SPECIES
C    AX - MAXIMUM BIOMASS
C    VH - MINIMUM BIOMASS
C    SUBROUTINE ROMIGR ROUTES TOO HIGH AND OR TOO LOW CONCENTRATIONS
C    TO/FROM SURROUNDING POINTS; BOUNDARY POINTS ARE SET ONLY TO MAX OR
C    MIN VALUES
C    TEST ON MINIMUM EXCLUDED
     JEH=NE-1
     JEH=ME-1
51   DO 10 I=2,IEH
     DO 10 J=2,JEH
     IF(ISL(I,J))10,10,2
2    IF(EP(I,J)-AX)10,10,4
4    NA=I-1
     NL=I+1
     MA=J-1
     ML=J+1
     SK=0.
     SU=0.
     DO 5 N=NA,NL
5    M=MA,ML
```

```
      IF(ISL(N,M))5,5,6
6     SU=SU+EP(N,M)
5     SK=SK+1.
      CONTINUE
      SME=SU/SK
      IF(SME-AX)12,12,8
8     SHE=AX
      GO TO 12
12    DO 11 N=NA,NL
      DO 11 M=MA,ML
      IF(ISL(N,M))11,11,13
13    EP(N,M)=SME
11    CONTINUE
10    CONTINUE
      RETURN
      END

      SUBROUTINE PRIAFF(FLD,KM)
      DIMENSION FLD(24,24),PRF(24,24)
C     KM=0, FULL UNITS, INTEGER
C     KM=1, DIVIDE BY 10
C     KM=2, DIVIDE BY 100
C     KM=3, FULL UNITS, REAL
C     KM=4, FULL UNITS, PRINT 1 DECIMAL PLACE
C     OPEN (UNIT=3,ACCESS=SEQUENTIAL, BLOCKSIZE=2560, FILE=C:NEWSLOT
C     1FORMAT=FORMATTED,RECL=256,STATUS=UNKNOWN)
      IF (KM-1)  1,2,3
3     IF (KM-3)  20,21,8
1     WRITE(3,10)
10    FORMAT (/,5X,18HORIG.UNITS, INTEG.)
      WRITE (3,500)(N,N=1,24)
500   FORMAT (/,5X,24I5)
511   WRITE (3,511)(N,(FLD(N,M),M=1,24),N=1,24)
      FORMAT (/I3,2X,24I5)
      GO TO 100
2     WRITE (3,11)
11    FORMAT (/,5X,16HUNITS DIV. BY 10)
      WRITE (3,500)(N,N=1,24)
      DIL=10
4     DO 12 N=1,24
      DO 12 M=1,24
      PRF(N,M)=FLD(N,M)/DIL
12    CONTINUE
      WRITE (3,513)(N,(PRF(N,M),M=1,24),N=1,24)
512   FORMAT (/I3,2X,24F5.1)
513   FORMAT (/I3,2X,24F5.0)
      GO TO 100
8     WRITE (3,10)
      WRITE(3,500)(N,N=1,24)
      WRITE (3,512)(N,(FLD(N,M),M=1,24),N=1,24)
      GO TO 100
20    DIL=100.
      WRITE (3,25)
25    FORMAT(/,5X,17HUNITS DIV. BY 100)

      WRITE (3,500)(N,N=1,24)
      GO TO 4
21    WRITE (3,24)
24    FORMAT(/,5X,11HORIG. UNITS)
      WRITE (3,500)(N,N=1,24)
      WRITE (3,513)(N,(FLD(N,M),M=1,24),N=1,24)
100   RETURN
      END

      SUBROUTINE PREDAT
      COMMON ISL(24,24),SD(24,24),T1(24,24),T2(24,24),SA(24,24),
     1AM(24,24),HW(24,24),CW(24,24),PLK(24,24),FR(24,24),VR(24,24),
     2SQ(24,24),NE,ME,K,RA(24,24)
      COMMON/PLNK/ZOO(10),ZOM(10),ZOMS(10),ZKAP(10),ZKAS(10)
      COMMON/FCO/FCO(24,24,7),FC(14,7),GR(12),FR(12),SM(12),AT(12)
      COMMON/BIM/PB(24,24,3),SB(24,24,4),DB(24,24,5)
C     XXXXXXXXXXXXX
C     COMP. OF MIGRATION AND PREDATION
C     XXXXXXXXXXXXX
      HERRING
      J=1
C     SEPARATE SPECIES
      DO 41 N=1,NE
      DO 41 M=1,ME
      HW(N,M)=PB(N,M,1)
41    CONTINUE
      IF(K-2)65,50,42
42    IF(K-4)65,60,65
C     CALL MIGRATION SUBROUTINES
50    KE=1
      GO TO 62
60    KE=2
62    SS=18.
      DD=2000.
      UK=4.2
      VK=3.8
      CALL UNIMIG(ISL,SD,UR,VR,SS,DD,UK,VK,KE)
      RP=0.8
      CALL RANPOR(ISL,SD,HW,AW,CW,RP,SS,DD)
      CALL RANNAK(AW,ISL,UR,VR)
      DO 52 N=1,NE
      DO 52 M=1,ME
      HW(N,M)=AW(N,M)+CW(N,M)
52    CONTINUE
65    CALL SOOK2(HW,T1,J)
      DO 67 N=1,NE
      DO 67 M=1,ME
      PB(N,M,1)=HW(N,M)
67    CONTINUE
C     XXXXXXXXXXXXX
C     ATKA MACKEREL
      J=2
      DO 70 N=1,NE
```

```
      DO 70 M=1,ME
      HW(N,M)=PB(N,M,2)
   70 CONTINUE
C
Note: Species by species computations should continue.similar
to the computations of herring above. In some species it is
not necessary to call migration subroutines.

At the end of this subroutine (PREDAT) after the computation
crabs, the consumption of benthos over deep water is adjusted.
C
C     XXXXXXXXX
C     CRABS
      J=12
      DO 270 N=1,NE
      DO 270 M=1,ME
      HW(N,M)=DB(N,M,5)
  270 CONTINUE
      CALL SOOK2(HW,T1,J)
      DO 287 N=1,NE
      DO 287 M=1,ME
      DB(N,M,5)=HW(N,M)
  287 CONTINUE
C
C     XXXXXXXXX
      SQUIDS=14;AND SALMON=13
      CALL SOOK2(SA,T1,5)
      CALL SOOK2(SQ,T1,5)
C     ADJUSTING CONSUMPTION OF BENTHOS OVER DEEP WATER
C     4-SQUIDS, 6-BENTHOS, 7-PLANKTON, 1-PELAGIC, 2-SEMIPEL.
C     3-DEMERSAL.
      DO 302 N=1,NE
      DO 302 M=1,ME
      BN=FO(N,M,6)
      BR=FO(N,M,3)
  301 IF(SD(N,M)-450)302,301,301
      FO(N,M,4)=FO(N,M,4)+0.10*BN
      FO(N,M,7)=FO(N,M,7)+0.50*BN
      FO(N,M,1)=FO(N,M,1)+0.15*BN+0.15*BR
      FO(N,M,2)=FO(N,M,2)+0.25*BN+0.25*BR
      FO(N,M,6)=0.
      FO(N,M,3)=FO(N,M,3)*0.3
  302 CONTINUE
C     ADJUSTMENT OF SQUIDS OVER SHELF
      DO 310 N=1,NE
      DO 310 M=1,ME
  311 IF(N-8)313,311,311
  312 IF(SD(N,M)-80)312,310,310
      BN=FO(N,M,4)
      FO(N,M,1)=FO(N,M,1)+0.3*BN
      FO(N,M,2)=FO(N,M,2)+0.3*BN
      FO(N,M,4)=FO(N,M,4)*0.15
  313 CONTINUE
  310 CONTINUE
      RETURN
      END


      SUBROUTINE BIOBAL
      COMMON ISL(24,24),SD(24,24),T1(24,24),T2(24,24),SA(24,24),
     1AW(24,24),HW(24,24),CW(24,24),PLK(24,24),UR(24,24),VR(24,24),
     2SQ(24,24),NE,ME,K,RA(24,24)

      COMMON/PLNK/ZOO(10),ZOM(10),ZOMS(10),ZKAP(10),ZKAS(10)
      COMMON/FCO/FO(24,24,7),FC(14,7),GR(12),FR(12),SM(12),AT(12)
      COMMON/BIH/PB(24,24,3),SB(24,24,4),DB(24,24,5)
C     COMPUTATION OF CONSUMPTION
C     I-IS THE GROUP NUMBER,1-PELAGIC, 2-SEMIPELAGIC, 3-DEMERSAL
C     J-IS THE NUMBER OF SPECIES IN GROUP (3,4 OR 5)
C     JA IS THIS NUMBER OF SPECIES (3,4 OR 5)
C     K-SEASON
      AR=63.5*63.5
  303 I=1
      JA=3
C     PELAGIC GROUP
C     JC IS THE CONSECUTIVE NUMBER OF SPECIES IN GROUP
C     L-IS THE NUMBER OF SPECIES (1 TO 12)
C     XXXXXXXXXXX
C     HERRING
      JC=1
      L=1
      DO 350 N=1,NE
      DO 350 M=1,ME
      CW(N,M)=0.
  350 CONTINUE
      NA=1
  351 DO 360 N=1,NE
      DO 360 M=1,ME
      CW(N,M)=CW(N,M)+PB(N,M,NA)
  360 CONTINUE
      NA=NA+1
      IF(NA-3)351,351,361
C     AMOUNT OF SPECIES L CONSUMED (AT ANY GRIDPOINT)
  361 DO 365 N=1,NE
      DO 365 M=1,ME
      IF(CW(N,M)-10.)365,365,362
  362 AW(N,M)=(PB(N,M,JC)/CW(N,M))*FO(N,M,I)
      IF(ISL(N,M)-9)364,365,365
      SU=SU+(AW(N,M)*AR/1000000.)
  364 CONTINUE
  365 CONTINUE
      SUM=0.
      DO 354 N=1,NE
      DO 354 M=1,ME
      RA(N,M)=FO(N,M,1)
      IF(ISL(N,M)-9)353,352,352
  352 RA(N,M)=0.
      GOTO 354
  353 SUM=SUM+(RA(N,M)*AR/1000000.)
  354 CONTINUE
  355 FORMAT(1H1,5X,'CONS. OF PEL. SP.. SUM, KT=',F6.0/)
```

```
      DIMENSION HW(24,24),SAM(24,24)
C     ANOTHER OUTPUT SUBROUTINE
C     SP - LATEST SPECIES BIOMASS
C     CON - CONSUMPTION
C     IT - SPECIES ECOLOGICAL GROUP (1 TO 3)
C     JC - SPECIES NUMBER IN GROUP
C     L - SPECIES NUMBER (1-14)
C     K - SEASON
      NE=24
      ME=24
      AR=63.5*63.5
      SU=0.
      SUR=0.
      DO 810 N=1,NE
      DO 810 M=1,ME
      SU=SU+((HW(N,M)*AR)/1000000.)
810   CONTINUE
802   IF(IT-2) 820,830,840
820   DO 825 N=1,NE
      DO 825 M=1,ME
      SAM(N,M)=HW(N,M)-PB(N,M,JC)
      SUR=SUR+((SAM(N,M)*AR)/1000000.)
825   CONTINUE
      GOTO 850
830   DO 835 N=1,NE
      DO 835 M=1,ME
      SAM(N,M)=HW(N,M)-SB(N,M,JC)
      SUR=SUR+((SAM(N,M)*AR)/1000000.)
835   CONTINUE
      GOTO 850
840   DO 845 N=1,NE
      DO 845 M=1,ME
      SAM(N,M)=HW(N,M)-DB(N,M,JC)
      SUR=SUR+((SAM(N,M)*AR)/1000000.)
845   CONTINUE
821   FORMAT(1H1,5X,'SP=',I3,'  BIOMASS, S=',I3,'  SUM,KT= ',F6.0/)
     1          'SP=',I3,'  BIOMASS,SUM.KT= ',F6.0/)
                'SP=',I3,'  SOURCE-SINK, S= ',I3,' SUM, KT= ',F6.0/)
822   FORMAT(1H1,5X,'SP=',I3,'  BIOMASS, S=',I3,' SUM, KT= ',F6.0/)
850   WRITE(3,822)L,K,SU
      CALL PRIAFP(HW,1)
      WRITE(3,822)L,K,SUR
      CALL PRIAFP(SAM,3)
860   RETURN
      END
```

```
      WRITE (3,355)SUM
      CALL PRIAFP(RA,1)
366   FORMAT(//,5X,'CONS.  SP= ',I3,' SUM, KT='F6.0/)
      WRITE (3,366)L,SU
C     AW CONTAINS CONSUMPTION OF SPECIES JC (=HERE L)
C     SEPARATE SPECIES BIOMASS
      DO 370 N=1,NE
      DO 370 M=1,ME
      HW(N,M)=PB(N,M,JC)
370   CONTINUE
      CALL ASVSUR(HW,AW,T1,L)
      CALL PUUK(HW,RA,L,0.025,SU)
      WRITE (3,369) L,K,SU
369   FORMAT(//,'   CATCH,SPECIES',I3,' S= ',I3,' KT=',F6.0/)
C     COMPUTE SOURCE - SINK
      IF(K-1)374,378,377
377   IF(K-3)374,378,374
378   DO 371 N=1,NE
      DO 371 M=1,ME
      RA(N,M)=HW(N,M)-PB(N,M,1)
371   CONTINUE
      WRITE (3,372)K
372   FORMAT(1H1,' SOURCE-SINK, HERRING, SEASON ',I3)
      CALL PRIAFP(RA,3)
374   DO 375 N=1,NE
      DO 375 M=1,ME
      PB(N,M,JC)=HW(N,M)
375   CONTINUE
      SBI=0
      DO 373 N=1,NE
      DO 373 M=1,ME
      SBI=SBI+((HW(N,M)*AR)/1000000.)
373   CONTINUE
      WRITE (3,376)K,SBI
376   FORMAT(1H1,' HERRING, BIOMASS, SEASON ',I3,' KT= 'F6.0/)
      CALL PRIAFP(HW,1)
C     XXXXXXXXXXX
C     ATKA MACKEREL
      JC=2
      L=2
      DO 390 N=1,NE
      DO 390 M=1,ME
      IF(CW(N,M)-1.)390,390,386
386   AW(N,M)=(PB(N,M,JC)/CW(N,M))*FO(N,M,I)
390   CONTINUE
```

This subroutine continues species by species as herring above. Various outputs can be taken within it.

```
      RETURN
      END
```

```
      SUBROUTINE VALAND(HW,IT,JC,L,K)
      COMMON/BIH/PB(24,24,3),SB(24,24,4),DB(24,24,5)
```

Index